Reclaiming Cognition

Published in the UK by Imprint Academic
PO Box 1, Thorverton EX5 5YX, UK

Published in the USA by Imprint Academic
Philosophy Documentation Center, Bowling Green State University,
Bowling Green, OH 43403-0189, USA

ISBN 0 907845 06 1 (paperback)

ISSN 1355 8250 (*Journal of Consciousness Studies*, **6**, No.11–12, 1999)

British Library Cataloguing in Publication Data
A catalogue record for this book is available from the British Library
Library of Congress Card Number: 99–068337

Cover illustration: Nicholas Gilbert Scott
Cover design: J.K.B. Sutherland

Printed in Exeter UK by Short Run Press Ltd.

ALSO OF INTEREST FROM IMPRINT ACADEMIC
Full details on: http://www.imprint.co.uk

Series Editor: Professor J.A. Goguen
Department of Computer Science and Engineering
University of California, San Diego

Thomas Metzinger, ed.
Conscious Experience

Francisco Varela and Jonathan Shear, ed.
The View from Within:
First-person approaches to the study of consciousness

Shaun Gallagher and Jonathan Shear, ed.
Models of the Self

Joseph A. Goguen, ed.
Art and the Brain

Benjamin Libet, Anthony Freeman and J.K.B. Sutherland, ed.
The Volitional Brain: Towards a neuroscience of free will

Leonard D. Katz, ed.
Evolutionary Origins of Morality: Cross-disciplinary perspectives

Reclaiming Cognition

The Primacy of Action, Intention and Emotion

edited by
Rafael Núñez and Walter J. Freeman

IMPRINT ACADEMIC

Contents

Mathematics and Neurobiology

Philosophy of Action, Intention and Emotion

Contributors

Paul Cisek

Dept. de physiologie, Université de Montréal, 2960 Chemin de la tour, Montréal, (Québec) H3C 3J7, Canada

Andy Clark

Department of Philosophy, Washington University, Campus Box 1073, One Brookings Drive, St Louis, MO 63130-4899, USA

Walter J. Freeman

Department of Molecular and Cell Biology, LSA 129, University of California, Berkeley, CA 94720-3200, USA

Ravi V. Gomatam

Bhaktivedanta Institute, Juhu Road, Juhu, Bombay 400049, India *or* 2334 Stuart Street, Berkeley, CA 94705, USA

Brian Goodwin

Schumacher College, Dartington, Devon TQ9 6EA, UK

Valerie Gray Hardcastle

Department of Philosophy, Virginia Polytechnic Institute and State University, Blacksburg, VA 24061-0126, USA

Jana M. Iverson

Department of Psychology, University of Misouri, 210 McAlester Hall, Columbia, MO 65211, USA

Giuseppe Longo

CNRS and Dépt. De Mathématiques et Informatique, Ecole Normale Supérieure, Paris, France

J.S. Nicolis

Department of Electrical Engineering, University of Patras, Greece

Rafael Núñez
Institute of Cognitive Studies, University of California,
Berkeley, CA 94720, USA

Eleanor Rosch
Department of Psychology, University of California,
3210 Tolman Hall, Berkeley, CA 94720-1650, USA

Hilary Rose
Department of Sociology, The City University,
4 Lloyd Square, London WC1X 9BA, UK

Robert E. Shaw
Center for the Ecological Study of Perception and Action,
University of Connecticut, Stoors, CT 06269-1020, USA

Maxine Sheets-Johnstone
Box 722, Yachats, OR 97498, USA

Christine A. Skarda
1544 Campus Drive, Berkeley, CA 94708, USA

Esther Thelen
Department of Psychology, Indiana University

I. Tsuda
Department of Mathematics, Hokkaido University, Sapporo 060, Japan

M.T. Turvey
Center for the Ecological Study of Perception and Action,
University of Connecticut, Stoors, CT 06269-1020, USA

Walter J. Freeman and Rafael Núñez

Restoring to Cognition the Forgotten Primacy of Action, Intention and Emotion

Making sense of the mind is the human odyssey. Today, the cognitive sciences provide the vehicles and équipage. As do all culturally shaped activities, they manifest crystallized generalizations and ideological legacies, many of which go unquestioned for centuries. From time to time, these ideologies are successfully challenged, generating revisions and new forms of understanding. We believe that the cognitive sciences have reached a situation in which they have been frozen into one narrow form by the machine metaphor. There is a need to thaw that form and move from a reductionist, atemporal, disembodied, static, rationalist, emotion- and culture-free view, to fundamentally richer understandings that include the primacy of action, intention, emotion, culture, real-time constraints, real-world opportunities, and the peculiarities of living bodies. These essays constitute an array of moves in that direction.

Making Sense of our Minds: What Have We Inherited?

For millennia our ancestors, in trying to understand what it is to think and to know, have made sense of these processes in terms of spirit and soul (Durkheim, 1915; Barfield, 1965; Freeman, 1995; Núñez, 1995). Many of their questions are still with us. How do our senses convey the world to our inner spirits? How do we convert that input to knowledge, and ultimately to wisdom? How do we remember what we have done and become? What kinds of spirits exist in beings other than humans, such as animals, trees, rocks, mountains, stars, and other planets? How do we establish relations with them, so as to predict the future, invoke their help, placate their anger, or give thanks for their aid, with the help of images, icons, totems, invocations, songs, dances, instruments of observation? Where are the spirits located; in their world, or even in our own? Where do we go when we sleep? When we die? How do we protect our families from spirits who return from other worlds to take vengeance, or entice those who can guide us?

The questions have been revised in various ways throughout recorded history by new developments in the natural sciences. In the West, the questions and their answers were most strongly influenced by the emergence of material medicine in the

Journal of Consciousness Studies, **6**, No. 11–12, 1999, pp. ix–xix

age of Hippocrates (Clarke and O'Malley, 1968). Where in the body is the mind located, in the heart or in the head? How do the elements of which the world is made combine to form the four humours? How do these humours determine our personalities: sanguine, phlegmatic, melancholic, and choleric? How do imbalances in the humours lead to illness, and how can bleeding and purging restore the natural balance? In the East, a comparable system arose simultaneously and independently in China, leading to emergence of the medical systems of acupuncture and moxibustion (Mann, 1962), which were materialist, practical aspects of a deep and insightful mechanistic theory of mind, based in the concept of the flow of qi (energy), centred in the navel (this explains the shape they gave to Buddha's body; a big belly represented a man of substance). In western philosophy, the distinction between form and matter led the Greeks to the question: are there ideal forms for material objects made of atoms or the elemental substances (Lakoff and Johnson, 1998)? Can the forms exist independently in the material world, the spirit world, or in ourselves? Are they accessed by perception or by reason? Are gods spirits that inhabit animals, mountains, and other worlds?

Under the aegis of Augustine, the Platonic view held sway for nearly a thousand years, until the revolution wrought by Thomas Aquinas (1272). Whereas for Platonists the apprehension of forms was by the passive imprint of imperfect copies from the world on the senses, and the reconstruction of the ideal forms by reason, the Thomists followed Aristotle (Kenny, 1980) in proposing that perception occurs by action of the body into the world ('intendere'), followed by assimilation ('adequatio', to make equivalent, for example, to shape the hand in conformance with a cup so as to drink, not to incorporate the form of the cup into the mind by reason), but with a crucial difference (Freeman, 1999). Aristotelians proposed that the forms in the mind ('in-formation' as we would say) were derived through the senses from the forms of matter, whereas Thomists proposed that material objects were unique (we would say 'infinitely complex', ultimately not comprehensible, Kant's 'Ding-an-sich'), and that all knowledge of forms was created intentionally within minds by imagination through the processes of abstraction and generalization. Their view was in opposition to the Averroists' universal mind (Aquinas, 1272, part I, question 76, article 2, p. 389), in postulating that every mind is isolated from every other by this constraint (we would call it God's own 'firewall'), giving the gift of ultimate privacy. Recent studies in mesoscopic (Ingber, 1992; Imry, 1997) brain dynamics appear now to support this view (Freeman, 1991; 2000).

With the emergence of modern science, Descartes, Leibniz and Kant (1781) enacted a counter-revolution, the essence of which was the mathematization of the human mind (Lakoff and Núñez, 1997; Núñez and Lakoff, 1998), and the replacement of Thomist intentionality by representationalism. Putnam (1990) identified the two concepts ('representation — that is to say, intentionality — ' p.107) and wrote:

> Kant's purpose, unlike Berkeley's, was not to deny the reality of matter, but rather to deny that things in themselves are possible objects of knowledge. What we can know — and this is the idea that Kant himself regarded as a kind of Copernican Revolution in philosophy — is never the thing in itself, but always the thing as represented. And the representation is never a mere copy; it always is a joint product of our interaction with the external world and the active powers of the mind. The world as we know it bears the stamp of our own conceptual activity (p. 261).

This re-formulation led to various modern questions. What occurs in the senses upon the impact of matter and energy, causing the transfer of forms (information) into the brain? How is the information transmogrified to representations in perception? What are the mechanisms for learning, storing, and retrieving representations from memory? How are they selected by attention? How do brains assemble them into knowledge bases, and how do data bases lead to wisdom? What is consciousness, and what is it for? The act of perception in this view is the construction of representations as symbols of objects and events in the world, whether real or imaginary, for presentation to consciousness, and the operation of the mind is the manipulation of these symbols in accordance with the laws of logic, induction, statistical inference, and mathematical deduction. Imagination is replaced by the manipulation of representations, and intention is reduced to the 'aboutness' of the representations (Searle, 1983), i.e., how those in brains that can be related to the world outside minds, as distinct (according to Brentano, 1889) from those in extant machines that cannot.

Descriptions of mental phenomena from the scientific point of view have been continuously influenced by the world views and methodologies required for studies in the so-called 'hard sciences'. A particularly strong influence has been that of Isaac Newton's physics, which gave mathematical form to the Cartesian conception of reflection as the basis for understanding the hierarchy of reflexes in the nascent sciences of physiology and psychology (Prochaska, 1784). The flip side was the practice of astrology, which was the logical cognitive science of its day. Astrologers were more respected, better paid, and in some respects more successful than contemporary physicians at treating indispositions of the mind. Many astronomers made their living by casting horoscopes and teaching medical students to do so. When, in 1609, the Grand Duke Ferdinando lay ill, Galileo on call predicted from the stars that he would live many years more. Three weeks later he died (Sobel, 1999), and Galileo went on to social troubles of his own. Newton found alternative employment; perhaps he was concerned that the action-at-a-distance required by his gravitational theory might be confused with astrological influences of the stars. Today, many intractable problems stem from this platform: the mind–body split, linear causality, the subject–object and nature–nurture dichotomies, to mention a few (Núñez, 1997).

The Cartesian method reached an apogee of a sort when Whitehead and Russell (1910) published 'Principia Mathematica', in which they attempted to reduce all of symbolic logic and the quintessence of human reasoning to formal mathematics (Lakoff and Núñez, 2000). Even upon this culmination there was a series of death-blows through the works of Gödel (1930) and Wittgenstein (1922), whose work according to Kenny (1984) was presaged by Aquinas, as though the very clarity of the foundations of reason that Whitehead and Russell had brought to the enterprise had exposed the essential contradiction on which it was based: the impossibility of the mind understanding itself through purely logical analysis. But the irony is that the positivist and reductionist study of the mind gained an extraordinary popularity through a relatively recent doctrine called *Cognitivism*, a view that shaped the creation of a new field — *Cognitive Science* — and of its most hard core offspring: Artificial Intelligence. Some practitioners called this endeavour 'The Cognitive Revolution', though the real revolution was already over 300 years old.

'Cognitivism': The Legacy of What Was to be the One True Science of Cognition

The new field of Cognitive Science, which we here call Cognitivism, was meant to be interdisciplinary, receiving contributions from psychology, neuroscience, linguistics, analytic philosophy, and computer science. But because of historical accidents of technology and of the nature of the enterprise, the contributions of the latter played the dominant role (Varela, 1989). The cognitivist-oriented study of the mind emerged in a specific place and time in history (the eastern seaboard of North America in the 1940s) as a concrete proposal for overcoming many of the analytic problems left unsolved by the approaches of the preceding era, such as introspection, psychoanalysis, and behaviourism. The aim was to provide a precise, pervasive, and uniform paradigm and methodology for operationalizing and emulating the essential aspects of the mind (not merely behaviour) in an objective, transparent, and controllable manner. The mind as a rational calculating device — an idea already posed by Leibniz and Descartes and expounded by Craik (1943) and Turing (1950) — served as a framework, and electronics provided the key tool for making the enterprise flourish: the digital computer (Berkeley, 1949), an artifact that could develop and operate an amazing variety of algorithms in software.

Technology, of course, was not the only factor. Cognitivism was built on a well-established doctrine — *substance dualism*, which postulates an essential separation between mind and body — and on a novel one — *functionalism*, which defines the mind as a set of mechanisms that can perform functions independently of the physical platform on which it is implemented (Block, 1980). The digital computer was the perfect platform on which these doctrines could be realized by positing 'a level of analysis wholly separate from the biological or neurological, on the one hand, and the sociological or cultural, on the other' (Gardner, 1985, p. 6). This reduced level of analysis focussed on the individual mind as a passive input–output device that processed information, and it characterized reasoning as the logical manipulation of arbitrary symbols. Thus technology and the philosophic *Zeitgeist* gave rise to a powerful and inviting metaphor, the legacy of which deeply influences us still today: the metaphor of the mind as a computer. Ironically, this happened despite the caveat, published posthumously shortly after the dawn of Cognitivism by John von Neumann (1958), chief architect of games theory, self-reproducing machines, and the programmable digital computer:

> Thus the outward forms of *our* mathematics are not absolutely relevant from the point of view of evaluating what the mathematical or logical language *truly* used by the central nervous system is. . . . It is characterized by less logical and arithmetical depth than what we are normally used to. . . . Whatever the system is, it cannot fail to differ considerably from what we *consciously* and *explicitly* consider as mathematics (pp. 81–2).

But von Neumann's conclusion was ignored. The Cognitive Revolution was already in full sway, so that all aspects of the mind were seen as software that happened in humans to be implemented in wetware having unreliable neurons, and that might be vastly improved if implemented instead in hardware with reliable transistors. As Gardner said, ' . . . not only are computers indispensable for carrying out studies of various sorts, but, more crucially, the computer also serves as the most viable model of how the human mind functions' (Gardner, 1985, p. 6). As a consequence, the entire domain of the science of cognition became narrowly defined as 'the study of intelligence and intelligent systems, with particular reference to intelligent behaviour as computation' (Simon and Kaplan, 1989, p. 1). This reductionist characterization became synonymous with 'Cognitive Science'. Cognitivism had not only *defined* Cognitive Science, but also it had *prescribed* how to conceive and carry out any true science of cognition.

The Cognitive Revolution was launched with great optimism amid predictions of devices for general problem solving, speech recognition, language translation, reading cursive script and photographs, and making free-roving household robots. (For details about these and other predictions and how they failed, see Dreyfus, 1992, and Winograd and Flores, 1986). This seductive way of looking at the mind engendered new journals, departments, and academic societies to exploit the new concepts and their related technology. As a result, the mind as computer metaphor provided not only a redefinition of fundamental concepts such as reasoning, thought, perception, knowledge, and learning, but also what later became an entire conceptual apparatus for how to understand people, neurons, languages, brains, and experiences in classrooms, in terms of information-processing: algorithms, subroutines, formal logic modules, content-addressable memories for storage, pattern completion by retrieval, and so on. The computer-oriented vision has enthralled (in the sense of 'enslaved') in various ways the minds of two generations of scientists. New approaches were parallel distributed processing, artificial neural networks and other forms of connectionism, hybrid systems, and more recently, neurocomputation. Although Cognitivism in the form of 'strong AI' is not as widely endorsed as it once was (Searle, 1992), its legacy is pervasive. What we observe today are approaches to understanding cognition and the mind that retain a substantial dose of residual Cognitivism hidden behind the attractive possibilities of the ever-developing computer technology. Together they constitute what we here call *Neo-Cognitivism*, comprising more subtle and updated forms of the doctrine which carry the heavy and reductionist legacy of the Cognitive Revolution and its seventeenth-century forebears.

Reclaiming What Is Primary

In this volume, we collect papers from a variety of disciplines with the goal of calling to attention the urgent need to reconsider the study of the mind. The aim is to bring to the fore some issues down-played under the legacy of Cognitivism, and which today are sustained by Neo-Cognitivism. The authors reinstate some inherent aspects of the mind: action in the world, intention, emotion, bodily grounded experiences, the human brain as an organ of social action, and so on. Because of the variety of the backgrounds of the contributors, the collection of papers included here is heterogeneous. The common thread is a determination to take full advantage of the social and biological contexts of the human animal, and to develop conceptual frameworks and methodologies freed from the assumptions of Cognitivism as outlined above.

Even under the dominance of Cognitivism a plurality of alternatives has continued to flourish, like mammals under the dinosaurs, and these provide the roots of many of the essays in this volume. Some authors have followed the route of Gestalt psychologists (for example, Koffka, 1935; Köhler, 1940) and their successor J.J. Gibson (1979) in proposing that internal representations are not static, localized entities, but are fluid, distributed structures in brains that are derived by 'in-forming'. That is, their contents are derived by actions of the body as the agent of the mind, which extract information from objects in the world through their relations to the agent in the form of affordances. These properties belonging to objects afford an agent the opportunities to use them to achieve its individual goals, through which the agent establishes relations with those facets of the world with which it can interact and from which it can learn. By virtue of these relations, the mind is not restricted to the brain or body but extends into the world (Clark, 1997), and the mind is a seamless fabric of inner and outer experience (Jarvilehto, 1998), perhaps having an ancestral affinity to the Averroist view. Others hold that the concept of internal representation is a mis-leading fiction, or perhaps even an outright category mistake (Ryle, 1949; Freeman, 1997; Núñez *et al.*, 1999). They follow the lead of Pragmatists such as John Dewey (1914), who conceived of action 'into the stimulus', not reaction and representation, and phenomenologists such as Merleau-Ponty (1945), who conceived of 'the inten-tional arc' as the means for the mind to exert 'maximum grip' on the world without need for representation. Yet others follow the lead of Aquinas, who distinguished between the infinite complexity of the material world and the inner world of imagina-tion, abstraction and generalization (Copleston, 1955; Freeman, 1999). He dismissed Plato and his concept of ideal forms, thereby avoiding what would become explicit as the Cartesian subject–object apposition, because he denied that minds could maintain images (representations) of material objects. The mind consisted of knowledge in action (Aquinas, 1272, part I, question 85, article 2, p. 454; Basti and Perrone, 1993) which it expressed into the world through its agent, the body, by shaping energy and matter for purposes of grasping the nature of the world, of other intentional beings, and of God, processes that we now call intention and social communication. These patterns of energy and material objects, which we call external representations, do not exist in brains, either as forms of neurons or patterns of neural activity. They exist only in the world outside mind. Aquinas' doctrine had contributed to the growth of the middle class, and became a parent of the Scientific Revolution in the Renaissance, and it, or rather its late and somewhat decadent residue, was the system from which Descartes took his departure: discarding intentionality and the phantasms of the imagination, re-introducing Platonic forms and the certitudes of mathematics, and treating the soul as the pilot of the corporeal boat, instead of the integrated function of the body engaged by intention with the world. When seeking the origins and resolu-tions of the nagging deficiencies in 'Cartesian Theater' (Dennett, 1991), there is no better place to find perspective for a fresh start than to read his 'Treatise on Man' (part I, questions 75–102) and digest the system of thought that had prevailed before Des-cartes (Barfield, 1965, pp. 85–91).

Developmental psychologists, drawing on work such as that of Myrtle McGraw and John Dewey (Dalton, 1999) in North America and Jean Piaget (1930) in Europe, and in opposition to Cognitivism, particularly Chomskian linguistics (see Piattelli-Palmarini, 1980), have continued to incorporate the biological bases of human and animal cognition and the ways in which they emerged through phylogen-etic evolution (Herrick, 1926; Maturana and Varela, 1987). Writers here follow the lead of non-linear dynamicists (Haken, 1983) in describing how brains largely direct and control their own ontogenetic evolution (Ashby, 1960; Walter, 1963; Clark, 1997; Hendriks-Jansen, 1996). These processes invoke circular causality by which bodily actions under brain control modify the brain through sensation, perception and learning, thereby incorporating the world through experience, not through information in the form of external symbols and internal representations. The somatomotor dialectic unfolds throughout infancy and childhood (Piaget 1930; 1967; Thelen and Smith, 1994), as subjects learn to control their bodies and direct them to the achievement of goals that emerge through internal creative dynamics (Freeman, 2000), and are not imposed by observers as ideal targets, that are to be approached by reduction of mean squared errors of the differences between sets of *a priori* and *a posteriori* numbers. Other than the acquisition of skills in music, language and abstract reasoning, similar ontogenetic processes occur in other mammals and possibly all animals. More generally, the array of species in the animal kingdom offers an incredibly rich reservoir of examples of cognition that is almost entirely inaccessible to Cognitivists and Neo-Cognitivists. Moreover, even after language emerges in humans, the study of colourful constructions that seem to defy logic, yet have compelling utility in social communication, exposes to view the rich alternatives to traditional logic that are characteristic of human mentation (Piattelli-Palmarini, 1980; Rosch, 1978; Lakoff, 1987; Johnson, 1987; McNeill, 1992). Some of these impressive mechanisms of everyday human cognition are analysed in this volume through the study of spontaneous gestures, conceptual systems, unconscious and effortless inference-making, metaphorical thinking, speech–gesture coordination, and natural language understanding. The arguments propose new forms of understanding human semantics and the nature of concepts, revealing the primary role played by bodily grounded experiences in making meaning and abstraction possible. These studies also show that meaning and concepts are socially and historically mediated, but unlike what many post-modern philosophers suggest, they are not the result of arbitrary social conventions. They are indeed realized through non-arbitrary, species-specific, bodily grounded experiences that are at the basis of consensual spaces and inter-subjectivity (Núñez, 1997). These results show the fundamental and intimate co-definition of minds and bodies, providing thus fruitful alternatives to the restrictions imposed by linear causality, the subject–object dichotomy, and the mind–body split.

Cognitivists have been known to deride researchers who investigate the carbon-based platform that evolution used for the creation of intelligence as 'hydrocarbon chauvinists' and their position as 'neuromachismo', for example using the phrase, 'you don't need feathers in order to fly'. They speculate about zombies — bodies without minds — and disembodied minds — brain-in-a-vat, brain transplants, humanoids, and cyberspace intellects, all of which for neurobiologists are science fiction of a rather low class — on the dualist and functionalist premise that minds are independent of the matter in which they are deployed. Considering that these views are anti-biological, the manner in which neurobiologists have been subverted by Cognitivists is most extraordinary. Fifty years ago, the excitabilities of cortical neurons were described in terms of their receptor fields. After the take-over of neurobiology by neuroengineers, the same neurons became 'feature detectors' (Lettvin *et al.*, 1959), whose activities were summed as a proposed solution to the 'binding problem' that this move created *de novo*. This change in terminology based in a recurring category error (Freeman, 1997) that has had the far-reaching effect of treating neurons as vectors of information that represent to other parts of brains the objects whose reflected or emitted energies are statistically linked to their firing rates (for example, Barlow, 1972; Abeles, 1991). Contrary to widespread beliefs among computer scientists and Cognitivists, action potentials are not binary digits, and neurons do not perform Boolean algebra. Their impulses are electrochemical waves serving as carriers for pulse frequency modulation, by which analogue quantities are transmitted across immense distances without the attenuation and delay imposed by diffusion (Bloom and Lazerson, 1988; Freeman, 1992; Kandel *et al.*, 1993).

The distinction here lies between using mathematics to describe brain function and postulating that brain function is computational. As some of our contributors argue, mathematics (and its formal systems) is part of human conceptual systems, so that numbers (and mathematical formulae) exist in the minds of observers, and not as representations in the brains being observed (Longo, 1998; Lakoff and Núñez, 2000). The functions of a brain are recorded by measurements either of its electrochemical activities or of the bodily behaviours that by intention represent its internal states. Our essayists show that the numbers and other symbols resulting from the measurements can be modelled by mathematics to provide compelling insights into brain dynamics, as well as by natural languages and the use of metaphor. But this doesn't mean that the language of the brain is mathematics or logic, or that the theories are prescriptions for algorithms by which to construct autonomous robots or even less complex devices that can simulate or emulate the observed behaviours. For example, one may describe the operation of the lens of the eye by using the Fourier transform without proposing that the lens computes it. Newer forms of NeuroCognitivism have stepwise moved the computational assumptions from an overt CPU in one or another thalamocortical circuit (Baars, 1997; Wright and Liley, 1996; Taylor, 1997) towards finer structural levels such as neurons equivalent to DSPs and chips, sub-neuronal components (Alkon, 1992) like microsomes and RNA (Hydén, 1973) as the repository of memories, and microtubules and subsynaptic webs (Hameroff, 1987) conceived as quantum computing devices, for which neurons serve as amplifiers (Penrose, 1994). As our essayists show, these speculations are far from reality.

The contributions to this volume include the recognition and exploitation of the obvious phylogenetic fact that human brains have evolved primarily as organs of social organization (Freeman, 1995; Núñez, 1997). They include descriptions of the higher levels of brain function that are enabled by the tools of non-linear dynamics and theories of chaos and complexity. They invite reconsideration of the primacy of everyday unconscious cognition. They invoke a resurrection of the forgotten foundation of the science, industry, medicine and law of the latter half of the Middle Ages in the philosophy of Thomas Aquinas. These plural approaches simply bypass the old traps of the mind–body split, the artificial subject–object dichotomy, and the reductionist, static, algorithmic, and emotion-free views of cognition. They avoid the Platonic view that in order to study cognition and objective knowledge, one has to isolate reason from the other mental faculties, and, as does Damasio (1994), they treat emotional and intentional aspects of cognitive psychology as central rather than irrelevant and disruptive or at best secondary to rational discourse. They incorporate the cultural and social dimensions of minds (Vygotsky, 1986). Even a simple act of observation with its consequences is a prototype for all intentional behaviour. Here we have a fresh insight into the story of observation in quantum physics as a circumscribed yet highly instructive microcosm of the relation of the observer to the world. And this is not only in relation to the material systems being acted upon but equally to other observers and the social interactions that are required for validation, interpretation, and acceptance or rejection of the meanings that emerge, constituting the traffic of minds. Our authors transform familiar scenes into gardens of delight, emphasizing the social dimensions of mind, for example, those relating to gender and psychosocial development, thereby again dispelling the notions that cognition is a purely rational, abstract activity, and that mind is an ahistorical, culture-free, and disembodied logical system (Varela *et al.*, 1991). The various contributions show that one need not be a monist, dualist or functionalist to practice the science of the mind.

Our aim with this collection of essays is not to answer questions about the nature of mind and cognition, but to reopen for consideration some avenues of exploration that have been shouldered aside by classic and Neo-Cognitivism, and, rather than abandon the name of cognitive science as if it were a bone to a dog, to reassert its pertinence in a wide range of behavioural sciences. We humanists and scientists should not allow ourselves to be bullied and dispossessed by analysts and technocrats, who create intractable problems that may often be claimed to be fundamental, if no solutions can be found. Perhaps we can rescue them from the morass of their own making, but perhaps we have other more interesting things to do.

References

Abeles, M. (1991), *Corticonics: Neural Circuits of the Cerebral Cortex* (Cambridge: Cambridge University Press).

Alkon, D. (1992), *Memory's Voice. Deciphering the Mind–Brain Code* (New York: Harper Collins).

Aquinas, St. Thomas (1272), *Treatise on Man* in Summa Theologica. Translated by Fathers of the English Dominican Province. Revised by Daniel J Sullivan. Published by William Benton as Volume 19 in the Great Books Series (Chicago: Encyclopaedia Britannica Inc., 1952), Questions 75–102.

Ashby, W.R. (1952), *Design for a Brain* (New York: Wiley).

Baars, B.J. (1997), *In the Theater of Consciousness: The Workspace of the Mind* (New York: Oxford University Press).

Barfield, O. (1965), *Saving the Appearances: A Study in Idolatry* (New York: Harcourt Brace).

Barlow, H.B. (1972), 'Single units and sensation: A neuron doctrine for perceptual psychology?', *Perception*, 1, pp. 371–94.

Basti, G. and Perrone, A. (1993), 'Time and non-locality: From logical to metaphysical being. An Aristotelian-Thomistic approach', in *Studies in Science and Theology: Part 1. Origins, Time and Complexity*, ed. G.V. Coyne and K. Schmitz-Moorman (Rome: Labor et Fides).

Berkeley, E.C. (1949), *Giant brains; or, Machines that Think* (New York: Wiley).

Block, N. (1980), 'What is functionalism?', in *Readings in Philosophy of Psychology*, ed. N. Block (Cambridge MA: Harvard).

Bloom, F.E. and Lazerson, A. (1988), *Brain, Mind, and Behavior* (2nd ed.) (New York: Freeman).

Brentano, F.C. (1889), *The Origin of our Knowledge of Right and Wrong*, tr. R.M. Chisolm and E.H. Schneewind (New York: Humanities Press, 1969).

Clark, A. (1997), *Being There. Putting Brain, Body, and World Together Again* (Cambridge MA: MIT Press).

Clarke, E. and O'Malley, C.D. (1968), *The Human Brain and Spinal Cord: A Historical Study Illustrated by Writings From Antiquity to the 20th Century* (Los Angeles: University of California Press).

Copleston, F.C. (1955), *Aquinas* (Harmondsworth: Penguin).

Craik, K. (1943), *The Nature of Explanation* (Cambridge: Cambridge University Press).

Dalton, T. (1999), 'The ontogeny of consciousness: John Dewey and Myrtle McGraw's contribution to a science of mind', *Journal of Consciousness Studies*, 6 (10), pp. 3–26.

Damasio, A.R. (1994), *Descartes' Error: Emotion, Reason, and the Human Brain* (New York: Grosset/Putnam).

Dennett, D.C. (1991), *Consciousness Explained* (Boston MA: Little, Brown and Co.).

Dewey, J. (1914), 'Psychological doctrine in philosophical teaching', *Journal of Philosophy*, 11, pp. 505–12.

Dreyfus, H. (1979), *What Computers Can't Do: The Limits of Artificial Intelligence* (New York: Harper Colophon).

Durkheim, E. (1915), *The Elementary Forms of the Religious Life: A Study in Religious Sociology*, tr. J.W. Swain (New York: Macmillan, 1926).

Freeman, W.J. (1991), 'The physiology of perception', *Scientific American*, 264, pp. 78–85.

Freeman, W.J. (1992), 'Tutorial in neurobiology', *International Journal of Bifurcation and Chaos*, 2, pp. 451–82.

Freeman, W.J. (1995), *Societies of Brains* (Mahwah NJ: Lawrence Erlbaum Associates).

Freeman, W.J. (1997), 'Three centuries of category errors in studies of the neural basis of consciousness and intentionality', *Neural Networks*, 10, pp. 1175–83.

Freeman, W.J. (1999), *How Brains Make Up Their Minds* (London: Weidenfeld and Nicolson).

Freeman, W.J. (2000), *Neurodynamics: An Exploration of Mesoscopic Brain Dynamics* (London: Springer-Verlag).

Gardner, H. (1985), *The Mind's New Science* (New York: Basic Books).

Gibson, J.J. (1979), *The Ecological Approach to Visual Perception* (Boston: Houghton Mifflin).

Griffin, D. (1992), *Animal Minds* (Chicago IL: University of Chicago Press).

Gödel, K. (1930/1962), *On Formally Undecidable Propositions of Principia Mathematica and Related Systems*, tr. B. Meltzer (New York: Basic Books).

Haken, H. (1983), *Synergetics: An Introduction* (Berlin: Springer).

Hameroff, S.R. (1987), *Ultimate Computing: Biomolecular Consciousness and Nanotechnology* (Amsterdam: North-Holland).

Hendriks-Jansen, H. (1996), *Catching Ourselves in the Act: Situated Activity, Interactive Emergence, Evolution, and Human Thought* (Cambridge, MA: MIT Press).

Herrick, C.J. (1926), *Brains of Rats and Men* (Chicago IL: University of Chicago Press).

Hydén, H. (1973), 'Macromolecules and behavior', Lectures and proceedings of the international symposium held at the University of Birmingham Medical School. (Baltimore MD: University Park Press).

Imry, Y. (1997), *Introduction to Mesoscopic Physics* (New York: Oxford University Press).

Ingber, L. (1992), 'Generic mesoscopic neural networks based on statistical mechanics of Neocortical interactions', *Physical Review A*, 45, pp. R2183–6.

Jarvilehto, T. (1998), 'Efferent influences on receptors in knowledge formation', *Psycholoquy* 98.9.41

Johnson, M. (1987), *The Body in the Mind: The Bodily Basis of Meaning, Imagination and Reason* (Chicago: University of Chicago Press).

Kandel, E.R., Schwartz, J.H. and Jessel, T.M. (1993), *Principles of Neural Science*, 3rd ed. (New York: Elsevier).

Kant, I. (1781), *Kritik der reinen Vernunft*, ed. W. von Weischedel (Frankfurt am Main: Suhrkamp Verlag, 1974).

Kenny, A.J.P. (1980), *Aquinas* (New York: Hill and Wang).

Kenny, A. (1984), 'Aquinas: Intentionality', in *Philosophy Through Its Past*, ed. T. Honderich (London: Penguin).

Koffka, K. (1935), *Principles of Gestalt Psychology* (New York: Harcourt Brace).

Köhler, W. (1940), *Dynamics in Psychology* (New York: Grove Press).

Lakoff, G. (1987), *Women, Fire, and Dangerous Things: What Categories Reveal about the Mind* (Chicago IL: University of Chicago Press).

Lakoff, G. and Johnson, M. (1999), *Philosophy in the Flesh* (New York: Basic Books).

Lakoff, G. and Núñez, R. (1997), 'The metaphorical structure of mathematics: sketching out cognitive foundations for a mind-based mathematics', in *Mathematical Reasoning: Analogies, Metaphors, and Images*, ed. L. English (Hillsdale, NJ: Erlbaum).

Lakoff, G. and Núñez, R. (2000), *Where Mathematics Comes From: How the Embodied Mind Creates Mathematics* (New York: Basic Books).

Lettvin, J.Y., Maturana, H.R., McCulloch, W.S. and Pitts, W.H. (1959), 'What the frog's eye tells the frog's brain', *Proceedings of the Institute of Radio Engineers*, 47, pp. 1940–51.

Longo, G. (1998), 'The mathematical continuum: From intuition to logic', in *Naturalizing phenomenology: Issues in contemporary phenomenology and cognitive science*, ed. J. Petitot *et al.* (Stanford: Stanford University Press).

Maturana, H. and Varela, F. (1987), *The Tree of Knowledge: The Biological Roots of Human Understanding* (Boston: New Science Library).

Mann, F. (1962), *Acupuncture; the Ancient Chinese Art of Healing*, foreword by Aldous Huxley (London: W. Heinemann Medical Books).

McNeill, D. (1992), *Hand and Mind: What Gestures Reveal About Thought* (Chicago: University Press).

Merleau-Ponty, M. (1945/1962), *Phenomenology of Perception*, translated by C. Smith (New York: Humanities Press).

Núñez, R. (1995), 'What brain for God's-eye? Biological naturalism, ontological objectivism, and Searle', *Journal of Consciousness Studies*, **2** (2), pp. 149–66.

Núñez, R. (1997), 'Eating soup with chopsticks: Dogmas, difficulties, and alternatives in the study of conscious experience', *Journal of Consciousness Studies*, **4** (2), pp. 143–66.

Núñez, R., Edwards, L. and Matos, J.F. (1999), 'Embodied cognition as grounding for situatedness and context in mathematics education', *Educational Studies in Mathematics*, **39** (1–3), pp. 45–65.

Núñez, R. and Lakoff, G. (1998), 'What did Weierstrass really define? The cognitive structure of natural and ε–δ continuity', *Mathematical Cognition*, **4** (2), pp. 85–101.

Penrose, R. (1994), *Shadows of the Mind* (Oxford: Oxford University Press).

Piaget, J. (1930), *The Child's Conception of Physical Causality* (New York: Harcourt, Brace).

Piaget, J. (1967), *Biologie et Connaissance* (Paris: Gallimard).

Piattelli-Palmarini, M. (Ed.) (1980) *Language and Learning: The Debate between Jean Piaget and Noam Chomsky* (Cambridge MA: Harvard University Press).

Prochaska, G. (1784/1851), *Principles of Physiology*, tr. T. Laycock (London, Sydenham Society) from *Adnotationum academicarum* (Prague: Wolfgang Gerle).

Putnam, H. (1990), *Realism With a Human Face* (Cambridge MA: Harvard University Press).

Rosch, E. (1978), 'Principles of categorization', in *Cognition and Categorization*, ed .E. Rosch and B. B. Lloyd (Hillsdale, NJ: Erlbaum).

Ryle, G. (1949), *The Concept of Mind* (New York: Barnes and Noble).

Searle, J.R. (1983), *Intentionality, An Essay in the Philosophy of Mind* (New York: Cambridge University Press).

Searle, J.R. (1992), *The Rediscovery of Mind* (Cambridge MA: MIT Press).

Simon, H. and Kaplan, C. (1989), 'Foundations of cognitive science', in *Foundations of Cognitive Science*, ed. M. Posner (Cambridge, MA: MIT Press).

Sobel, D. (1999), *Galileo's Daughter* (New York: Walker).

Taylor, J.G. (1997), 'Neural networks for consciousness', *Neural Networks*, **10**, pp. 1207–25.

Thelen, E, and Smith, L.B. (1994), *A Dynamic Systems Approach to the Development of Cognition and Action* (Cambridge MA: MIT Press).

Turing, A.M. (1950), 'Computing machines and intelligence', *Mind*, **59**, pp. 459–76.

Varela, F. (1989), *Connaître les Sciences Cognitives: Tendances et Perspectives* (Paris: Seuil).

Varela, F., Thompson, E. and Rosch, E. (1991), *The Embodied Mind: Cognitive Science and Human Experience* (Cambridge, MA: MIT Press).

von Neumann, J. (1958), *The Computer and the Brain* (New Haven CT: Yale University Press).

Vygotski, L. (1986), *Thought and Language* (Cambridge, MA: MIT Press).

Walter, W.G. (1963), *The Living Brain* (New York: Norton).

Whitehead, A.N. and Russell, B. (1910), *Principia Mathematica*, paperback edition (Cambridge: Cambridge University Press, 1967).

Winograd, T. and Flores, F. (1986), *Understanding Computers and Cognition: A New Foundation for Design* (New Jersey: Ablex).

Wittgenstein, L. (1922), *Tractatus Logico-Philosophicus*, introduction by Bertrand Russell (New York: Harcourt, Brace).

Wright, J.J. and Liley, D.T.J. (1996), 'Dynamics of the brain at global and microscopic scales: Neural networks and the EEG', *Behavioral and Brain Sciences*, **19**, pp. 285–95.

Andy Clark

Visual Awareness and Visuomotor Action

Recent work in 'embodied, embedded' cognitive science links mental contents to large-scale distributed effects: dynamic patterns implicating elements of (what are traditionally seen as) sensing, reasoning and acting. Central to this approach is an idea of biological cognition as profoundly 'action-oriented' — geared not to the creation of rich, passive inner models of the world, but to the cheap and efficient production of real-world action in real-world context. A case in point is Hurley's (1998) account of the profound role of motor output in fixing the contents of conscious visual awareness — an account that also emphasizes distributed vehicles and long-range dynamical loops. Such stories can seem dramatically opposed to accounts, such as Milner and Goodale (1995), that stress relatively local mechanisms and that posit firm divisions between processes of visual awareness and of visuomotor action. But such accounts, I argue, can be deeply complementary and together illustrate an important lesson. The lesson is that cognition may be embodied and action-oriented in two distinct — but complementary — ways. There is a way of being embodied and action-oriented that implies being closely geared to the fine-grained control of low level effectors (hands, arms, legs and so on). And there is a way of being embodied and action-oriented that implies being closely geared to gross motor intentions, current goals, and schematic motor plans. Human cognition, I suggest, is embodied and action-oriented in both these ways. But the neural systems involved, and the size and scope of the key dynamic loops, may be quite different in each case.

I: **Local** *versus* **Highly Interactive Explanations**

Science in general, and cognitive science in particular, seems currently torn between two superficially competing kinds of explanation.[1] The struggle is between explanations that highlight distributed complexity and large-scale, non-linear, interactive effects and those that highlight smaller circuits; between — to invoke an old but persistent dichotomy — holism and reductionism. In the next section, I'll show how this

---

[1] For a nice — and also ultimately reconciliatory — account, see Bechtel and Richardson (1992), as well as Fontana and Ballati (1999).

Journal of Consciousness Studies, **6**, No. 11–12, 1999, pp. 1–18

general tension plays out in the specific arena of (accounts of) visual awareness. But it is useful, I think, to begin with a rough sketch of the larger issues.

Thus consider the most general claim of what has come to be known as 'embodied, embedded cognitive science'.[2] The claim is that the essence of adaptive intelligence often lies in complex, non-linear processes that span multiple inner systems and include, as active contributors, aspects of body, action and world. A frequent corollary of this claim is that there are no neat dividing lines between perception, cognition and action; that much of what looks like action plays fundamental cognitive and computational roles; and that perceptual processing phases gradually into cognition and abstract reason. To get an explanatory grip on such world — and action — exploiting systems, it has been argued,[3] we must attend heavily to extended feedback and feedforward loops whose (often continuous) activity underpins adaptive intelligence.

Such an expanded focus does indeed seem helpful when confronting several aspects of biological intelligence. I shall give one example, and merely gesture at several others. The example concerns the production of rhythmic motor actions, such as walking, rowing, certain kinds of coordinated finger motion, and even human speech. There is now a large and compelling literature[4] that shows how such activity arises from the complex interactions of neural resources, bodily bio-mechanics and (sometimes) external environmental structure. Such accounts reject the once-popular (e.g. Wallace, 1981; Gibbon et al., 1984) view of 'centralized control' or 'central pattern-generation', which 'assumes a central representation of the movement, including its form, amplitude and temporal characteristics, that is imposed on the periphery' (Hatsopoulos and Warren, 1996, p. 3). Instead of seeing rhythmic motor action as the muscular expression of an inner rhythmic command, theorists such as Kelso (1995), Thelen and Smith (1994), Turvey and Kugler (1987) and, more impressionistically, Bernstein (1967), see such actions as the tuned product of neural–bodily interactions. Such accounts need not — and often do not — deny the existence of something *like* a central pattern generator. But they deny that such a resource *determines* the rhythmic notions of the embodied agent:

> Any central timing process, such as a neural oscillator, is not acting as an extrinsic time-keeper to drive the peripheral segments but must be reciprocally modulated by information about the dynamics of the periphery. Preferred motor timing emerges from the interaction of central and peripheral components, rather than from being dictated by either (Hatsopoulos and Warren, 1996, p. 10).

The basic mechanism proposed by Hatsopoulos and others involves the use of proprioceptive feedback (from a musculoskeletal system with its own intrinsic, spring-like dynamics) to tune the neural resources to the resonant frequency determined by the peripheral biomechanics. This tuning, which can be experimentally demonstrated (Hatsopoulos, 1996; Hatsopoulos and Warren, 1996), allows the

[2] The focus of embodiment, in the recent tradition, probably began with Dreyfus (1979), Varela et al. (1991) and Johnson (1987). Historical precedents clearly include Merleau-Ponty (1942), Heidegger (1961/1927) and, to some extent, Gibson (1979). The specific phrase 'embodied, embedded' is due to Haugeland (1998).

[3] Haugeland (1998); Merleau-Ponty (1942); Hatsopoulos (1996); Chiel and Beer (1997); Beer (1995); Hurley (1998) — to name but a few.

[4] For a sampling, see Turvey and Kugler (1987); Kelso (1995); Thelen and Smith (1994); Port et al. (1995); Hatsopoulos and Warren (1996). Historical antecedents include Bernstein (1967).

system to cope fluently with physical changes (limb growth, muscle growth, etc.) and to find 'least-energy' solutions to the problem of generating rhythmic actions. Moreover, the very same mechanisms that allow a biological agent to thus couple its neural resources to properties of the bodily periphery also allow for coupling to the wider environment. Experiments in which a subject performs a task using a hand-held pendulum (Turvey and Kugler, 1987) show a selected frequency of oscillation that becomes matched to the resonant frequency of the whole wrist-pendulum system. The same effect will occur in a good golfer or racket sport player, or when rocking a car to get it out of the snow (Hatsopoulos and Warren, 1996, p. 12). In all these cases, the proprioceptive information couples the neural system to bodily and/or environmental resources in a way that creates 'a larger autonomous dynamical system' (Hatsopoulos and Warren, 1996, p. 12). This effect is even reflected, it now seems, in the receptive field size of cells involved in somatosensory and visual processing. Iwamura (1998) cites Iriki *et al.* (1996) as showing that after a monkey repeatedly used a tool (a rake) for the retrieval of food 'the visual RF[5] [of some cells in the arm/hand region of monkey postcentral gyrus] became elongated along the axis of the tool, as if the image of the tool was incorporated into that of the hand' (Iwamura, 1998, p. 525).

Other examples of an embodied, environmentally embedded perspective include Beer's (1995) work on leg control in a simple robot 'insect', Webb's (1994) work on phonotaxis (sound-dependent tracking) in both real and robot crickets, and Thelen and Smith's (1994) work on reaching and stepping in human infants. Chiel and Beer (1997) offer a useful review of the literature on the importance of brain/body interactions, concluding that 'adaptive behaviour is the result of the continuous interaction between the nervous system, the body and the environment' and that 'one cannot assign credit for adaptive behavior to any one piece of this coupled system' (p. 555).

It can often seem, however, as if the embodied approach really pays dividends only for relatively low-level, motoric aspects of adaptive behaviour. And without a doubt, the most radical versions of the embodied approach do face special challenges as we ascend to the levels of reasoned thought, imagination and off-line planning and rehearsal (for a critical but sympathetic account, see Clark, 1999). It seems clear, however, that the broad idea of neural systems becoming deeply geared to the exploitable presence of bodily or external structure is highly applicable to certain aspects of 'advanced reason' — we can see the biological brain as coupled, via the body, to all kinds of technological and artifactual resources (pens, paper, PCs), such that (to paraphrase Chiel and Beer) 'one cannot assign credit for intellectual success to any one piece of this complex system'.[6]

In many cases, however, there is only a surface opposition between such a distributed, interactive perspective and more traditional (localist, internalist) concerns. There may, for example, be a perfectly good localist story to tell about how certain aspects of individual brain function provide the pattern-completing substrate necessary for the larger person-plus-pen-and-paper system to then function as an integrated, extended numerical computation device.

---

[5] Receptive Field.

[6] For such arguments, see Hutchins (1995); Clark (1995); Clark and Chalmers, (1998); Dennett (1995), chapters 11 and 12.

To set up a genuine *opposition* between the two visions, certain conditions need to be met. First, we need to be convinced that we really confront two different stories about the *same phenomenon* (a condition I'll call 'same target'). Second, we need to be convinced that the interactive story, if it is to be preferred, is also in some important but hard-to-pin-down sense *deep* (a condition I'll call 'deep embodiment').

In the rhythmic motion case just described, these conditions seem to be met. The target phenomenon was rhythmic motion generation. One (localist, internalist) story explained this as the effect of a central pattern generator whose own frequency of oscillation determined a temporally matched oscillation in, say, a limb. The other (interactionist) story denied the presence of such a closely matched inner resource, and explained the very same target motion via a complex process of reciprocal modulation in which the inner resources sought a kind of energy efficient compromise with the intrinsic dynamics of the limb and musculo-skeletal system. The 'embodied' account is here a genuine 'same target' competitor. But is it deep? By this I mean, does the stress on interaction illuminate the phenomenon in a truly revealing way? Consider a scenario in which interaction might matter yet (intuitively) not in a deep way. Imagine a central pattern generator with a small number of pre-determined settings (frequencies of oscillation), and suppose, in addition, that certain instances of proprioceptive feedback determine which of these (three or four) frequencies is generated. In such a case, interaction and feedback *matters*. But it matters only as a kind of input to a central system whose operation still fixes the target behaviour. We might thus respond by, in essence, re-parsing the original target into two semi-independent components: a peripheral, dynamic loop, and a central system which, once informed, does the real work. Contrast this with the actual story, in which the target behaviour was revealed as a genuine product of complexly interacting dynamics and in which the detailed biomechanics of the periphery continuously and sensitively influences the central resource so as to yield a signal whose effect — in biomechanical context — is to match the preferred frequency of motion to the resonant frequency of whatever peripheral system (arm, arm-plus-golf club, etc.) is in play. When the interactions are this important, continuous, and subtle, there is — or so I claim — a very real sense in which no re-parsing can yield an internalist story which still *keeps sight* of the target phenomenon. There is, of course, an inner story to tell. But the inner story now fails to account for the subtlety, power and efficiency of the target phenomenon — a task that necessitates a more extended and interactive perspective.

Hopefully, then, we now have at least a rough sense of the kind of case in which an embodied, embedded story is a *genuine* and *deep* competitor to some localist/internalist alternative. On, then, to our target case: perceptual (and especially visual) awareness.

II: Two Takes on Visual Awareness: The Ventral Stream *versus* Escher Spaghetti

The recent debate concerning visual awareness looks superficially similar to the one concerning rhythmic motion. Here, too, we find both localist accounts and more 'embodied' alternatives that stress larger-scale dynamic loops. A compelling example of a recognizably localist account is Milner and Goodale's (1995; 1998) account of the ventral stream correlates (see below) of visual awareness. A good example of a

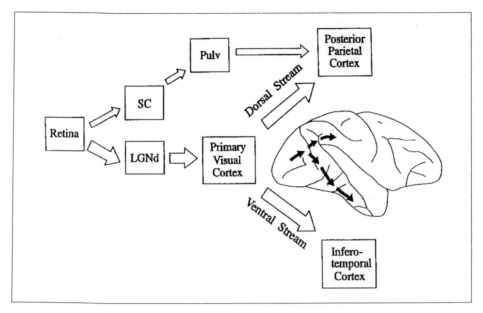

Figure 1

Major routes whereby retinal input reaches the dorsal and ventral streams. The diagram of the macaque brain (right hemisphere) on the right of the figure shows the approximate routes of the cortico-cortical projections from the primary visual cortex to the posterior parietal and the inferotemporal cortex respectively. LGNd, lateral geniculate nucleus, pars dorsalis; Pulv, pulvinar; SC, superior colliculus. (From Milner and Goodale, 1995, by permission of the authors and of Oxford University Press.)

dynamic, interactionist account is Susan Hurley's (1998) depiction of conscious visual experience as dependent on a complex web of relations between perception and action. Brief sketches seem in order.

Milner and Goodale (1995) suggest that visual awareness depends on the activity of selected parts of the visual processing system. By 'visual awareness' they mean the capacity to know and recognize objects by means of conscious, visual experience: a capacity that is taken to imply — in a normally-functioning language user — the ability to report that a visually presented object is a such-and-such, that is spatially oriented thus-and-so, and so on.

The specific claim is that these capacities of visual awareness (which they sometimes call capacities of visual *perception* and contrast with capacities for visually-guided *action*) depend on a specific visual processing stream — the ventral stream — that is said to operate semi-independently of the processing stream (the dorsal stream) that guides fine-tuned motor action in the here-and-now. The ventral stream structures that are thus claimed to support awareness include areas (V2, V3, V4, TEO) projecting to the inferior temporal cortex (IT), while the dorsal stream story implicates areas projecting to the posterior parietal (PP) cortex (see fig. 1). Regarding these two streams the claim is that:

> The visuomotor modules in the primate parietal lobe function quite independently from the occipitotemporal mechanisms generating perception-based knowledge [visual awareness] of the world (Milner and Goodale, 1998, section 4).

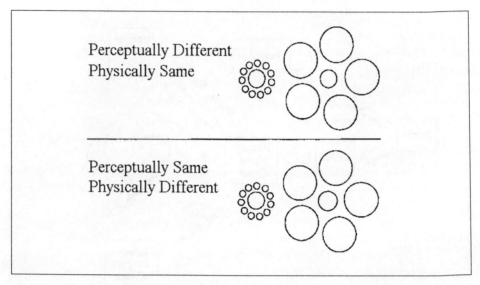

Figure 2

Diagram showing the 'Titchener circles' illusion. In the top figure the two central discs are of the same actual size, but appear different; in the bottom figure, the disc surrounded by an annulus of large circles has been made somewhat larger in size in order to appear approximately equal in size to the other central disc. (From Milner and Goodale, 1995, by permission of the authors and of Oxford University Press).

Evidence for this deep dissociation comes in three main varieties: deficit data concerning patients with damage to areas in either the dorsal or ventral streams; performance data from normal human subjects; and computational conjectures concerning the inability of a single encoding to efficiently support both visual form recognition and visuomotor action.

The relevant deficit data concerns two classes of patients: visual form agnosics and optic ataxics. The visual form agnosic DF (who suffered from carbon monoxide poisoning causing ventral occipital damage) is unable to judge (unable to report on, and reports no visual experience of) the shape and orientation of visually presented items. But she retains significant visuomotor skills, and is able to catch a ball or stick with correct hand orientation, to place her flattened hand through a visually presented 'letterbox' and so on (for a full account, see Milner and Goodale, 1995, ch. 5). Optic ataxics, by contrast, have damage to the dorsal stream and, although they are visually aware of (and can report on) shape and orientation, they are unable to fluently reach for and grasp the objects they so clearly see. As Gazzaniga (1998, p. 109) has it, 'it is as though they cannot use the spatial information inherent in any visual scene.'

Moving to unimpaired, normal agents, Milner and Goodale appeal to various experimental results displaying a conflict between verbal judgments, based on visual experience, and the visual knowledge that is manifest in non-verbal actions. A neat example concerns the 'Ebbinghaus' or 'Tichener Circles' illusion (see fig. 2), in which visual awareness (visual experience) delivers an illusory content which nonetheless fails to inform subsequent fine-tuned motor action. To show this, Aglioti *et al.* (1995) set up a physical version of the illusion, using poker ships as circles. Subjects were told to pick up a specific disc if they saw the two discs as differing in size. By choosing a specific disc, subjects showed they were susceptible to the illusion, since the discs were actually identical in size. Nonetheless, by using infra-red light emitting diodes attached to finger and thumb, the act of picking up the chosen disc was shown to be finely calibrated in advance of actually touching the disc, displaying a pre-formed precision (thumb and forefinger) grip closely tuned to the actual (non-illusory) size! The explanation, according to Milner and Goodale (1995, p. 168) is that this precision grip was calculated, using non-conscious visual information, by the dorsal stream, and only the conscious, ventral stream was 'fooled'.

The deep reason for such functional compartmentalization, Milner and Goodale conjecture, involves the very different computational demands of visuomotor guidance and object recognition. The former requires precise knowledge of spatial location and orientation, and must be constantly and egocentrically updated to reflect real-world motion and relative location. The latter requires us to identify something as *the same* thing, irrespective of motion and current spatial orientation, and demands only as much spatial sensitivity as is necessary to support conscious object recognition and reasoning. The dorsal stream is thus said to be specialized for fluent motor interaction while the ventral stream deals with enduring object properties and subserves explicit recognition and semantic recall. As a kind of corollary, the ventral stream must take over whenever the real-world object is not present-at-hand: actions in respect of imagined or recalled objects are under ventral stream control, and this is reflected in grosser kinematics of grasp and sometimes (in the case of ventrally damaged subjects) total failure to perform (see Milner and Goodale, 1995, pp. 136–8).

Milner and Goodale thus offer an account of the physical correlates of visual awareness which stresses the importance of a closely circumscribed subset of neural mechanisms (V2, V3, V4, TEO and IT), and excludes the occipito-parietal structures implicated in fine-grained motor control. The upshot, provocatively expressed, is that 'what we think we "see" is not what guides our actions' (Milner and Goodale, 1995, p. 177).

The Milner and Goodale account belongs squarely in the tradition[7] of betting on relatively local neural structures and features as the likely physical correlates of visual awareness. The general alternative, as we noted, is to bet on models that stress co-ordination and interaction.[8] As an example — and one which now moves us all the way into the terrain of complex interactive dynamics — consider Susan Hurley's (1998) account in which the contents of perceptual experience are linked (in a deep, 'non-instrumental' way — see below) to motor outputs, and in which there is heavy stress on 'multiple channels of motor feedback, some of which go through states of the environment' (p. 328).

---

[7] In this vein, Koch and Braun recently wrote:

> We believe that further progress in understanding visual, and indeed all forms of, awareness will only come from precisely locating the NCC [neural correlates of consciousness] to specific brain areas and local populations (Koch and Braun, 1998, p. 170).

[8] Farah (1997) draws exactly this distinction, contrasting accounts which depict consciousness as 'the privileged role of particular brain systems' with ones which stress 'states of integration between brain systems' (pp. 203–15). As examples of such integrative accounts Farah cites Kinsbourne (1998), Crick and Koch (1990), and Damasio (1990). We might add Edelman (1992) and even Dennett's (1993) account of consciousness as 'cerebral celebrity'.

To lay the groundwork for such an account, it is necessary to recognize an important idea, central to the literature on embodied cognition, and increasingly reflected in
mainstream neuroscientific conjecture. It is the idea that we should, in many cases,
resist the temptation to think in terms of a simple linear flow in which the senses
deliver input which is progressively processed and refined until an output (usually a
motor action) is selected and the process repeats. Hurley calls this the Input-Output
picture and argues that it should be rejected in favour of accounts which give due
weight to the looping and temporally continuous nature of many processes of information flow and control. For example, there is evidence that a great deal of activity in
'early' visual processing areas is modulated by activity flowing back down from
'higher' areas. Experimenters have found top-down context effects on the receptive
fields of cells in V1, enhanced responses of cells in V1 and V2 to locations to which
monkeys are about to saccade, and so on (see Van Essen and De Yoe, 1995; Knierim
and Van Essen, 1992; Wurtz and Mohler, 1976). Moreover, as Churchland *et al.*
(1994, p. 43) point out, 'all cortical areas, from the lowest to the highest, have numerous projections to lower brain centers, including motor-relevant areas such as the
striatum, superior colliculus, and cerebellum'. Such back-projections allow highly
processed states — decisions to move, top-level semantic information, etc. — to loop
back to affect the low-level processing of current inputs. Such descending connectivity lies at the heart of Edelman's (1987) account of 'reentrant processing' in which
such downward pathways establish correlations between multiple cortical and
sub-cortical areas. Patterns of mutual influence between distinct areas, set up by the
use of re-entrant connectivity, allow activity in one site to become correlated with
activity at others, and these correlated patterns of activity (according to Edelman —
see also Sporns *et al.*, 1989) are themselves the distributed vehicles that carry information about high-level features. Most high-level cognitive capacities, on this
account, depend on the correlated activity of multiple neural areas, including motor
areas. Similar themes surface in the 'convergence zone' hypothesis of Damasio and
Damasio (1994). Convergence zones are local brain areas whose task is not to directly
encode information but to enable correlated activity to occur in multiple neural
sub-systems. A convergence zone is thus an area in which several long-range
corticocortical connections (feedback and feedforward) converge, enabling signals
from that area to simultaneously influence multiple distant regions of neural tissue.
And once again, various types of high-level knowledge are said to depend on various
kinds of correlated activity, orchestrated by a kind of hierarchy of convergence zones
(see e.g. Damasio and Damasio, 1994, p. 73).

Many theorists, it thus seems, are stressing the looping dynamics of inner processing, and describing cognitive mechanisms in which 'information flows back as it
flows up, and it flows more or less continuously' (Hardcastle, 1998, p. 341).[9] The
physical *vehicles* of certain kinds of higher level content, on these accounts, are often
extended dynamic loops connecting 'higher' to 'lower' brain areas, and encompassing both 'cognitive' and 'motor' systems.[10] It is an image that I think of as Escher
Spaghetti — not just multiple criss-crossing strands (ordinary spaghetti), but strands

---

[9] See also Clark (1997) on 'Continuous reciprocal causation' and Jarvilehto (1998).

[10] The scare quotes are included because these distinctions, in such accounts, are frequently themselves
called into question.

whose ends feed back into their own (and others) beginnings, making 'input' and 'output', and 'early' and 'late' into imprecise and misleading visions of complex recurrent and reentrant dynamics.[11]

Hurley (1998) is, in large part, a sustained application of these kinds of ideas to issues concerning perceptual awareness and its relation to motor output. The key claim is that such perception can depend *non-instrumentally* on motor output. Perception depends merely *instrumentally* on output when, for example, you turn your head and see something new. The new perceptual content is (trivially) made available by your action. The cases Hurley is chasing, however, are ones in which what matters is not this pragmatic effect of the current output (action command) on the next input, but a more profound effect of the output signal *itself* on perceived content: 'When perceptual content depends *non-instrumentally* on output, it does not do so via input, but directly' (Hurley, 1998, p. 342). One mark of genuine non-instrumental dependence, for Hurley, seems to be the capacity to vary the perceptual content by a change in the motor output signal *even though the current input remains fixed*. The goal is to capture 'the thought that perception and action are in some way co-constituted' (p. 342).

The very simplest example of the kind of thing Hurley has in mind is the case of the paralysed eye. According to Gallistel (1980), if someone with paralysed eye muscles tries to look (say) to the right, the eye does not respond (so there is no change in gross retinal image) but the world, as presented in visual experience, appears to jump to the right. Hurley concludes that since there is a change in perceptual (conscious) content, and no other apparent difference except the introduction of the motor signal, the difference in content is best explained by a non-instrumental role for the motor signal itself (1998, p. 372). Several other cases are discussed, including cases involving TVSS (tactile visual substitution systems), adaptation to left-right reversing goggles, cases of output neglect and post-commissurotomy experiments (Hurley, ch. 9). In the latter case, a patient with a severed corpus callosum attempts to perform a left-hand task relating to a card that is visually presented in the right visual field, while fixating a central point. The patient reports that, 'The image of the object, initially seen, was blotted out of awareness in the left hemisphere the instant a movement, initiated by the right hemisphere, had started' (Trevarthen, 1984, p. 333, quoted by Hurley, p. 355).

Hurley's interpretation is that 'the motor intention relating to the left hand brought sensory information from the left visual field to perceptual awareness' (p. 374), and adds that in this case 'not just the content of consciousness, but also the presence or absence of consciousness, may depend on relations of input to output' (p. 374). Once again, we seem to see the 'motor-dependent selection of which information is present to perceptual consciousness' (p. 365). From such cases, Hurley is led to conjecture that the perceptual content (the content available to perceptual awareness) depends on an entire 'feedback relationship between input and output' (p. 375).

In the case of the paralysed eye, this feedback loop must be wholly internal, perhaps involving the influence of efferent copy (functioning as a kind of 'virtual input') on the visual processing stream. But, as Hurley frequently stresses, feedback loops may also run through the external environment. This would be the case if, for example, perceptual content depended at times on 'patterns of afference that are a function of movement through the environment' (p. 416) — a view often stressed in ecological

---

[11] Susan Hurley (1998), p. 183, speaks of the 'twisted rope' metaphor of neural processing.

views of perception. Such world-involving feedback loops will count as issuing in non-instrumental dependence if (and only if) what fixes the perceptual content is not *just* the sequence of inputs dictated by the motions (as Gibson thought) but rather the *relation* between motion *signals* and resultant inputs (i.e. passive motion won't have the same effect). It is thus the putative involvement of the motor output signal *itself* (and not its effects on subsequent worldly input) in determining the contents of our perceptual experience that makes for the *non-instrumental* dependence of perception on motor output, and that allows, more generally, for a genuinely *deep* (see secton I above) dependence of perceptual content upon 'feedback loops with orbits of varying sizes . . . that can . . . in principle spread across internal and external boundaries' (p. 327). The full conjecture is thus that perceptual contents (including the contents of visual awareness) may sometimes depend on whole loops involving motor signals, body and world. It is a picture which, as Hurley (p. 20) notes, 'takes the notion of dis-tributed processing to its logical extreme': just as the physical bearer of a given con-tent might be a pattern of activation across a whole neural population, so too it might be a pattern of relations *between* neural populations, or between sensory inputs, motor outputs, and external states of affairs.

We seem to thus confront a genuine conflict. Milner and Goodale offer a localist, internalist account of visual awareness, which draws a firm distinction between mecha-nisms of visual awareness and mechanisms of visuomotor action. Hurley offers a highly interactionist picture in which the contents of visual awareness are *deeply* bound up with motor output and (perhaps) bodily and external states of affairs.

III: Two Ways to be Action-Oriented

Is the conflict depicted in the previous section real or merely superficial?[12] Hurley's story does indeed involve what we termed (in section I) a 'deep' role for large-scale dynamics. But is it really a 'same target' competitor? Hurley's principal conjecture, recall, is that the content of the visual experience would not be as it is were it not for the non-instrumental role of motor output: a conjecture echoed by Mandik (1999), for instance, who argues that the introspectible properties of visual experience (the qualia, on his definition) are:

> determined not only by the nature of the information transduced by the nerve endings in the sensory organs, but also by what type of subsequent motor activity that information is employed in (Mandik, 1999, p. 56).

But now consider Milner and Goodale's claim in just a little more detail. The ventral processing stream, they argue, is the physical locus of visual awareness, while the dorsal stream guides fine-grained visuomotor action in the here-and-now. Clearly, though, there must be interactions between the two.[13] I can choose to reach for an item

---

[12] Hurley herself addresses this question in a long footnote on pages 183–4 of Hurley (1998). In the foot-note Hurley accepts much of the Milner & Goodale story, but questions the 'overlay' which assimi-lates the dorsal/ventral systems to systems dedicated to action/conscious perception, pointing out the ventral functions 'include responses and hence actions-it's just that these are much more flexible and cognitively mediated' (*op cit*). The account I develop is compatible with this, but goes further in attempting to distinguish two classes of motor output signal, only one of which is plausibly implicated in the constitution of conscious visual awareness.

[13] Milner and Goodale comment, revealingly, that 'understanding these interactions would take us some way towards answering what is one of the central questions in modern neuroscience: how is sensory information transformed into purposeful acts' (Milner and Goodale, 1995, p. 202).

because it looks (phenomenally, in my visual experience) larger than another, or redder, or whatever. The two streams anatomically exhibit multiple kinds of looping cross-connectivity, and there is ample evidence of information transfer between the two. Milner and Goodale's suggestion (1995, pp. 201–4) is that the ventral stream — the one said to be subserving visual experience — helps select both the *type* of action to perform and the *object* upon which to perform it.

Thus recall the Titchener Circles experiment. Here, according to Milner and Goodale, the choice of *which* disc to pick up, and the choice of what *kind* of action to perform (picking it up and not, e.g., poking at it), both depend on conscious, ventral stream dominated processing. It is only the most fine-grained visuomotor performance (the specifics of the precision grip) that is then deferred to full dorsal stream control. Let's call the choice of targets for action and the choice of type of action 'gross motor selection', and let's call fine-grained visuomotor control 'fine motor tuning'. The possibility then remains open — or so it seems to me — that acts of gross motor selection might actually *influence* the conscious content of the visual experience even if fine motor tuning remains relatively insulated. It remains possible, for example, that had the gross motor command selected different action-types or different goal-objects, the visual experience would have differed, and done so non-instrumentally, i.e. without the need for any change in gross input. In short, just because *fine motor tuning* is insulated from perceptual experience, it doesn't follow that perceptual experience is independent of *all types* of motor output signal: in particular, it may not be independent of the grosser — and ventral stream involving — events of action-type and target-object selection.

The most general lesson here is that the notion of an inner state or process being 'action-oriented' is by no means as simple as it sounds. The kinds of encoding supported by the dorsal stream are, as Milner and Goodale point out, heavily geared to what we have called 'fine motor tuning'. They are geared to action in the sense of being geared to precise and egocentrically defined *movements*, which must themselves be geared to the real locations, sizes and orientations of objects in space. The kinds of encoding supported (in part) by the ventral stream are geared to action in quite a different sense. They are geared to *action* in the sense in which actions reflect the specific needs, purposes and intentions of the agent. What is required here is an encoding that captures how the surrounding world is in just those respects necessary for planning, choice and conscious reasoning. Milner and Goodale's prime conjecture is that the computational demands of the latter complex of activities (planning, choice and reason) are dramatically distinct from — and even inconsistent with — those of the former (fine motor control). For the reasoning and planning complex requires us to identify objects regardless of spatial orientation and absolute location, and regardless of the current disposition of our limbs and bodies; and it demands only rough and relative information about spatial matters (what is closer, further, what is next to what, and so on). In this vein, Goodale and Humphrey recently speculate that:

> ... if perceptual representations [the ones underlying conscious visual awareness] were to attempt to deliver the real metrics of all objects in the visual array, the computational load would be astronomical. The solution that perception [visual awareness] seems to have adopted is to use world-based co-ordinates — in which the real metric of that world need not be computed. Only the relative position, orientation, size and motion of objects is [here] of concern. For example, we can watch the same scene unfold on television or on

a movie screen without being confused by the enormous change in the co-ordinate frame. [But] as soon as we direct a motor act towards an object, an entirely different set of constraints applies. (Goodale and Humphrey, 1998, pp. 195–6.)

I have quoted this passage at length since it is, I think, very revealing. Visual awareness, if this story is correct, simply cannot *afford* to be action-oriented in the first of our two senses. It simply cannot afford to represent each and every aspect of the scene present in visual awareness in the precise and egocentrically defined co-ordinates required to support complex physical interactions with that very scene. But the pressure for computational economy may very well drive the systems underlying visual awareness to *be* action-oriented in the *second* sense we described. They are action-oriented insofar as they register just those aspects of the visual scene relevant to our current needs, purposes and motor plans.

There is, indeed, suggestive evidence that conscious visual awareness is action-oriented and motor-involving in this latter sense. Hurley's own example of the visual field effects in the post-commissurotomy patient suggest a system which (when functioning normally) ties visual awareness to the locations indicated by current motor plans. While the growing body of evidence concerning so-called change-blindness (Simons and Levin, 1997) seems to show that conscious visual awareness is not in the business of building and maintaining a detailed inner model of the visual scene. (In a typical change-blindness experiment, elements of a visually present scene are altered while the subject saccades around the display. Subjects prove remarkably insensitive to these changes, leading theorists — such as Churchland *et al.* (1994) — to conclude that the feeling we have of perceiving a richly detailed scene is due to our capacity to repeatedly scan and foveate the visual scene so as to extract task-relevant information just as it is needed; but not a moment before). For some lively demonstrations, see Ballard *et al.* (1997).

Another factor that probably contributes to the sparse yet gross-action-oriented nature of the contents of visual awareness is the surprisingly tight relation between visual awareness and certain processes of *attention*. Thus consider the fascinating claim that 'there is no conscious perception at all in the absence of attention' (Mack and Rock, 1998, p. 227). Mack and Rock are driven to this claim by multiple experimental routes, including (especially) the observation that in the presence of an attention-diverting stimulus, a stimulus which would otherwise be seen and reported is very often not consciously perceived at all: an effect which is actually *magnified* if the critical (nondiverting) stimulus is foveally presented. For example, if the critical stimulus was a small, black square presented while the viewer was asked to report the longer arm of a briefly presented cross, the square was not consciously perceived in 25% of cases in which the critical stimulus was parafoveally presented, and in a full 75% of cases where it was foveally presented! By contrast, in experiments with no diverting stimulus or in conditions of divided attention, most subjects could perceive the square. Mack and Rock dub this phenomenon 'inattentional blindness' and conclude, quite generally, that it is only once attention is engaged by an object that the object becomes 'visible' to conscious awareness (see e.g. Mack and Rock, pp. 227–8). This may seem impossible — how can attention be drawn to that which is not yet perceived? But it makes good sense once we distinguish nonconscious sensory pickup from conscious perceptual awareness. Indeed, it is well-known — and their own subsequent experiments (see e.g. Mack and Rock 1998, pp. 232–5) also reflect this — that stimuli that do not make it into conscious awareness may nonetheless be highly processed, and this information can be used to guide behaviour. Indeed, it is often necessary for a stimulus to be highly processed (yielding an implicit grasp of abstract meaning and significance for the agent) *before* attentive processes render it consciously available (see Mack and Rock, p. 229).[14]

The link between visual awareness and attention is, I believe,[15] deep and significant. It fits nicely, moreover, with the compelling idea (Baars, 1988; Dennett, 1991; Chalmers, 1998) that conscious awareness is bound up with processes (attention being a strong contender) that make information widely available in the brain. The present suggestion, however, is much simpler. If (for whatever reason) attention plays some role in determining the contents of visual awareness, perhaps gross motor signals could affect conscious visual experience *by affecting or helping to engage processes of attention.* Thus, recall the case of the post-commisuratomy patient. Perhaps what matters here is not the gross motor signal *itself* (the intention to reach with the left hand) so much as the motor-dependent disposition of *attention.* Where Hurley depicts the case as one in which 'the side of motor activity seems to be determining the side of visual awareness' (Hurley, 1998, p. 355), we might wonder whether the gross motor intention is more perspicuously seen as engaging an attentional mechanism which in turn, and crucially, modulates conscious awareness.

It is worth pausing to clarify this suggestion.[16] The idea here is *not* that the fine-grained motor signals (the province, let us assume, of the dorsal stream) influence ventral stream activity and thus conscious awareness. Rather, it is that the gross motor *intention* (in this case, simply to move the left hand) results in a disposition of attention that directly influences conscious awareness. Such influence seems both intuitively familiar and neuroscientifically plausible. There is plenty of evidence, for example, that the deliberate assignment of attention (even covert attention, with no associated eye movement) modulates the receptive field properties of cells in V1, V2, V4 and MT (see e.g. Motter, 1993; Assad and Maunsell, 1995). And PET studies show that shifts of covert attention intentionally directed to specific aspects of a stimulus such as its colour, shape and spatial location, results in increased activity in the neuronal groups specialized for processing that type of information (Corbetta *et al.*, 1991; 1993; Haxby *et al.*, 1993). This kind of evidence is thus distinct from (but compatible with) the idea, embraced by Milner and Goodale, of other processes of selective attention operating entirely *within* the dorsal stream and having no influence on the contents of conscious awareness — see their endorsement of Rizzolatti *et al.*'s (1994) 'premotor' theory of selective spatial attention in Milner and Goodale (1995), pp. 183–8.[17] The present suggestion, to recap, is that gross motor intentions engage

---

[14] Note that the claim is that attention is *necessary* for a stimulus to become consciously perceived. It need not — and probably should not — be counted as sufficient.

[15] Thanks to Jesse Prinz for helping to convince me of this.

[16] Thanks to an anonymous referee for pressing me on this point.

[17] This kind of attention-invoking story is also clearly compatible with Milner and Goodale's further suggestion that conscious awareness of visual content requires both ventral stream coding *and* a sharpening effect due to 'spatial gating processes known to be active during selective attention' (Milner and Goodale, 1998, p. 7).

attentional mechanisms which in turn modify the response characteristics of neuronal groups implicated in conscious awareness.

It is not clear (to me) whether Hurley should regard this as a significant revision. In one sense it would seem to be, since it suggests that the motor signal does not itself help *constitute* the state of visual awareness — rather, it is now depicted as engaging neural mechanisms of attention which *do* seem to play a constitutive role. Against this, however, we must set the fact that what Hurley really aims to show is just the non-instrumental dependence of perception on motor output. If this is defined simply as motor output making a difference, independent of any change in gross sensory input, to the contents of perceptual awareness, the case stands regardless. It does not matter whether the non-instrumental dependence is direct or indirect (going *via* its effects on attentional mechanisms).

It is important in addressing such issues (and thanks to Susan Hurley, in a personal communication, for clarifying this) to distinguish between arguments that attempt to show that motor output makes a constitutive contribution to the fixation of conscious contents and ones that attempt to show that the physical vehicles of the contents (roughly, the most restricted set of inner states sufficient to cause the experience) are *themselves* whole, extended dynamical loops including motor output circuitry. Hurley (1998) attempts both kinds of argument, but merely establishing a non-instrumental role for motor outputs in content fixation cannot itself warrant any conclusions about extended physical vehicles. The present suggestion (concerning the attention-engaging role of gross motor attentions) is best seen as affecting only the issues concerning conscious content fixation. In Hurley's own terms, the idea would be that instead of the gross motor intentions making a constitutive contribution to conscious contents, they affect what she terms the 'borderland' of attentional mechanisms. It is the borderland activity itself which then makes the decisive contribution to content.

The main moral, however, is independent of these speculations about a bridging role for attention. It is that there is plenty of evidence that visual awareness is action-oriented and non-instrumentally motor-sensitive, but that this is fully compatible with Milner and Goodale's depiction of the dorsal stream as a semi-insulated system for visuo-motor action and of the ventral stream as a semi-insulated system for visual awareness. For the sense in which visual awareness is tied up with action and motor commands is quite distinct from the sense in which the non-conscious dorsal stream 'takes care' of action. What influences visual awareness, I have suggested, is a kind of schematic intentional version of motor control: one involving broad motor plans, projects and intentions. While what falls to the dorsal stream is the fine-tuned implementation of these plans. Both systems are thus profoundly motor-oriented, but in different and complementary ways. This complementarity is nicely captured by Goodale himself, who recently suggests that:

> [the] interplay between a 'smart' but metrically-challenged ventral stream and a 'dumb' but metrically-accurate dorsal stream is reminiscent of . . . what engineers call teleassistance — where a human operator looks at a scene, say the surface of a hostile planet, makes a decision that a particular rock needs to be examined, and then sends a command to pick up the rock to a semi-autonomous robot on the planet's surface (Goodale, 1998, p. 491).

Notice, finally, that none of this need imply a return to the old notion of a central executive. What matters is *not* that we identify the full intelligence of the agent with the ventral stream (which would, I think, be a serious mistake: the tele-assistance metaphor misleads in this respect). Rather, what matters is that we recognize the computationally efficient division of labour achieved by using a semi-insulated system for fine-tuned visuomotor action, and a semi-insulated system for visual awareness.

Recognition of such neural division of labour need not (and should not) blind us to the equal importance of large-scale distributed dynamics. It is perfectly possible, for instance, to hold that visual awareness itself depends on much more than mere ventral stream activity. It could involve, for example, complex recurrent dynamics linking multiple cortical and sub-cortical sites. Complex dynamic loops, as suggested by theorists such as Skarda and Freeman (1987); Freeman (1991; 1995); Edelman (1987) and Damasio (1994), may act as the neural vehicles of many kinds of mental content, and these loops may effectively combine different features into meaningful packages geared to an animal's current needs and goals. Visual awareness may thus reflect current context, projects and memories as well as relevant aspects of ongoing sensory input. What it need not (and probably should not) reflect is the finest-grained detail of visuomotor action control.

IV: Conclusions: Balancing Intimacy and Estrangement

The cognitive scientific understanding of perception, awareness and action is, I conclude, likely to turn on the appreciation of an especially complex kind of dynamic balance. It is a balance, as I have argued elsewhere (Clark, 1997, ch. 7), between large-scale, multi-component dynamics and pockets of local order and specialization: a balance between *intimacy* and close cooperation on the one hand, and *estrangement* and semi-autonomous specialization on the other. In considering the relations between perception, awareness, and action, we must do simultaneous justice to processes and dynamics of both kinds, and (hardest of all) to the ways in which the two harmonize and interact.

In the specific case we considered — the role of embodied action in visual awareness — there is convincing evidence that perception and action are both intricately intertwined and multiply dissociated! There is clear evidence of fine-tuned action-oriented coding in the dorsal stream. But there is also suggestive evidence that this whole visual stream operates semi-autonomously from the ventrally-dominated systems underlying major aspects of visual awareness. Within the ventral stream itself, however, we find *another* kind of interpenetration of perception and action: the kind stressed by Hurley and involving the influence of gross motor intentions and schematic action plans on conscious visual content. Finally, we must also consider how best to conceptualize the way these two semi-autonomous systems work in harmony so as to yield useful visual awareness of the very world in which we move and act.

The case of visual perception and action thus presents a much more complex problem than the case of rhythmic motion with which we began. In the rhythmic motion case, there was a clean 'same-target' conflict between a centralist, localist model of a specific phenomenon and a large-scale, interactionist, dynamical alternative. In the case of visual perception and action, there are multiple phenomena presenting very different explanatory targets, and at least two quite distinct ways in which visual and motor elements may be said to combine and interact.

Let me end, then, by relating all this to some of the bigger issues implied by the overarching theme of 'reclaiming cognition'. Recent years have indeed, or so it seems to me, been marked by the emergence of a new kind of science of the mind — one that places embodiment and action at the forefront, and that recognizes the crucial role of distributed dynamics (rather than static symbolic structures) in underpinning human thought and reason. But as with all dramatic shifts in emphasis, there is a concurrent danger — the danger of letting the pendulum swing too far in the opposite direction. For a mature science of the mind needs, somehow, to do *simultaneous* justice to the emergent unity (courtesy of looping webs of dynamical influence) and the frequent sub-systemic estrangement (courtesy of computationally efficient pockets of specialization and insulated functioning) characteristic of biological brains and natural intelligence. Current cognitive scientific research still tends to be drawn to one or other of these poles, oscillating between stress on complex dynamical intimacy and recognition of significant specialization and dissociation. Nature, as ever, contrives to have it both ways and all at once. Finding the models, metaphors and analytic tools necessary to describe and comprehend this unique balancing act is vital if cognition is to be truly *reclaimed*, rather than simply buffeted by another academic tug of war.

Acknowledgements

Thanks to: Susan Hurley, Brian Cantwell Smith, David Chalmers, Jesse Prinz, Josefa Toribio, an anonymous referee for *JCS* and all the seminar participants at the Washington University School of Medicine, and at the 1st Gulbenkian Symposium on Cognitive Neuroscience, Lisbon, July 1998.

References

Aglioti, S., Goodale, M. and DeSouza, J. (1995), 'Size contrast illusions deceive the eye but not the hand', *Current Biology*, **5**, pp. 679–85.

Assad, J.A. and Maunsall, J.H.R. (1995), 'Neuronal correlatives of inferred motion in primate posterior parietal cortex', *Nature*, **375**, pp. 518–21.

Baars, B. (1988), *A Cognitive Theory of Consciousness* (Cambridge: Cambridge University Press).

Ballard, D., Hayhoe, M. *et al.* (1997), 'Deictic codes for the embodiment of cognition', *Behavioral and Brain Sciences*, **20** (4), pp. 723–42.

Bechtel, W. and Richardson, R. (1992), *Discovering Complexity: De-Composition and Localization as Scientific Research Strategies* (Princeton, NJ: Princeton University Press).

Beer, R. (1995). 'A dynamical systems perspective on environment agent interactions', *Artificial Intelligence*, **72**, pp. 173–215.

Bernstein, N. (1967), *The Co-ordination and Regulation of Movements* (London: Pergamon Press).

Chalmers, D. (1996), *The Conscious Mind* (New York, Oxford University Press).

Chalmers, D. (1998), 'On the search for the neural correlate of consciousness', in *Towards a Science of Consciousness II*, ed. S. Hameroff, A. Kaszniak and A. Scott (Cambridge, MA: MIT Press).

Chiel, H. and Beer, R. (1997), 'The brain has a body: adaptive behavior emerges from interactions of nervous system, body and environment', *Trends in Neurosciences*, **20** (12), pp. 553–7.

Churchland, P., Ramachandran, V. and Sejnowski, T.J. (1994), 'A critique of pure vision', in *Large-Scale Neuronal Theories of the Brain*, ed. C. Koch and J. Davis (Cambridge, MA: MIT Press).

Clark, A. (1995), 'I am John's brain', *Journal of Consciousness Studies*, **2** (2), pp. 144–8.

Clark, A. (1997), *Being There: Putting Brain, Body and World Together Again* (Cambridge, MA: MIT Press).

Clark, A. (1999), 'An Embodied Cognitive Science?' *Trends in Cognitive Sciences*, **3** (9), pp. 345–51.

Clark, A. and Chalmers, D.J. (1998), 'The extended mind', *Analysis*, **58**, pp. 7–19.

Corbetta, M., Miezin, F.M. *et al.* (1991), 'Selective and divided attention during visual discriminations of shape, color and speed', *Journal of Neuroscience*, **11** (8) pp. 2383–402.

Corbetta, M., Miezin, F.M. *et al.* (1993), 'A PET study of visuospatial attention', *Journal of Neuroscience*, **13** (6) pp. 1202–26.

Crick, F. and Koch, C. (1990), 'Function of the thalamic reticular complex: The searchlight hypothesis', *Seminars in the Neurosciences*, **2**, pp. 263–75.

Crick, F. and Koch, C. (1997), 'Towards a neurobiological theory of consciousness', in *The Nature of Consciousness*, ed. N. Block, O. Flanagan and G. Güzeldere (Cambridge, MA: MIT Press).

Damasio, A. (1990), 'Synchronous activation in multiple cortical regions: A mechanism for recall', *Seminars in the Neurosciences*, **2**, pp. 287–96.

Damasio, A. and Damasio, H. (1994), 'Cortical systems for retrieval of concrete knowledge: The convergence zone framework', in *Large-Scale Neuronal Theories of the Brain*, ed. C. Koch and J. Davis (Cambridge, MA: MIT Press).

Dennett, D. (1991), *Consciousness Explained* (New York: Little Brown & Co.).

Dennett, D. (1993), 'The message is: There is no medium', *Phenomenological Research*, **53**, pp. 919–31.

Dennett, D. (1995), *Darwin's Dangerous Idea* (New York: Simon & Schuster).

Dreyfus, H. (1979), *What Computers Can't Do* (New York: Harper & Row).

Edelman, G. (1987), *Neural Darwinism: The Theory of Neuronal Group Selection* (New York: Basic Books).

Edelman, G. (1992), *Bright Air, Brilliant Fire: On the Matter of Mind* (New York: Basic Books).

Farah, M.J. (1997), 'Visual perception and visual awareness after brain damage: A tutorial overview', in *The Nature of Consciousness*, ed. N. Block, O. Flanagan and G. Güzeldere (Cambridge, MA: MIT Press).

Fontana, W. and Ballati, S. (1999), 'Complexity', *Complexity*, **4** (3), pp. 14–16.

Freeman, W.J. (1991), 'The physiology of perception', *Scientific American*, **204**, pp. 78–85.

Freeman, W.J. (1995), *Societies of Brains: A Study in the Neuroscience of Love and Hate* (Hillsdale, NJ: Erlbaum).

Gallistel, C.R. (1980), *The Organization of Action : A New Synthesis* (Hillsdale, NJ, and New York: L. Erlbaum Associates).

Gazzaniga, M. (1998), *The Mind's Past* (University of California Press).

Gibbon, J., Church, R. *et al.* (1984), 'Scalar timing in memory', in *Timing and Time Perception*, ed. J. Gibbon and L. Allan (New York: New York Academy of Sciences).

Gibson, J.J. (1979), *The Ecological Approach to Visual Perception* (Boston, MA: Houghton-Mifflin).

Goodale, M. (1998), 'Visuomotor control: Where does vision end and action begin?', *Current Biology*, **8**, pp. R489–R491.

Goodale, M. and Humphrey, G. (1998), 'The objects of action and perception', *Cognition*, **67**, pp. 181–207.

Hardcastle, V. (1998), 'The puzzle of attention, the importance of metaphors', *Philosophical Psychology*, **11** (3), pp. 331–52.

Hatsopoulos, N. (1996), 'Coupling the neural and physical dynamics in rhythmic movements', *Neural Computation*, **8**, pp. 567–81.

Hatsopoulos, N. and Warren, W. (1996), 'Resonance tuning in rhythmic arm movements', *Journal of Motor Behavior*, **28** (1), pp. 3–14.

Haugeland, J. (1998), 'Mind embodied and embedded', in *Having Thought* (Cambridge, MA: MIT Press).

Haxby, J.V., Grady, C.L., Horawitz, B., Salerno, J., Ungerleider, L., Mishkin, M. and Shapiro, M.B. (1993), 'Dissociation of object and spatial visual processing pathways in human extrastriate cortex', in *Functional Organization of Human Visual Cortex*, ed. B. Gulyas, D. Orroson and P.E. Roland (Oxford: Pergamon Press).

Heidegger, M. (1961/1927), *Being and Time* (New York: Harper and Row).

Hurley, S. (1998), *Consciousness in Action* (Cambridge, MA: Harvard University Press).

Hutchins, E. (1995), *Cognition in the Wild* (Cambridge, MA: MIT Press).

Iriki, A., Tamaka, M. and Iwamura, Y. (1996), 'Coding of modified body schema during tool use by macaque postcentral neurons', *Neuroreport*, **7**, pp. 2325–30.

Iwamura, Y. (1998), 'Hierarchical somatosensory processing', *Current Opinion in Neurobiology*, **8**, pp. 522–8.

Jarvilehto, T. (1998), 'Efferent influences on receptors in knowledge formation', *Psycoloquy*, **9** (41); ftp://ftp.princeton.edu/pub/harnad/Psycoloquy/1998.volume.9/

Johnson, M. (1987), *The Body in the Mind : The Bodily Basis of Meaning, Imagination, and Reason* (Chicago: University of Chicago Press).

Kelso, J.A.S. (1995), *Dynamic Patterns : The Self-Organization of Brain and Behavior* (Cambridge, MA: MIT Press).

Kinsbourne, M. (1988), 'Integrated field theory of consciousness', in *Consciousness in Contemporary Science*, ed. J. Marcel and E. Disiach (Oxford: Clarendon Press).

Knierim, J. and Van Essen, D. (1992), 'Visual cortex: Cartography, connectivity and concurrent processing', *Current Opinion in Neurobiology*, **2**, pp. 150–5.

Koch, C. and Braun, J. (1998), 'Towards the neuronal correlate of visual awareness', in *Findings and Current Opinion in Cognitive Neuroscience*, ed. L. Squire and S. Kosslyn (Cambridge, MA: MIT Press).

Mack, A. and Rock, I. (1998), *Inattentional Blindness* (Cambridge, MA: MIT Press).

Mandik, P. (1999), 'Qualia, space and control', *Philosophical Psychology*, **12** (1), pp. 47–60.

Merleau-Ponty, M. (1942), *La Structure du Comportment* (France, Presses Universitaites de France).

Milner, A. and Goodale, M. (1995), *The Visual Brain in Action* (Oxford: Oxford University Press).

Milner, D. and Goodale, M. (1998), 'The visual brain in action', *Psyche*, **4** (12), Oct. 1998.

Motter, B.C. (1993), 'Social attention produces spatially selective processing in visual cortical areas V1, V2 & V4 in the presence of competing stimuli', *Journal of Neurophysiology*, **70**, pp. 909–19.

Port, R., Cummins, F. and McCauley, J. (1995), 'Naive time, temporal patterns and human audition', in *Mind As Motion*, ed. R. Port and T. Van Gelder (Cambridge, MA: MIT Press).

Rizzolatti, G., Riggio, L. *et al.* (1994), 'Space and selective attention', in *Attention and Performance* [**vol XV**, pp. 231–66], ed. C. Umilta and M. Moscovitch (Cambridge, MA: MIT Press).

Simons and Levin (1997), 'Change blindness', *Trends in Cognitive Sciences*, **1** (7), pp. 261–7.

Skarda, C. and Freeman. W. (1987), 'How brains make chaos in order to make sense of the world', *Behavioral and Brain Sciences*, **10**, pp. 161–95.

Sporns, O., Gally, J.A., *et al.* (1989), 'Reentrant signaling among simulated neuronal groups leads to coherency in the oscillatory activity', *Proc. Natl. Acad. Sci. USA*, **86**, pp. 7265–9.

Thelen, E. and Smith, L. (1994), *A Dynamic Systems Approach to the Development of Cognition and Action* (Cambridge, MA: MIT Press).

Trevarthen, C. (1984), 'Biodynamic structures, cognitive correlates of motive sets, and the development of motives in infants', in *Cognition and Motor Processes*, ed. W. Prinz and A. Sanders (Berlin: Springer-Verlag).

Turvey, M.T. and Kugler, P.N. (1987), *Information, Natural Laws, and Self-Assembly of Rhythmic Movement* (Hillsdale, NJ: L. Erlbaum Associates).

Van Essen, D. and DeYoe, E. (1995), 'Concurrent processing in the primate visual cortex', in *The Cognitive Neurosciences*, ed. M. Gazzaniga (Cambridge, MA: MIT Press).

Varela, F., Thompson, E. and Rosch, E. (1991), *The Embodied Mind* (Cambridge, MA: MIT Press).

Wallace, S.A. (1981), 'An impulse-timing theory for reciprocal control of muscular activity in rapid, discrete movements', *Journal of Motor Behavior*, **13**, pp. 144–60.

Webb, B. (1994), 'Robotic experiments in cricket phonotaxis', in *From Animals to Animats 3*, ed. D. Cliff, P. Husbands, J.A. Meyer and S. Wilson (Cambridge, MA: MIT Press).

Wurz, R. and Mohler, C. (1976), 'Enhancement of visual response in monkey striate cortex and frontal eye fields', *Journal of Neurophysiology*, **39**, pp. 766–72.

Jana M. Iverson and Esther Thelen

Hand, Mouth and Brain

The Dynamic Emergence of Speech and Gesture

Introduction

The past fifteen years have seen a resurgence of interest in ideas of embodiment, the claim that bodily experiences play an integral role in human cognition (e.g., Clark, 1997; Johnson, 1987; Sheets-Johnstone, 1990; Varela *et al.*, 1991). The notion that mind arises from having a body that interacts with the environment in particular ways stands in stark contrast to the predominant view since the 'cognitive revolution' of the post-war years. Using the computer as a metaphor for describing the structure of the mind, this cognitivist tradition has viewed thought as a product of abstract mental symbols and the rules by which they are mentally manipulated.

The fundamental difference between the embodiment and cognitivist perspectives lies in the role ascribed to the body, its characteristics, and its interactions with the environment. From a cognitivist point of view, the body is an output device that merely executes commands generated by symbol manipulation in the mind; the properties and activities of the body are irrelevant. From an embodiment perspective, however, cognition depends crucially on having a body with particular perceptual and motor capabilities and the types of experiences that such a body affords. In other words, cognition is a product of the body and the ways in which it moves through and interacts with the world.

In this chapter, we examine the embodiment of one foundational aspect of human cognition, language, through its bodily association with the gestures that accompany its expression in speech. Gesture is a universal feature of human communication. Gestures are produced by all speakers in every culture (although the extent and typology of gesturing may differ). They are tightly timed with speech (McNeill, 1992). Gestures convey important communicative information to the listener, but even blind speakers gesture while talking to blind listeners (Iverson and Goldin-Meadow, 1998), so the mutual co-occurrence of speech and gesture reflects a deep association between the two modes that transcends the intentions of the speaker to communicate. Indeed, we believe that this linkage of the vocal expression of language and the arm movements produced with it are a manifestation of the embodiment of thought: that human mental activities arise through bodily interactions with the world and remain linked with them throughout the lifespan. In particular, we

Journal of Consciousness Studies, **6**, No. 11–12, 1999, pp. 19–40

propose that speech and gesture have their developmental origins in early hand–mouth linkages, such that as oral activities become gradually used for meaningful speech, these linkages are maintained and strengthened. Both hand and mouth are tightly coupled in the mutual cognitive activity of language. In short, it is the initial *sensorimotor linkages* of these systems that form the bases for their later cognitive interdependence.

Our defence of this proposition proceeds in this way. First, we show from extensive neurophysiological and neuropsychological evidence that, in adults, language and movement are very closely related in the brain. The question then becomes: How did they get that way? To answer this question, we invoke principles of dynamic coordination to show how two mutually active systems can influence and entrain one another. We then apply these principles to the early development of the speech–gesture system. We argue from developmental evidence that the motor actions of hand and mouth are present from birth and evolve in a mutually interactive fashion during the first year. We demonstrate how, as infants learn language, the changing thresholds and activation of hands and mouth for communication lead to the tight, synchronous speech–gesture coupling seen in adults. Finally, we speculate on this developmental story for the understanding of embodied cognition.

Before we embark on the details of our proposition, we review the current thinking about the relations between speech and gesture.

The Relationship Between Gesture and Speech

Currently, there are three competing views of the relationship between gesture and speech. The first of these posits that gesture and speech are separate communication systems, and that any existing links between the two modes are the result of the cognitive and productive demands of speech expression (e.g., Butterworth and Beattie, 1978; Hadar, 1989; Hadar *et al.*, 1998; Levelt *et al.*, 1985). According to this view, gesture functions as an auxiliary 'support system' whose primary role is to compensate for speech when verbal expression is temporarily disrupted (e.g., by coughing) or unavailable (e.g., when the speaker is unable to put thoughts into words). Importantly, any feedback links between speech and gesture are unidirectional, moving uniquely from speech to gesture. The production of gesture is thus assumed to have no effect on speech production or the cognitive processes that guide it.

The second view, recently articulated by Robert Krauss and colleagues (e.g., Krauss, 1998; Krauss and Hadar,1999, Rauscher *et al.*, 1996), differs from the first in that it assumes the existence of reciprocal links between gesture and speech. However, these links are located at a specific point in the process of speech production: the phonological encoding stage (cf. Levelt, 1989), or the moment at which a word form must be retrieved from lexical memory. Krauss and colleagues have argued that when speakers encounter difficulty in lexical retrieval, the production of gestures activates spatio-dynamic features of the concept in question. This in turn activates the lexical affiliate of that concept in memory and leads to successful articulation of the word. In other words, while gesture and speech are viewed as a linked system, the connection is highly limited in scope, with gesture influencing speech processing to the extent that it provides for cross-modal activation of concepts at a moment of difficulty in word form retrieval.

The third view of the gesture–speech relationship has been put forth by David McNeill (1992). In McNeill's view, gesture and speech form a single system of communication based on a common underlying thought process. Gesture and speech are tightly connected to one another, and there are links between gesture and speech throughout the process of speech production, occurring at the levels of discourse, syntax, semantics, and prosody. From this perspective, gesture and speech co-occur during production because they are linked to one another and to the same underlying thought processes (even though each modality may express a different aspect of that thought). Any disruption in the process of speech production should therefore have an effect on gesture, and vice versa.

In this chapter, we are inspired by and expand upon this third view, that articulated by McNeill. In particular, we begin by reviewing evidence from studies of normal adults and those with brain injuries and neurological disorders indicating that these two modalities are indeed linked in all aspects of language production. We then ask the developmental question: Where did these links come from?

Neurophysiological Links Between Language and Movement

Four lines of research from neurophysiology and neuropsychology provide converging evidence of links between language and movement at the neural level. These studies have revealed that: a) some language and motor functions share underlying brain mechanisms; b) brain regions typically associated with motor functions (e.g., motor cortex, premotor area, cerebellum) are involved in language tasks; c) classical 'language areas' (e.g., Broca's area) are activated during motor tasks; and d) patterns of breakdown and recovery in certain language and motor functions appear to be closely linked in some patient populations. We review each of these lines of work in turn.

Common brain mechanisms for language and motor functions

Studies employing electrical stimulation mapping techniques have indicated that some language and motor functions may share common mechanisms in certain brain regions. In these studies, electrodes are inserted into the cortex of a patient under local anaesthesia. A small amount of electrical current is delivered to each electrode in turn, and the effects of this stimulation on the patient's behaviour are measured.

Results from a series of studies conducted by Ojemann and colleagues (see Ojemann, 1984, for a review) point to a common brain mechanism for sequential movement and language that appears to be located in the lateral perisylvian cortex of the dominant hemisphere. Ojemann and colleagues reported that stimulation of this region resulted in two distinct patterns of change in motor and language functions. The first of these occurred primarily at sites at the posterior end of the inferior frontal gyrus, where stimulation disrupted imitation of any type of orofacial movement and also inhibited speech production. The second occurred at sites more widely distributed throughout the perisylvian cortex, where stimulation disrupted mimicry of sequences of orofacial movements (but not single movements) and evoked disturbances in naming or reading. However, recent verbal memory was not hampered by stimulation at any of these sites. This is important because it suggests that the observed disturbances in language production and movement were specific effects of stimulation at these sites, and not simply the product of global perceptual or attentional disruptions that might be a general consequence of the stimulation procedure.

Such observations suggest that there may be a mechanism common to language and sequential motor tasks located in this area. One candidate for a mechanism underlying language production and motor sequencing is precise timing, which is essential for the kinds of rapid movements that are involved in both motor sequencing and language production (Ojemann, 1984). This is a particularly appealing notion if we view gesture production as a motor sequencing task that co-occurs with speech production. A common timing mechanism for language and movement could account for the fact that gesture and speech are tightly linked in time, with the stroke of the gesture being executed in synchrony with the semantically co-expressive word or phrase (McNeill, 1992).

Not only is there some evidence for common mechanisms for speech production and sequential movement, but there is also some indication that the hands and arms and the vocal tract may be represented in neighbouring sites in certain brain regions. Fried *et al.* (1991) used electrical stimulation to map the functional organization of the supplementary motor cortex (SMA) in a group of patients preparing to undergo neurosurgery for chronic epilepsy. The obtained patterns of somatotopic organization indicated that sites where stimulation elicited movements of the hands and arms were adjacent to sites where stimulation resulted in speech disruption. Thus, in one patient, stimulation at one site in the left SMA was followed by the patient's report of a strong urge to raise the right elbow. Application of a slightly more intense current at the same site elicited abduction of the right arm, but no speech difficulties. At an adjacent site (approximately 1 cm away), however, stimulation elicited speech arrest in the form of hesitation during a naming task, but no arm movement.

Interestingly, in transitional areas between neighbouring somatotopic representations, stimulation often elicited complex movements involving body regions represented in these adjacent regions. For instance, at a site that appeared to mark a transitional area between the hand/arm and speech representations described above, stimulation elicited both speech arrest and finger flexion of the right hand.

These results raise the possibility that the tight temporal co-occurrence between gesture and language may be the product of spreading levels of activation in neighbouring areas, such that when the portion of the region associated with speech production is activated, activity spreads to the neighbouring site associated with movement of the hand and arm. Patterns of co-activation may be influenced by a common precise timing mechanism in the lateral perisylvian cortex, resulting in the production of gestures that are highly synchronous with co-occurring speech.

Motor areas are involved in language tasks

Additional evidence for neurophysiological connections between language and movement comes from work demonstrating that brain regions traditionally known as 'motor areas' become active in language tasks that do not explicitly involve speech production. In the motor cortex, for example, there are high levels of EEG activity when adults are asked to read words silently from a video screen. Interestingly, patterns of activity are particularly high when the target words are verbs (Pulvermüller, et al., 1996).

Premotor regions are also closely involved. For instance, when Grabowski *et al.* (1998) examined patterns of PET activity in a task involving retrieval of words from various conceptual categories (e.g., animals, tools, persons), they found high levels of activity in the left premotor area, but only when the words to be retrieved were tool names. One interpretation for this pattern of findings is that verbs and tool names have a strong motoric component that is stored with the semantic features of the word, and that the motor affiliate of such words becomes activated during lexical processing and retrieval.

Even the cerebellum, the portion of the brain most closely identified with movement, participates in language functions. Petersen *et al.* (1989) presented a group of normal, right-handed adults with two word production tasks: a) a simple task, in which participants were only asked to repeat a visually-presented word; and b) a complex task, in which participants viewed a word, had to think of a different word associated with the use of the presented word, and then say the associated word (e.g., saying 'sew' when the presented word is 'needle'). While both tasks require a similar vocal response (i.e., saying a word), the complex task also required participants to generate a word association. To identify cerebellar areas that were active during word association, a method of subtractive data analysis was employed, in which motor activation obtained in the simple task was subtracted from activation patterns obtained in the complex task.

The simple task (saying a visually-presented word) activated an area in the superior anterior lobe of the cerebellum. Interestingly, this area is just lateral to those activated by movements of the fingers. The word association task, however, activated an entirely different area, the inferior lateral cerebellum. Significant activation was found in this area even after subtracting away the motor activity generated by word production. Moreover, activation of the inferior lateral cerebellum was localized to the right hemisphere, the side that projects to the left hemisphere and was dominant for language in these participants.

These findings point strongly to connections between the cerebellum and classical 'language areas' such as Broca's area. Indeed, such connections have been identified anatomically (Leiner *et al.*, 1989; 1993). This cerebro–cerebellar loop consists primarily of cerebellar output connections, which are projected to the reticular formation and the thalamus. Via the thalamus, these cerebellar projections can reach areas of the frontal lobe. The loop is completed by projections from prefrontal areas that are sent back to the cerebellum. Within this pathway, there are additional connections between cerebellar regions and cortical areas that have been implicated in language processes. For example, the dentate nucleus of the cerebellum projects through the medial thalamus and into Broca's area. In addition, signals can be transmitted via Türck's bundle from an area of temporal cortex known to be involved in language to the pontine nuclei and then to the cerebellum.

Language areas are involved in motor tasks

In addition to evidence pointing to motor area involvement in language tasks, there is now growing indication that language areas are activated during motor tasks in which linguistic mediation (i.e., using language to guide movements) is unlikely. While there are many so-called 'language areas' distributed throughout the brain, we focus our review here specifically on studies that have examined activity in Broca's area, which is perhaps the best known of these sites.

The question of whether brain areas activated by motor tasks overlap with those activated during language tasks was addressed in an fMRI study conducted by Erhard *et al.* (1996). Twelve healthy, right-handed participants performed a series of motor tasks (random tongue movement, toe movement, complex instruction-guided finger tapping, and copying of displayed hand shapes) and a language task. As expected, there was activation throughout Broca's area during the language task. The striking finding was that portions of Broca's area were also activated during each of the motor tasks, particularly the two tasks involving hand movement (see also Bonda *et al.*, 1994).

Perhaps even more impressive, however, is that Broca's area is even activated when individuals think about moving their hands. Krams *et al.* (1998) looked at changes in cerebral blood flow patterns that occurred when healthy adults were asked to copy sequenced finger movements. Participants either executed the movements immediately, experienced a short delay prior to movement execution, or simply prepared the movements without executing them. There was a significant change in blood flow in Broca's area (specifically in Brodmann's area 44) in the two conditions involving a relatively extended period of movement preparation (the delayed execution and prepare-only conditions) relative to the immediate movement condition. In other words, merely planning a sequenced hand movement was sufficient to activate a portion of Broca's area.

Thus, in addition to its well-documented role in language processing and production, Broca's area appears to be involved in some motor activities related to the extremities and facial areas. In our view, this is important because it points to a possible neural substrate for the link between gesture and speech. Specifically, Broca's area appears to play a critical role in the generation of coherent sequences of body movements. Such a mechanism (along with others controlling precise timing of the sort described above) may well be involved in the co-production of speech and gesture, which requires the generation of sequential movements that are precisely timed with one another.

Evidence from special populations

The notion that gesture and speech co-production may draw on common brain mechanisms is further supported by studies of patients with a variety of different linguistic and motor impairments. Here we review evidence suggesting that some motor functions (particularly movement sequencing abilities) tend to be compromised when language is impaired; that gesture production can improve language skills in aphasic patients; and that gestures are produced even when there is damage to motor systems and proprioceptive and spatial position feedback are lost.

In a classic study of motor functioning in patients with left- and right-hemisphere injury, Kimura and Archibald (1974) reported that relative to right-hemisphere patients, adults with left-hemisphere damage performed significantly worse on a task involving copying of meaningless hand movements (e.g., closed fist, thump sideways on table; open hand, slap palm down on table) and on a traditional test of apraxia requiring demonstration of the use of familiar objects (e.g., show how to use a cup) and production of familiar gestures on verbal command (e.g., show how to wave goodbye). Additional analyses revealed that the poorer performance of the left-hemisphere group could not be explained by general difficulties with hand movement or the presence of linguistic deficits in these patients.

It may be the case, therefore, that the speech disturbances and movement difficulties manifested by left-hemisphere patients in this study are the product of a more general impairment in the type of motor sequencing involved in speech and gesture production. Additional support for this conclusion is provided by analyses of patterns of spontaneous speech and gesture production in patients with left-hemisphere damage (Pedelty, 1987). Specifically, patients with Broca's aphasia (generally a product of damage to anterior portions of the language-dominant hemisphere) exhibit parallel interferences in speech and gesture. The speech of Broca's aphasics tends to be agrammatic, consisting largely of content-bearing 'open-class' words and relatively lacking in grammatical functors (articles, prepositions, and other structural words). With regard to gesture production, Broca's aphasics produce many imagistic iconic gestures, which convey pictorial content (e.g., holding the arms out and extended slightly to the sides, conveying information about the size of a box) and relatively few of the fluid, hand-waving gestures that are often used to mark relationships within a conversation (e.g., the rhythmic beats of the hand that are observed at the moment in which new information is introduced into a conversation). Thus, when language breaks down in aphasia, parallel deficits are found in gesture.

That impairment in motor sequencing may be a more general feature of language disturbance is suggested by work examining the motor skills of children with specific language impairment (i.e., impaired language skills in the face of normal cognitive abilities and hearing). Hill (1998) tested children with specific language impairment (SLI) on a standard motor development battery and the familiar and unfamiliar hand movement tasks developed by Kimura and Archibald (1974) described above. She reported two striking findings.

First, despite the fact that children with SLI did not have any documented motor difficulties and were not selected for the study on the basis of their motor development battery scores, over half of the children (11 of 19) obtained scores that fell within the range for a group of children with developmental coordination disorder (DCD, a diagnosis characterized by movement difficulties out of proportion with the child's general level of development). Normally, 6% of the population of children between the ages of 5 and 11 years are diagnosed with DCD (American Psychiatric Association, 1994). Second, children with SLI scored significantly worse than age-matched peers and like children with DCD on the two tests of representational gesture imitation (with and without objects). This pattern was apparent in the performance of every child in the SLI group, even those who scored within the normal range on the movement battery.

The fact that motor functions related to the production of gesture are impaired when language is compromised is consistent with two recent studies suggesting that some language functions in aphasic patients may be improved by gesture production and training. The principal hypothesis of these studies was that if the output systems of speech and gesture are overlaid on the same 'general cognitive/movement cerebral systems', then gesturing should help stimulate the verbal articulatory system.

In one study, Hanlon et al. (1990) examined the effects of gesture production on performance in a confrontation naming task in patients with severe aphasia following left hemisphere damage. Patients were presented with black and white photos of common objects and asked to try to name the objects while either pointing at the picture or making a fist. They found that pointing with the right hand significantly improved performance, compared to fisting the right hand or pointing with the left hand. This suggests that functional activation of the right arm in the production of communicative gestures may facilitate activity in left hemisphere areas involved in the naming task, which may in turn result in improved naming performance.

This type of gestural activity also appears to have an effect on language functions that lasts beyond a single session in the laboratory. In a study of a single patient with nonfluent aphasia, Pashek (1997) employed a training procedure over multiple sessions to provide extensive practice with naming line drawings of gesturable objects and actions (e.g., a comb, a cigarette, scissors, to knock). Some of the stimuli were presented with verbal plus gestural training (i.e., oral repetition together with production of an associated gesture with either the right or the left hand), while others were associated with verbal-only training (i.e., oral repetition alone). The issues of interest were how naming performance would compare over time for verbal plus gesture versus verbal-only items, and whether the effects of training would be retained over time.

At baseline sessions prior to the beginning of training, accuracy was consistently poor across items, with the patient naming approximately 30% of the items correctly. By the fourth training session, however, performance had improved substantially for items associated with gestures (85% and 70% correct for left- and right- hand gestures respectively), while accuracy for verbal-only items was 50%. By the end of the training period, accuracy was quite high for verbal-plus-gesture targets (90% and 85% for left-hand and right-hand targets respectively), but had dropped down to initial levels for verbal-only items. What is perhaps most impressive is that gains made in naming for verbal-plus-gesture targets were retained for six months post-training.

In short, the finding that gesturing stimulates language functions associated with naming tasks (e.g., word retrieval, verbal articulation) is consistent with the hypothesis that the output systems of speech and gesture may draw on underlying brain mechanisms common to both language and motor functions. Further support for this view comes from a recent case study of spontaneous gesture production by a single patient who, as a young adult, suffered an infection that led to the loss of all proprioceptive feedback and spatial position sense from the neck down (Cole et al., 1998). Movements requiring precision and maintenance of postural stability were effortful for this patient, and thus one might expect to find a total absence of gesture under these extreme conditions.

Contrary to this expectation, the patient produced gestures, and continued to do so even when he could not see his hands and make use of visual feedback to control their movement. Moreover, these gestures were tightly synchronized with speech, even when visual feedback was not available. Despite the fact that movements requiring spatial accuracy were virtually impossible for this patient, he was able to use space to differentiate meanings conveyed in gesture (e.g., a movement executed on the right side to represent one meaning, another on the left for a contrasting meaning).

These observations are striking because they indicate not only that gestures can occur in the absence of visual monitoring and proprioceptive feedback, but also that the gesture-speech relationship remains temporally and semantically intact even when other types of motor activities (e.g., walking, reaching) have been severely disrupted. This is consistent with the notion that the speech-gesture system is controlled by common brain mechanisms. Thus, even when there is damage to motor control systems, gesture may remain relatively spared because it is controlled at least in part by systems related to language that are distinct from traditional 'motor areas'.

In summary, a body of evidence from electrical stimulation, neuroimaging, and behavioural studies of healthy adults and patient populations is consistent with the view that gesture and speech form a tightly coupled system. Tasks requiring precisely timed movements of the vocal tract and hands and arms appear to share common brain mechanisms; classical 'language areas' in the brain are activated during motor tasks, and vice versa; subtle motor deficits, particularly in the production of sequential movement, co-exist with language breakdown and disorder; and gesture production appears to have a facilitating effect on language recovery. The strength of the coupling between gesture and speech is further underscored by preliminary findings indicating that spontaneous gesture production occurs even in the face of damage to brain regions involved in motor control.

There is thus compelling neurophysiological evidence suggesting that in adults, gesture and speech are inextricably linked in the brain. In the next section of this chapter, we argue that the foundations of these linkages are in place from birth, likely with phylogenetic origins. Furthermore, the gesture–speech system in adults can be understood as the product of the mutual, interacting development of these two systems over the first few years of life. We view this developmental pathway from the perspective of dynamic systems theory, and in particular, the principles derived for understanding the coordination of human movement. Our assumption here, based on the evidence we presented above, is that mouth and hand are two related movement systems that start out coordinated with one another and remain so, although the nature of the coordination changes. Thus, contemporary formulations of such coordination can be applied. We discuss limb coordination and then suggest that the same principles apply to the heterogeneous systems of mouth and hand.

The Dynamics of Motor Coordination

One of the central issues in understanding human movement is the question of coordinating the limbs and body to perform adaptive actions. How do people and other animals so precisely move their limbs in time and space to walk, run, or manipulate objects? Dynamic systems theory in motor control was initially formulated to address the problem of coordination of the limbs as a special case of the more general issue of coordination in complex systems (e.g. Kugler and Turvey, 1987). The principle tenet of a dynamic systems approach is that in such complex, heterogeneous systems (such as moving animals), the individual parts cooperate to form patterns, which exist in space and time. This cooperativity occurs without any 'executive' direction, but rather strictly as a function of the coherence of the parts under certain energetic constraints. Many such self-organized patterns occur in nature in physical and biological systems, with no 'cognitive' intervention (see Kelso, 1995).

The most well-studied phenomena are issues of coordination in rhythmic limb movements in humans and other animals. Rhythmic movements are universal in animal movement, primarily for locomotion, but also in humans for tool use, music, (and in speech and gesture!). The contemporary dynamic view rests heavily on earlier work by the physiologist von Holst, who studied locomotion in fish and insects. In particular, von Holst described the actions of fish fins as individual oscillators that were, however, coupled to one another. Von Holst enumerated several principles of this interlimb coordination:

(1) Each fin had a preferred frequency when acting alone.

(2) Sometimes the oscillation of one fin could be detected in the oscillation of another. This is the *superposition* effect.

(3) Each fin tries to draw the other fins to its characteristic oscillation. This is the *magnet* effect, and it results in a cooperative tempo, often a balance between the two competing tempos.

(4) Each fin tries to maintain its preferred frequency when participating in a coupling, leading to variations around the mean cooperative tempo. The *maintenance* effect illustrates the dynamic nature of the coupling of several oscillators: there is a tension between maintaining the preferred frequency and the strength of the entrainment to other oscillators.

These principles were best illustrated in human limb movements by experiments done by Kugler and Turvey (1987) over a decade ago. They examined the entrainment of rhythmical arm movements when they experimentally changed the arms' natural frequencies. Under normal circumstances, it is very natural for people to flex and extend their arms rhythmically about the elbows, either in phase or alternating. The comfortable frequency that people choose for the movement of the combined limbs is very similar to the natural frequencies that people find comfortable when swinging one arm alone. Kugler and Turvey asked people to do this simple movement while holding weighted pendula. People swinging heavy weights preferred a lower oscillation rate than those holding light weights. What happens when people are holding a heavy weight with one hand and a light weight with the other and they are asked to find a common rhythmical coordination pattern? The solution is just what von Holst predicted: they find a compromise frequency that is neither as fast as the light arm nor as slow as the heavy one. In short, these two coupled oscillators, represented by the two arms, mutually influenced one other to produce a single coordinated behaviour, synchronous in time and space.

We have shown that, in terms of their control, the speech articulators and the hands and arms are closely related. We suggest that, indeed, the systems activating mouth and arms can mutually influence and entrain one another, much as has been amply demonstrated for limb systems alone. Furthermore, we propose that these entrainments are dynamic and flexible such that activation of one system can have various effects on the other — tight temporal synchrony, or more loosely coupled influence — according to von Holst's principles above. We believe that this conceptualization of mutually influential systems can help explain the linkage between speech and gesture.

To begin to understand the initial hand–mouth linkages and the subsequent developmental changes that we describe, we propose a simple, qualitative model. Two concepts are critical: the notions of the *thresholds* for eliciting vocal and manual behaviours, and their *relative activation strengths*, and in particular, their ability to pull in and entrain the activity of the complementary system. The threshold for a behaviour measures its ease of performance: in naturally occurring behaviours seen in infants, a good measure is how frequently they are performed. Behaviours with a low threshold for performance are seen frequently and under different task contexts. Behaviours with a high threshold, in contrast, are effortful and less frequently produced. Thus, we assume that first gestures and words have a high threshold (as do first appearances of any new skill). One effect of repeated practice is to lower the threshold for performance, to make that behaviour available at different times and in different and variable contexts.

Activation is the relative strength of the behaviour once the threshold is reached. Because a great deal of effort is required in order for relatively novel, unpracticed forms of behaviour to emerge, we assume that new behaviours have relatively low levels of activation. In contrast, more established, well-practiced behaviours can be said to have relatively higher levels of activation; that is, they are strong, stable skills. A critical assumption is that the dynamic coupling of two effector systems — either limbs or limbs and oral structures — requires relatively high levels of activation in order for mutual entrainment to occur.

The Development of the Coupled Speech–Gesture System

We now put these ideas of coupled oscillators, thresholds, and activation together to describe the ontogeny of oral and limb movements leading to gesture and speech coupling. We propose a dynamic developmental progression characterized by four phases: 1) *initial linkages*: hand and mouth activity are loosely coupled from birth; 2) *emerging control*: increasing adaptive use of hands and mouth, especially marked by rhythmical, sometimes coordinated, activities in both manual and vocal modalities; 3) *flexible couplings*: emergence of coupled, but not synchronous gesture and speech; 4) *synchronous coupling*: more adult-like, precisely-timed coupling of gesture and speech. This progression is summarized in Table 1 and Figure 1. We now turn to a description of how the initial biases that link hand and mouth become progressively elaborated as language and gestural communication emerge and relate these changes to the model outlined here.

The early oral–manual system

Connections between the oral and manual systems are in place from birth (cf. Table 1). This link is initially apparent in the Babkin reflex: newborns react to pressure applied to the palm by opening their mouths. Moreover, coordination between oral and manual actions is extremely common in infants' spontaneous movements. For instance, newborns frequently bring their hands to the facial area, contact the mouth, and introduce the fingers for sucking, often maintaining hand–mouth contact for extended periods of time. Hand-to-mouth behaviour in young infants looks goal-directed. Infants bring their hands to the mouth in the absence of prior facial contact. They open their mouths in 'anticipation' of the arrival of the hand. The trajectory followed by the hand en route to the mouth varies widely from bout to bout, suggesting that they can attain mouth contact from many different starting positions (Butterworth and Hopkins, 1988).

Hand-to-mouth behaviour continues to be an important action throughout the first year, but the behaviour shifts in function. As soon as infants are able to grasp and hold objects placed in their hands, usually at two months, they bring these objects to their

Developmental period	Evidence	Oral (speech)/manual linkages	
Newborn: Initial linkages	Oral: Sucking, crying, vegetative sounds Manual: Hand to mouth/ reflexive grasping, spontaneous movements, no ability to reach	Babkin reflex Spontaneous hand/mouth coordination	Hand and mouth are mutually activated.
Six to eight months: Emerging control	Oral: cooing, sound play, reduplicative babbling Manual: Onset of reaching, rhythmical waving and banging, manual babbling	Onsets of rhythmical vocal and manual babbling, rhythmical arm movements coincide	Rhythmical activities in arms and hands and in speech articulators mutually entrained.
Nine to 14 months: Emergence of gestures and words	Oral: variegated babbling, onset of first words Manual: onset of first gestures, fine motor control in fingers improves	Communicative gestures precede first words; gesture use predicts first words. Gestures and speech have different referents. When gestures and speech co-occur, they are sequential.	Threshold for gestural activation lower than for speech. No simultaneous coactivation of speech and gesture because threshold for both is high, but entrainment activation is low.
16 to 18 months: Emergence of synchronous speech and gesture	More communication Increasing vocabulary Continued fine motor improvement	Onset of meaningful, synchronous word + gesture combinations	Practice with communication lowers thresholds and increases entrainment activation, leading to synchrony.

Table 1. Developmental progression of oral–manual linkages during the first two years.

mouths and explore them orally (Lew and Butterworth, 1997; Rochat, 1989). Indeed, when infants learn to reach out and grab objects on their own, they invariably bring these objects to their mouths, a behaviour that continues throughout the first year.

These hand–mouth linkages are also apparent in communicative settings. Fogel and Hannan (1985) observed a group of infants between the ages of 9 and 15 weeks during face-to-face interaction with their mothers and found systematic relationships between certain types of hand actions and oral activity. In particular, extensions of the index finger were especially likely to co-occur with either vocalization or mouthing movements.

Taken together, these observations suggest that discrete manual actions and oral or vocal activity are linked from birth and continue to be coupled in the first months of life, well before the emergence of first gestures and words. In terms of our dynamic model, they further suggest that thresholds for hand–mouth activity are relatively low and activation is high in the first months (cf. Figure 1). Instances of hand–mouth

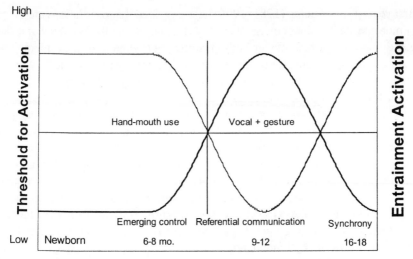

Figure 1. Threshold and entrainment activation levels in the oral–manual system during the first two years.

Initially, activation (depicted in the thick line) is high and the threshold (depicted in the thin line) is low and remain so until the emergence of referential communication. At this point, communication is a novel and effortful skill, and thus the threshold is raised and activation becomes relatively weak. As children practice their communicative skills, the threshold is lowered and the level of activation increases.

contact and co-occurrences of hand movements with vocalizations are seen frequently; and infants tend to spend a substantial portion of their time with their hands in their mouths (Vereijken *et al.*, 1999). In short, there appears to be some degree of co-activation of the hands and mouth from very early in life, such that co-occurring manual and oral behaviours form a central part of the young infant's behavioural repertoire.

It is tempting to speculate that these initial hand–mouth linkages are phylogenetically established, possibly as the result of mechanisms that link manipulation and feeding. In non-human primates, for instance, manual dexterity is associated with food-gathering processes such as opening seeds, fishing for termites, and using tools to break shells and husks. This suggests that the speech–gesture linkages may be using brain systems that long predate human language and, indeed, may have evolved for completely different functional purposes. In this way, we echo Bates *et al.*'s (1979) claim that 'Language is a new machine that nature made from old parts'.

Reorganization of the oral–manual system and emerging control

At around three or four months of age, infants show increasing adaptive control of both the hand–arm and the oral–vocal systems (cf. Table 1). Visually-elicited reaching and grabbing objects emerges at this time, as does the ability to produce differentiated vowel sounds and cooing vocalizations, especially during social interactions. Although the manual and vocal systems appear to be developing relatively independently, there are also indications of continued coupling, and indeed, mutual influence.

Manual–vocal coupling is best evidenced in the production of *rhythmical movements* in both effectors. Rhythmicity is highly characteristic of emerging skills during

the first year of life. Thelen (1981; 1996) has suggested that oscillations are the product of motor systems under emergent control; that is, when infants attain some degree
of intentional control of limbs or body postures, but when their movements are not
fully goal-corrected. Thus, for instance, infants commonly rock to and fro when they
can assume a hands-and-knees posture, but before they can coordinate their four
limbs for forward propulsion in creeping.

Rhythmical movements of the arms and hands — waving, swaying, banging —
indeed increase dramatically between the ages of 26 and 28 weeks. This is several
months after infants first reach, but before they develop differentiated use of arms and
hands for manipulation. The emergence of canonical babbling (i.e., when babies
begin to produce strings of reduplicated syllables, such as 'gagaga' or 'bababa'),
which is also a rhythmic behaviour, occurs at about the same age, averaging about 27
weeks (e.g., Oller and Eilers, 1988). Most importantly, findings from two studies suggest that there is a close temporal relationship between the onset of babbling and
changes in patterns of rhythmic hand activity.

The extent to which the emergence of babbling and changes in rhythmic hand and
arm movements are temporally related was specifically addressed in a cross-sectional
study of repetitive arm activity in infants who had either not yet begun to babble or
who had been babbling for varying lengths of time (Locke et al., 1995). Infants were
given a rattle to shake in either their right or left hand, and the overall frequency of
shakes per second was recorded. Results indicated that rate of shaking was relatively
low among prebabblers, increased substantially among infants who had just begun to
babble, and then declined somewhat (but remained above that for prebabblers)
among infants who had been babbling for longer periods of time. Importantly, the frequency of shakes was consistently higher for the right relative to the left hand across
all infants, regardless of amount of babbling experience, suggesting that the sharp
increase observed among new babblers cannot be explained simply by heightened
arousal levels in this group of infants.

The assumption that the oral and manual articulators are tightly linked from birth
and remain so also explains the recently described phenomenon of 'manual babbling'
observed in both deaf and hearing infants. First described by Petitto and Marentette
(1991), manual babbles are gestures that are neither communicative nor meaningful
and tend to consist of more than one movement cycle. For instance, a child might
extend the first finger of the left hand and repeatedly contact the palm of the right
hand, while at the same time giving no indication that the movement is meaningful or
directed toward a specific addressee. In other words, although the form of the movement may be gesture-like, the apparent absence of meaning and communicative
intent make it difficult to classify as a communicative gesture. Petitto and Marentette
interpreted manual babbling in deaf infants as a manual analogue of vocal babbling
that is evidence of a dedicated, amodal language acquisition faculty.

More recently, however, Meier and Willerman (1995) recorded instances of manual babbling in hearing infants with no exposure to sign language. In a longitudinal
study of two hearing infants, they reported that manual babbling co-existed with
vocal babbling, and that manual babbles accounted for a majority of the children's
manual output between the ages of 7 and 8.75 months. Based on these findings, they
concluded that the manual babbling may not necessarily be an indication of a
'brain-based language capacity', but rather one of a class of rhythmic behaviours that

emerge during the transition to more differentiated motor control. Thus, the period between about 6 and 9 months is one in which rhythmical and repetitive movements abound as transient patterns consistent with emergent fine motor control in both mouth and hand.

As we discussed earlier, it is well-known that biological oscillators tend to interact and entrain one another. Assuming an initial linkage of the mouth and hand subserved by the same brain systems, it is plausible that rhythmicity in the two effector systems is mutually influential. In our model, we see that this is also a period of low thresholds and high activation for rhythmical vocal and manual behaviours (cf. Figure 1); they are relatively common and often performed (see Thelen, 1979; Oller and Eilers, 1988, for frequencies). Given the combination of low thresholds and relatively high activation, our dynamic prediction is that these two systems should mutually entrain. Indeed, in a recent longitudinal study of Japanese infants, Ejiri (1998) reported that approximately 40% of all rhythmic manual activity co-occurred with babbling, and that 75% of all babbling co-occurred with rhythmic manual activity. Additional evidence of mutual entrainment comes from the finding that the average syllable length in bouts of babbling accompanied by hand activity was significantly longer than that in bouts that did not co-occur with manual actions (Ejiri and Masataka, 1999). This difference is illustrative of a principle of coupled oscillators that we described earlier, namely that two motor systems (in this case, the hands and the jaw) mutually influence one another and ultimately settle on a 'compromise' frequency at which they entrain to produce a coordinated behaviour.

We may speculate further that the development of vocal babbling is actually facilitated by early rhythmical limb movements. Infants have a long history of producing rhythmic arm, leg, and torso movements prior to the onset of canonical babbling. It is possible that production of repetitive, rhythmically-organized movements gradually entrains vocal activity, leading eventually to the production of the mandibular oscillations that comprise babbling. In short, to the extent that manual and vocal babbling are indicative of increased control over the manual and oral articulators, they may be transitional behaviours in the development of the speech–gesture system. The repetition of babbling activity in both modalities may then allow the child to gain further control over the oral and manual articulators, control that is clearly necessary for the production of first words and gestures.

Learning to talk and to gesture: the period of flexible coupling

In the last few months of the first year, infants' manual and vocal behaviours change (cf. Table 1). Banging and waving decrease, and infants use their hands for more finely differentiated manipulation. Likewise, babbling gives way to words and word-like productions. This period also sees the emergence of communicative gestures such as pointing, showing, and requesting. As production of communicative gestures increases, beginning between the ages of 10 and 11 months and continuing through the first few months of the second year, manual babbles tend to decline (Meier and Willerman, 1995). Rhythmic repetition thus gives way to more articulated control and more directed communication.

Importantly, however, during this third transition, communication by gesture is predominant, while verbal communication tends to lag behind. For example, children often produce their first gestures several weeks before they say their first words (e.g.,

Bates *et al.*, 1979; Caselli, 1990). In addition, gestures often outnumber words in the communicative repertoires of individual children at this stage, and many children show a strong preference for gestural over verbal communication in their spontaneous interactions (Iverson *et al.*, 1994).

What does this transition tell us about the organization of the coupled speech–gesture system? In our model, this period is characterized by asymmetry in the relative control and activation of these effectors. At this time, gestures become increasingly frequent, while speech develops somewhat more slowly and is more effortful. Thus, relative to speech, for which thresholds are high and activation is relatively weak, well-practiced manual activities and gesture have lower thresholds and higher activation (cf. Figure 1). In short, we believe that the threshold for communication in the manual mode is much lower in late infancy than in the vocal mode, likely because control of the hands and arms is more advanced than that of the vocal articulators. If infants are motivated to communicate, it is simply easier for them to use movements that have been well-practiced in the service of object exploration. This is a well-known phenomenon of motor control, namely that stable and well-patterned movements are preferred over newer and less well-established coordinations (Zanone and Kelso, 1991).

Although gestures often play a predominant role in children's early production, the two systems are still tightly linked. Several observations provide support for this claim. First, gesture production can predict impending change in speech. Thus, for example, Bates *et al.* (1979) reported that gesture production was positively related to gains in language development between 9 and 13 months. In other words, children who made the most extensive use of gesture were also those who exhibited the most precocious language development.

Second, work by Acredolo and Goodwyn suggests that teaching typically-developing, hearing children to gesture has positive effects on language development (e.g., Goodwyn and Acredolo, 1993; 1998). These researchers asked parents to teach their infants a small set of communicative gestures and encourage them to use these gestures in daily interactions. Longitudinal data on the attainment of early language milestones indicated that these children produced their first symbols (i.e., a word or gesture that is used reliably to 'stand for' a referent, independent of context or proximity to the referent) and attained the five-symbol milestone approximately one month earlier than groups of children who had received no training or who had been taught a small set of words respectively.

A final piece of evidence comes from studies indicating that continued delay in the development of productive language may be predicted from gesture production. This research involves examining patterns of speech and gesture production in young children who are 'late talkers', a group generally characterized by delayed acquisition of productive vocabulary in the absence of hearing loss, mental retardation, behavioural disturbances, or known forms of neurological impairment. Thal and Tobias (1992) analysed late talkers' spontaneous speech and gesture production at an initial visit and at a follow-up one year later. By the follow-up visit, some of the children originally identified as late talkers had caught up with their peers in terms of their language production abilities (the 'late bloomers'), while others continued to show delays in expressive language (the 'truly delayed' children).

These investigators found that the late bloomers and the truly delayed children could be reliably distinguished from one another on the basis of their communicative gesture production at the initial visit. Thus, late bloomers produced significantly more communicative gestures than did truly delayed children, who looked more like a group of younger children matched on the basis of productive language. That reduced use of communicative gestures is related to delayed language development is further indicative of the coupled nature of the gesture–speech system; that is, when functioning in one part of the system is compromised, functioning in other components may also be disrupted.

The emergence of synchronous speech and gesture

As we mentioned earlier, gestures and speech are very tightly coupled in adults. When people talk and gesture at the same time, the 'stroke' (or active phase) of the gesture is timed precisely with the word or phrase it accompanies. This timing relationship is so strong that when a speaker stutters, the gesture tends to be held motionless until the speech is recovered (Mayberry et al., 1998).

This tight temporal link appears to develop during the initial period of gestures and first meaningful words (cf. Table 1). Butcher and Goldin-Meadow (in press) described patterns of relative timing of words and gestures in infants as they acquired early language. They found that at first, gestures tended to be produced without speech or with meaningless speech. Even when gesture and vocal utterances occurred together, they were not tightly linked in time. The gesture and the word or vocalization occurred sequentially, not simultaneously. Adult-like gesture–speech synchrony emerged rather dramatically when infants began to combine meaningful words with gestures.

This change in the timing relationship between gesture and speech can be accounted for in our model in the following way. During the time when infants are just beginning to acquire many new words, speech requires concentration and effort, much like the early stages of any skill learning. As infants practice their new vocal skills, thresholds decrease, and activation for words becomes very high (cf. Figure 1). Since the level of activation generated by words is well beyond that required to reach threshold, it has the effect of capturing gesture and activating it simultaneously. The behavioural result of this co-activation is a word–gesture combination in which the two elements are fully synchronous. In other words, as words are practiced, they are able to activate the gesture system sufficiently to form synchronous couplings, and thus the two motor systems become entrained. Thus, by the time infants are combining meaningful words with single gestures, speech is synchronously coupled with gesture, a coupling that remains tight throughout life.

Thus, we see the intensive period of word learning as the point at which initial oral/manual linkages are consolidated into a new organization that couples the emergent gesture system with the emergent speech system. It is the dynamics of change in the effort required for these early skills that provides the 'spill-over' activation needed to link the two effector modes in a common communicative intent.

Speculations on an Embodied Cognition

We have shown with converging evidence that systems of movement for mouth and for hand cannot be separated from one another, and that they are intimately linked in the production of language, the pinnacle of human cognition. We invoked concepts of coupled oscillators that hold that related systems can interact, mutually activate and entrain one another. We further speculated that during development, with initial linkages established phylogenetically through feeding systems, it is manual activity that acts as a magnet. Through rhythmical activity, and later through gesture, the arms gradually entrain the activity of the vocal apparatus. This mutual activation increases as vocal communication through words and phrases becomes more practiced, leading to tight synchrony of speech and gesture in common communicative intent.

What does this mean for a theory of embodiment? As language develops, its expression through speech is continually accompanied by movement, such that vocal behaviour is tightly intertwined with hand and arm activity. These movements co-occur with the communicative intent that produces them. Thus, every communicative act, either by speech or gesture is remembered as an ensemble, which includes the proprioceptive consequences of that movement. As utterances become common and frequently practiced, the motor repertoire that is mapped with the growing language competence also becomes strengthened. The initial biases to move hand and mouth together thereby cascade into a single coupled, communicative system, where the mental aspects of the expression are manifest in movement.

The speech–language–gesture system is a particularly compelling example, we believe, not only of the sensorimotor origins of thought, but also of its continued embodiment throughout life. In this chapter we showed that a model based on notions of coupled oscillators can explain the changing patterns of coordination of mouth and hand in the first years and their link with emergent language. The critical point for embodiment is that such coupling demands that the systems involved in speech, gesture, and language are represented in the brain in commensurate codes. That is, the representations of the mental aspects of language must be able to mesh seamlessly with those involved in the control of movements. Traditionally, language is viewed as symbolic and discrete, represented by lexical items and grammatical rules. Perception and action, on the other hand, are subsymbolic and better described in the analogic realm of dynamics. Yet the fact that gesture shares a semantic and communicative burden with speech as well as a tight temporal coupling means that they must also share a common, integrative mechanism. Indeed our speculative model offers a mechanism — dynamics — by which these seemingly incommensurate codes may be reconciled. We suggest that just as these aspects are linked initially, when language emerges, so they remain coupled throughout life.[1]

The issue of commensurate codes is equally relevant for all aspects of human cognition. Cognition — remembering, planning, deciding, and rehearsing — is abstract and mental and often couched in terms of concepts and symbols. Again, perception

---

[1] For a recent review of dynamic approaches in cognitive science, see Beer (in press). Efforts to cast mental events in the language of dynamics or connectionist networks have increased in the last decade. See, for instance, Elman's (1995) work on connectionist models of language, Thelen *et al.* (in press) for a dynamic model of Piaget's A-not-B error, and Schöner *et al.* (1995) and Pfeiffer and Scheier (1999) for work in autonomous robots.

and action deals with the here-and-now and seems not to require an elaborate representational structure. But if we consider how real people behave in everyday life, it is impossible to draw a line between the two modes of processing. Sometimes people engage in contemplation, problem solving, and day-dreaming, where mental activity is predominant. At other times, the situation demands being closely 'clamped' to the environment (Glenberg, 1997). But most often, people shift rapidly and seamlessly between the two types of engagement: having moments of thought interspersed with nearly continual on-line activity. For instance, imagine driving down the highway mentally reviewing your next lecture when a deer darts in front of the car. In a split second, you become totally and completely 'clamped' to the immediate situation. Or consider the ability to mentally rehearse an unfamiliar route or a difficult and new motor skill and then carry it out (Jeannerod, 1997). Here again, it is difficult to imagine how this integration of thought and perception–action could be accomplished if their mental currencies are fundamentally incompatible. The issue, therefore, is not how to transgress a divide between cognition and action or between body and mind. The integration is seamless in both directions. Rather, the critical dimension is the balance between on- and off-line, or the relative strength of the coupling between the mental dynamics and those of the body and the environment. Put another way, people are adept at shifting the relative dominance of the immediate input versus the relative strength of the remembered input as the context demands.

This formulation also recasts the developmental issue. Traditionally, cognitive development is construed as moving from purely sensorimotor processing to that which is more conceptual and abstract. Gaining the ability to process 'off-line' is indeed a tremendous developmental advance, moving infants from being dominated by the immediate input to the ability to hold aspects of the environment in memory and using those stored memories to plan actions. But children must also learn to perform well on-line and, most importantly, to rapidly and appropriately switch between these modes of functioning. In terms of a dynamic model, this means tight coupling when the occasion warrants, but also great flexibility to modulate that coupling when the situation demands different skills. Thus, development is not so much saying 'bye-bye' to being in the world as learning to use cumulative experiences to adaptively act in the world.

In sum, our argument for embodiment rests on the necessity for compatible dynamics so that perception, action, and cognition can be mutually and flexibly coupled. Such dynamic mutuality means that activity in any component of the system can potentially entrain activity in any other component, as illustrated by von Holst's principles. We suggested that in the development of communication, rhythmic manual activity captured and entrained the coordination of the oral system and that both were linked with emerging speech and language. In that all cognition grows from perception and action and remains tied to it, body, world and mind are always united by these common dynamics. Action influences thought as much as thought motivates action.

Acknowledgment

Jana M. Iverson was supported by NIH Training Grant HD07475 and Esther Thelen by an NIMH Research Scientist Award.

References

American Psychiatric Association (1994), *Diagnostic and Statistical Manual of Mental Disorders (4th Edition)* (Washington, D.C.: American Psychiatric Association).

Bates, E., Benigni, L., Bretherton, I., Camaioni, L., and Volterra, V. (1979), *The Emergence of Symbols: Cognition and Communication in Infancy* (New York: Academic Press).

Beer, R.D. (in press), *Dynamic Approaches to Cognitive Science. Trends in Cognitive Science*

Bonda, E., Petrides, M., Frey, S., and Evans, A.C. (1994), 'Frontal cortex involvement in organized sequences of hand movements: Evidence from a positron emission tomography study', *Society for Neurosciences Abstracts*, **20**, p. 353.

Butcher, C.M. and Goldin-Meadow, S. (in press), 'Gesture and the transition from one- to two-word speech: When hand and mouth come together', to appear in *Language and Gesture: Window into Thought and Action*, ed. D. McNeill (Cambridge: Cambridge University Press).

Butterworth, B. and Beattie, G. (1978), 'Gestures and silence as indicators of planning in speech', in *Recent Advances in the Psychology of Language: Formal and Experimental Approaches*, ed. R. Campbell and P.T. Smith (London: Plenum).

Butterworth, G. and Hopkins, B. (1988), 'Hand–mouth coordination in the new-born baby', *British Journal of Developmental Psychology*, **6**, pp. 303–13.

Caselli, M.C. (1990), 'Communicative gestures and first words', in *From Gesture to Language in Hearing and Deaf Children*, ed. V. Volterra and C.J. Erting (New York: Springer-Verlag).

Clark, A. (1997), *Being There: Putting Brain, Body and World Together Again* (Cambridge, MA: MIT Press).

Cole, J., Gallagher, S., McNeill, D., Duncan, S.D., Furuyama, N., and McCullough, K.-E. (1998), 'Gestures after total deafferentation of the bodily and spatial senses', in *Oralité et gestualité: Communication multi-modale, interaction*, ed. Santi *et al.*, (Paris: L'Harmattan), pp. 65–9.

Ejiri, K. (1998), 'Synchronization between preverbal vocal behavior and motor action in early infancy. I. Its developmental change', *Japanese Journal of Psychology*, **68**, pp. 433–40.

Ejiri, K. and Masataka, N. (1999), 'Synchronization between preverbal vocal behavior and motor action in early infancy. II. An acoustical examination of the functional significance of the synchronization', *Japanese Journal of Psychology*, **69**, pp. 433–40.

Elman, J.L. (1995), 'Language as a dynamical system', in *Mind as Motion*, ed. R.F. Port and T. van Gelder (Cambridge MA: MIT Press).

Erhard, P., Kato, T., Strupp, J.P., Andersen, P., Adriany, G., Strick, P.L., and Ugurbill, K. (1996), 'Functional mapping of motor in and near Broca's area', *Neuroimage*, **3**, S367.

Fogel, A., and Hannan, T.E. (1985), 'Manual actions of nine- to fifteen-week-old human infants during face-to-face interactions with their mothers', *Child Development*, **56**, pp. 1271–9.

Fried, I., Katz, A., McCarthy, G., Sass, K.J., Williamson, P., Spencer, S.S., and Spencer, D.D. (1991), 'Functional organization of human supplementary motor cortex studied by electrical stimulation', *Journal of Neuroscience*, **11**, pp. 3656–66.

Glenberg, A.M. (1997), 'What memory is for', *Behavioral and Brain Sciences*, **20**, pp. 1–56.

Goodwyn, S.W., and Acredolo, L.P. (1993), 'Symbolic gesture versus word: Is there a modality advantage for the onset of symbol use?', *Child Development*, **64**, pp. 688–701.

Goodwyn, S. and Acredolo, L.P. (1998), 'Encouraging symbolic gestures: A new perspective on the relationship between gesture and speech', in *The Nature and Functions of Gesture in Children's Communications. New Directions for Child Development, no. 79*, ed. J.M. Iverson and S. Goldin-Meadow (San Francisco: Jossey Bass).

Grabowski, T.J., Damasio, H., and Damasio, A.R. (1998), 'Premotor and prefrontal correlates of category-related lexical retrieval', *Neuroimage*, **7**, pp. 232–43.

Hadar, U. (1989), 'Two types of gesture and their role in speech production', *Journal of Language and Social Psychology*, **8**, pp. 221–8.

Hadar, U., Wenkert-Olenik, D., Krauss, R., and Soroker, N. (1998), 'Gesture and the processing of speech: Neuropsychological evidence', *Brain and Language*, **62**, pp. 107–26.

Hanlon, R.E., Brown, J.W., and Gerstman, L.J. (1990), 'Enhancement of naming in nonfluent aphasia through gesture', *Brain and Language*, **38**, pp. 298–314.

Hill, E.L. (1998), 'A dyspraxic deficit in specific language impairment and developmental coordination disorder? Evidence from hand and arm movements', *Developmental Medicine and Child Neurology*, **40**, pp. 388–95.

Iverson, J.M., Capirci, O. and Caselli, M.C. (1994), 'From communication to language in two modalities', *Cognitive Development*, **9**, pp. 23–43.

Iverson, J.M. and Goldin-Meadow, S. (1998), 'Why people gesture when they speak', *Nature*, **396**, p. 228.

Jeannerod, M. (1997), *The Cognitive Neuroscience of Action* (Oxford: Blackwell).

Johnson, M. (1987), *The Body in the Mind* (Chicago: The University of Chicago Press).

Kelso, J.A.S. (1995), *Dynamic Patterns: The Self-Organization of Brain and Behavior* (Cambridge, MA: MIT Press).

Kimura, D. and Archibald, Y. (1974), 'Motor functions of the left hemisphere', *Brain*, **97**, pp. 337–50.

Krams, M., Rushworth, M.S.F., Deiber, M.-P., Frackowiak, R.S.J., and Passingham, R.E. (1998), 'The preparation, execution, and suppression of copied movements in the human brain', *Experimental Brain Research*, **120**, pp. 386–98.

Krauss, R.M. (1998), 'Why do we gesture when we speak?', *Current Directions in Psychological Science*, **7**, pp. 54–60.

Krauss, R.M. and Hadar, U. (1999), 'The role of speech-related arm/hand gestures in word retrieval', in *Gesture, Speech, and Sign*, ed. L.S. Messing and R. Campbell (Oxford: Oxford University Press).

Kugler, P.N. and Turvey, M.T. (1987), *Information, Natural Law, and the Self-Assembly of Rhythmic Movement* (Hillsdale, NJ: Erlbaum).

Leiner, H.C., Leiner, A.L., and Dow, R.S. (1989), 'Reapprasing the cerebellum: What does the hindbrain contribute to the forebrain?', *Behavioral Neuroscience*, **103**, pp. 998–1008.

Leiner, H.C., Leiner, A.L., and Dow, R.S. (1993), 'Cognitive and language functions of the human cerebellum', *Trends in Neuroscience*, **16**, pp. 444–7.

Levelt, W.J.M. (1989), *Speaking* (Cambridge, MA: MIT Press).

Levelt, W.J.M., Richardson, G., and La Heij, W. (1985), 'Pointing and voicing in deictic expressions', *Journal of Memory and Language*, **24**, pp. 133–64.

Lew, A.R. and Butterworth, G. (1997), 'The development of hand–mouth coordination in 2- to 5-month-old infants: Similarities with reaching and grasping', *Infant Behavior and Development*, **20**, pp. 59–69.

Locke, J.L., Bekken, K.E., McMinn-Larson, L., and Wein, D. (1995), 'Emergent control of manual and vocal–motor activity in relation to the development of speech', *Brain and Language*, **51**, pp. 498–508.

Mayberry, R.I., Jacques, J., and DeDe, G. (1998), 'What stuttering reveals about the development of the gesture–speech relationship', in *The Nature and Functions of Gesture in Children's Communication. New Directions for Child Development, no. 79*, ed. J.M. Iverson and S. Goldin-Meadow (San Francisco: Jossey-Bass).

McNeill, D. (1992), *Hand and Mind: What Gestures Reveal About Thought* (University of Chicago Press).

Meier, R.P. and Willerman, R. (1995), 'Prelinguistic gesture in hearing and deaf infants', in *Language, Gesture, and Space*, ed. K. Emmorey and J. Reilly (Hillsdale, NJ: Erlbaum).

Oller, D.K. and Eilers, R.E. (1988), 'The role of audition in infant babbling', *Child Development*, **59**, pp. 441–66.

Ojemann, G.A. (1984), 'Common cortical and thalamic mechanisms for language and motor functions', *American Journal of Physiology*, **246** (Regulatory Integrative and Comparative Physiology 15), R901–R903.

Pashek, G. (1997), 'A case study of gesturally cued naming in aphasia: Dominant versus non-dominant hand training', *Journal of Communication Disorders*, **30**, pp. 349–66.

Pedelty, L.L. (1987), 'Gesture in aphasia'. Unpublished doctoral dissertation, The University of Chicago.

Petersen, S.E., Fox, P.T., Posner, M.I., Mintun, M. and Raichle, M.E. (1989), 'Positron emission tomographic studies of the processing of single words', *Journal of Cognitive Neuroscience*, **1**, pp. 153–70.

Petitto, L.A. and Marentette, P. (1991), 'Babbling in the manual mode: Evidence for the ontogeny of language', *Science*, **251**, pp. 1493–6.

Pfeiffer, R. and Scheier, C. (1999), *Understanding Intelligence* (Cambridge MA: MIT Press).

Pulvermüller, F., Preissl, H., Lutzenberger, W., and Birbaumer, N. (1996), 'Brain rhythms of language: Nouns versus verbs', *European Journal of Neuroscience*, **8**, pp. 937–41.

Rauscher, F.H., Krauss, R.M., and Chen, Y. (1996), 'Gesture, speech, and lexical access: The role of lexical movements in speech production', *Psychological Science*, **7**, pp. 226–31.

Rochat, P. (1989), 'Object manipulations and exploration in 2- to 5-month-old infants', *Developmental Psychology*, **25**, pp. 871–84.

Schöner, G., Dose, M., and Engels, C. (1995), 'Dynamics of behavior: theory and applications for autonomous robot architectures', *Robotics and Autonomous Systems*, **16**, pp. 213–45.

Sheets-Johnstone, M. (1990), *The Roots of Thinking* (Philadelphia: Temple University Press).

Thal, D.J. and Tobias, S. (1992), 'Communicative gestures in children with delayed onset of oral expressive vocabulary', *Journal of Speech and Hearing Research*, **35**, pp. 1281–9.

Thelen, E. (1979), 'Rhythmical stereotypies in normal human infants', *Animal Behaviour*, **27**, pp. 699–715.

Thelen, E. (1981), 'Kicking, rocking, and waving: Contextual analyses of rhythmical stereotypies in normal human infants', *Animal Behaviour*, **29**, pp. 3–11.

Thelen, E. (1996), 'Normal infant stereotypes: A dynamic systems approach', in *Stereotyped Movements*, ed. R.L. Sprague and K. Newell (Washington, DC: American Psychological Association).

Thelen, E., Schöner, G., Scheier, C., and Smith, L.B. (in press), 'The dynamics of embodiment: A field theory of infant perseverative reaching', *Behavioral and Brain Sciences*.

Varela, F. J., Thompson, E., Rosch, E. (1991), *The Embodied Mind* (Cambridge, MA: MIT Press).

Vereijken, B., Spencer, J.P., Diedrich, F.J., and Thelen, E. (1999), 'A dynamic systems study of posture and the emergence of manual skills', manuscript under review.

Zanone, P.G. and Kelso, J.A.S. (1991), 'Experimental studies of behavioral attractors and their evolution in learning', in *Tutorials in Motor Neuroscience*, ed. J. Requin & G.E. Stelmach (Dordrechts: Kluwer).

Rafael Núñez

Could the Future Taste Purple?
Reclaiming Mind, Body and Cognition

*This article examines the primacy of real-world bodily experience for understanding the human mind. I defend the idea that the peculiarities of the living human brain and body, and the bodily experiences they sustain, are essential ingredients of human sense-making and conceptual systems. Conceptual systems are created, brought forth, understood and sustained, through very specific cognitive mechanisms ultimately grounded in bodily experience. They don't have a transcendental abstract logic independent of the species-specific bodily features. To defend this position, I focus on a case study: the fundamental concept of **time flow**. Using tools of cognitive linguistics, I analyse the foundations of this concept, as it is manifested naturally in everyday language. I show that there is a precise conceptual metaphor (mapping) whose inferential structure gives an account of a huge variety of linguistic expressions, semantic contents, and unconscious spontaneous gestures: Time Events Are Things In Space. I discuss various special cases of this conceptual metaphor. This mapping grounds its source domain (space) in specific spatial bodily experiences and projects its inferential structure onto a target domain (time) making inferences in that domain possible. This mechanism allows us to unconsciously, effortlessly, and precisely understand (and make inferences with) expressions such as 'the year 2000 is **approaching**' or 'the days **ahead** of us'. The general form of the mapping seems to be universal. The analysis raises important issues which demand a deeper and richer understanding of cognition and the mind: a view that sees the mind as **fully embodied**. In order to avoid misunderstandings with a general (and somewhat vague) notion of 'embodiment' which has become fashionable in contemporary cognitive science, I describe what I mean by 'full embodiment': an embodied–oriented approach that has an explicit commitment to **all** of cognition, not just to low-level aspects of cognition such as sensory-motor activity or locomotion (lower levels of commitment). I take embodiment to be a living phenomenon in which the primacy of bodily grounded experience (e.g., motion, intention, emotion) is inherently part of the very subject matter of the study of the mind.*

The Colleague Who Had a Question For You . . .

Let's say that you are a scientist. Your subject matter is the study of the mind, the human mind. More specifically, you study conceptual systems and how humans make basic effortless inferences in everyday life. One day a colleague comes in and tells you that she went out and did a field study. She observed people in everyday

Journal of Consciousness Studies, **6**, No. 11–12, 1999, pp. 41–60

conversations, talking, making gestures, making jokes. She also observed TV commercials, scientists giving talks, and priests celebrating ceremonies, and studied how ideas are expressed in printed material such as technical books, newspapers, holy texts, and commercial advertisements. Among many things, your colleague observed that no matter what the field, and irrespective of whether it was oral or written English, people use expressions such as:

> *Faster* than ever we are *approaching* the end of the millennium; he finally *left* his sad past *behind*; the winter hasn't *arrived* yet; he is organizing a *retro*spective of Hitchcock's movies; the days *ahead* of us are promising; the concert took place the day *before* yesterday; Christmas is *gone*; it started all the *way back* in the thirties; the millennium bug will bother us well *beyond* the year 2000; so *far* we have been lucky. And so on.

Your colleague points out to you that these expressions, although making use of completely different words, being about different subjects, and being observed in different contexts, do have something in common: they all serve to express ideas about *time* in terms of objects, positions and movements in *space* (see Figure 1). What is interesting, your colleague adds, is that people seemed to use these expressions in an absolutely natural way, with no effort, and without even being aware of the fact that they were talking (or reading) about time events in terms of objects in space. Listeners engaged in conversations seemed to follow what was said in an equally natural effortless manner. Moreover, and interestingly enough, people seemed to make quick, precise, and effortless inferences when expressions like those were made. People immediately understood that the days *ahead* referred to days *in the future; that those days have not occurred yet; that those days will occur earlier than days that are even further ahead*, and so on. People never seemed to be puzzled or confused with expressions such as 'we are *approaching* the *end* of the millennium'. Somehow, they implicitly understood that *approaching* implies that the given time (the end of the millennium) is a moment that has not occurred yet, and that it may occur sometime relatively soon. Completely intrigued with these observations, your colleague asks you a very direct and simple question: *How is it that human beings understand so effortlessly and unconsciously ideas, experiences, and inferences about **time**, while talking about **space**? How can that be?*

So, here you are — you the scientist — someone who is trying to understand the human mind, the conceptual systems, and the inferential mechanisms that human animals take for granted and that make the most basic details of our everyday life so unbelievably livable . . . to the point that we don't even notice them! So, what do you say?

Serious business

My point with this little story is to make clear that these questions — although they may seem at first trivial and anecdotal — are far from being obvious, and that they should be taken very seriously in the various fields of cognitive science[1] and the study of the mind. In fact, there are more interesting details that make these observations

---

[1] By cognitive science, I simply mean the scientific study of cognition in a large sense (i.e., including various aspects of the mind). That is, it is the study of a particular subject matter — cognition (and the mind) — through the explicit use of the scientific methodology. From this perspective, nothing says that cognitive science *is* (or should be) about computation, or that it necessarily makes use of computer technology and computer-based concepts in studying the subject matter.

Figure 1

Human everyday language requires the capacity to make an impressive ammount of unconscious, effortless, and precise inferences in real time. For example, when looking at this cartoon, we 'naturally' understand the events in time in terms of things or locations in space. The cognitive mechanisms that make these everyday phenomena possible, are structured by fundamental bodily-grounded experiences. Their study reveals the embodied nature of the mind. (Reprinted with permission. KAL, *Cartoonists & Writers Syndicate*, 1999. www.cartoonists.com)

extremely relevant. When producing speech, people usually generate an impressive amount of spontaneous gestures, bodily postures, and facial expressions. More precisely, people produce — in a *perfectly synchronized* manner — spontaneous gestures which somehow match the meaning, timing, and form of the oral expressions used (McNeill, 1992; Iverson and Thelen, 1999). For instance, with a hand or a finger, people point towards something in their backs at the very moment when they say 'all the way *back* in the thirties'. Or they show something in front of them when saying 'the days *ahead* of us'. Therefore, bodily actions (i.e., spontaneous gestures) and speech, not only are coherent, but occur with an impressive synchronicity with speech. But there is more. In everyday conversations, people make the most amazing inferences in a matter of milliseconds. For example, consider the following questions.

What does it mean to say that 'Christmas is *gone*'? After all, Christmas is a social (and commercial) event. As such, it does not move anywhere. So, gone where? In what space did it *move*? *From* where *to* where? Going through what locations? Similarly, if two people are sitting in a pub drinking beer, why should they say '*faster* than ever we are *approaching* the end of the millennium', if they are just there, simply sitting, statically, drinking beer? How can they *approach* anything at all — much less a 'moment' such as 'the end of the millennium'? From where are they *approaching* it? *Faster* than what? How is it that people simply go about in their everyday conversations deeply understanding all these expressions, with *no effort at all*, often *not even being aware* of them, and what is more, making quite sophisticated inferences about the structure of temporal experiences?

Towards reclaiming cognition

In the following pages, I will analyse this intriguing phenomenon of human everyday conversations. Through the analysis of this time–space case study, I will try to give support to the main goal of the present volume. That is, to show that the scientific study of the mind needs to reconsider its very subject matter in a broader, deeper, and richer manner. I will argue that the questions above can be answered only in a limited manner from the perspective provided by traditional mainstream cognitive science. That is, when one approaches these questions with a dualistic and functionalistic view of cognition, where one sees cognition as a purely abstract rule-driven information-processing phenomenon, inherently separated from the nature of the body of the living human animal, and the bodily experiences it sustains (Freeman & Núñez, this volume). Through the time–space case study, I will defend the idea that cognition — and the study of the mind — needs to be reconsidered. It needs to be redefined, reclaimed in order to be understood in a more appropriate way than has been done by mainstream cognitive science and its various computer-metaphors for the mind.

In the rest of the article, I intend to accomplish several things. First, I will analyse the above time–space case through current work done in the emerging field of cognitive linguistics, focussing on the understanding of the notion of *time flow*. Second, this analysis will raise important issues which, I will argue, demand a deeper and richer understanding of cognition and the mind: a view that sees the mind as *fully embodied*. Third, I shall describe what I mean by (full) embodiment, and avoid misunderstandings with a general (and somewhat overused) notion of 'embodiment' which has become fashionable in contemporary cognitive science. I will take embodiment to be a living phenomenon in which the primacy of bodily grounded experience is inherently part of the very subject matter of the study of the mind. Finally, I will close defending the idea that in order to meet the foundational demands imposed by this view, we need to free ourselves from several taken-for-granted harmful dogmas that impede a clear understanding of important mental phenomena, and which have been at the core of much of mainstream cognitive science and philosophy of mind.

How do we Conceptualize Time Events? A View from Cognitive Linguistics

In recent years, the emerging field of cognitive linguistics has made some interesting contributions. Among others, it has confirmed that an important amount of abstract thought is unconscious (i.e., it happens below the level of awareness and therefore is

often beyond introspection), and it has shown that concepts are systematically organized through everyday cognitive mechanisms such as *conceptual mappings*. The
most well known conceptual mappings are *conceptual metaphors* (Lakoff and Johnson, 1980; see Lakoff, 1993, for a general overview) and *conceptual blends* (Turner
and Fauconnier, 1995; Fauconnier, 1997; Fauconnier and Turner, 1998). For the purpose of our case study, I will focus only on the first one.

A conceptual metaphor is a cognitive mechanism that allows us to make precise
inferences in one domain of experience (*target domain*) based on the inferences that
hold in another domain (*source domain*). Through this mechanism, the target domain
is understood, often unconsciously, in terms of the inferential structure that holds in
the source domain. One shouldn't get the idea that 'metaphor' here is a mere figure of
speech used by poets or politicians to illustrate an idea for aesthetic or manipulative
purposes (respectively). In fact, a conceptual metaphor, as understood in cognitive
linguistics, does not belong to the realm of words but to the realm of thought. And this
is very important to keep in mind: a conceptual metaphor is a cognitive mechanism,
an *inference-preserving cross-domain mapping*.

A key concept in this theory is that the 'projections' from source to target domain
are not arbitrary, and that they can be studied empirically and stated precisely. They
are not arbitrary, because they are motivated (in general) by our bodily grounded
experience, which is biologically constrained. For example, underlying expressions
like 'She greeted me *warmly*' or 'send her *warm* hellos', there is a conceptual metaphor which allows us to conceptualize Affection in terms of bodily grounded thermic
experiences: Warmth. This mapping is not a mere arbitrary social convention. It is
based on a (human) invariant, which is the shared experience of the correlation
between the bodily sensation of warmth and affection from the most early days of our
ontogeny.

Research in contemporary conceptual metaphor theory has shown that there is an
extensive conventional system of conceptual metaphors in every human conceptual
system. The empirical evidence comes from a variety of sources, including, among
others, psycholinguistic experiments (Gibbs, 1994), generalizations over inference
patterns (Lakoff, 1987), historical semantic change (Sweetser, 1990), and the study
of spontaneous gestures (McNeill, 1992). When combined in an appropriate way,
these conceptual mappings sustain even the most sophisticated forms of abstract
thinking, such as the conceptual apparatus underlying the whole edifice of mathematics
(Lakoff and Núñez, 1997; 2000; Núñez and Lakoff, 1998).

Among the hundreds of conceptual metaphors that have been studied in depth in
the last decade, there is the one concerning our understanding of time in terms of
motion in space, originally described as the Time Passing Is Motion metaphor
(Lakoff, 1993). Today we know that there are different forms of this mapping.[2] Particularly relevant is the distinction between *time-based* metaphors and *ego-based*
metaphors. Both of them are present in our everyday language, but they work in
rather different manners. The reason why they are called time- and ego-based is
because the former works in terms of a metaphorical 'orientation' applied to events

---

[2] Following a convention in Cognitive Linguistics, capitals here serve to denote the name of the conceptual mappings as such (e.g. Time Passing Is Motion). Particular instances of mappings, called metaphorical expressions (e.g., 'summer is approaching'), are not written with capitals.

(or times), and the latter applied to a bodily orientation of the speaker or his/her audience.[3] For the purpose of this article, I will call Time Events Are Things In Space the general conceptual mapping under which time-based and ego-based metaphors can be classified. Let's analyse them in more detail.

1. Time-based metaphor

This is a relatively simple conceptual metaphor. Its structure is the following.

Nature: Time is understood in terms of things (objects in a sequence) and motion in space.

Background conditions:
- There is a sequence of objects which:
  - may move horizontally (as a whole), and
  - in the direction of one of its extremes.

Mapping:

SOURCE DOMAIN Space		TARGET DOMAIN Time
Things	→	Times
Sequence of objects	→	Chronological order of times
Horizontal movement of the entire sequence in one direction	→	Passing of time
Things oriented with their fronts[4] in their direction of motion	→	Times oriented with their fronts in their direction of motion
An object *A* in front (behind) of an object *B* in the sequence	→	A time *A* occurs earlier (later) than a time *B*

Entailments:
- If time *B* follows time *A* (in the sequence), then time *B* occurs later than time *A* (it is in the future relative to time *A*).
- Transitivity properties applying to relative positions in the sequence in the source domain are preserved by the mapping so they are available in the target domain. For example, if event *C* is behind (in the future relative to) an event *B*, and event *B* is behind (in the future relative to) *A*, then event *C* is behind (in the future relative to) event *A*.
- Since the sequence of objects is one-dimensional, time is one-dimensional.

---

[3] The term 'ego' here is used in the sense found in the technical literature in linguistics (see, for example, Dunkel, 1983). This shouldn't be confused with the term 'ego' commonly used in psychology which designates a complex and dynamic integrative apparatus relating with the notion of self.

[4] Notice that in the source domain, 'front' is already a metaphorical front brought in from another conceptual mapping. A mapping that allows us to ascribe an orientation to objects relative to their normal direction of motion (as in the *front* of the car). Furthermore, the same mechanism allows us to ascribe a metaphorical orientation to objects which don't have inherent orientation, such as a cube: We can unmistakenly conceptualize the 'front' side of a cube sliding along a flat surface.

This time-based conceptual metaphor accounts for a variety of linguistic expressions (and their semantic entailments) such as,

> In the *preceding* session ... ; in the days *following* next Wednesday ... ; the day *before* yesterday ... ; Greenwich Mean Time is lagging *behind* the scientific standard.

2. Ego-based metaphor

This is a relatively complex conceptual metaphor which has two layers of encompassing inferential structure: a basic static one and a dynamic one. The complete dynamic mapping, in turn, manifests itself in two distinct forms depending on the nature of the moving entity. The basic (non-dynamic structure) is the following.

Ego-based metaphor: Basic static structure

Nature: Time is understood in terms of things (entities and locations) in space.

Background conditions:

- There is a landscape, and a canonical observer.

Mapping (basic static structure):

SOURCE DOMAIN Horizontal Uni-dimensional Space		TARGET DOMAIN Time
Things	→	Times
Order of things in a horizontal one-dimensional landscape	→	Chronological order of times
Things in front of the observer	→	Future times
Things behind the observer	→	Past times
Things at the location of the observer	→	Present times

Entailments:

- Transitivity properties applying to relative positions in the source domain are preserved by the mapping so they are available in the target domain. For example, if event A is *further away* ahead relative to the observer's orientation (in the future) than event B, and event B is further away (in the future) than C, then event A is further away ahead (in the future) than event C.

The basic static structure of the conceptual metaphor provides a rather rich inferential structure based on relations between positions in space. This fundamental partial mapping (without motion) accounts for an important number of linguistic expressions and their semantic entailments:

> The end of the world is *near*; I'm looking *ahead* to the winter holidays; It occurred in the *remote* past; The days *ahead* of us. . .; Way *back* in the sixties; It will happen some day in the *distant* future; Election day *is here*. The wedding is still *far away*.

Ego-based metaphor: Additional dynamic structure

Nature: Time is understood in terms of things (entities and locations) and motion in space.

Background conditions:

- There is a landscape, and a canonical observer which may move.
- The observer and the thing(s) in the landscape don't move simultaneously.
- When one entity is moving (thing or observer), the other is stationary; the stationary entity is the deictic centre.

Mapping (additional dynamic structure):

SOURCE DOMAIN Horizontal Uni-dimensional Space		TARGET DOMAIN Time
Relative motion (of the things with respect to the observer) along a one-dimensional landscape	→	The passing of time

Entailments:

- Since motion is continuous and one-dimensional, the passage of time is continuous and one-dimensional.

When (relative) movement is incorporated into the general mapping, new precise inferential properties emerge. As indicated above, one of the background conditions of this conceptual mapping establishes that the canonical observer and the thing(s) in the landscape don't move simultaneously. Either the observer moves while things are stationary, or things move while the observer is static. These two possibilities sustain two specific forms of the more general conceptual mappings observed in everyday language. These special forms are the following:

2.1. Dynamic ego-based form 1: Time Passing Is Motion Of An Object

In this case, the observer is fixed and the times are entities moving with respect to the observer. These elements bring additional inferential structure to the general mapping.

SOURCE DOMAIN Horizontal Uni-dimensional Space		TARGET DOMAIN Time
Things moving horizontally with respect to the fixed observer (and with their fronts in their direction of motion)	→	Times

Entailments:

- The time passing the observer is the present time.
- Time has a velocity relative to the observer.
- If time B follows time A (in the movement towards the observer, or away from him/her), then time B occurs later than time A (time B is in the future relative to time A).

This form of the conceptual metaphor accounts for the linguistic form and the semantic entailments of expressions like:

The time to take a decision *has arrived* . . . ; The summer had long since *gone* when . . . ; Christmas is *coming* up on *us*. Time is *flying* by. The time *has passed* when . . . ; The end of the world is *approaching*. That time will never *come* . . .

2.2. Dynamic ego-based form 2: Time Passing Is Motion Over A Landscape

In this case, times are fixed locations and the observer moves with respect to time.

SOURCE DOMAIN Horizontal Uni-dimensional Space		TARGET DOMAIN Time
Fixed objects (or locations) with respect to which the observer moves	→	Times

Entailments:

- Time has an extension, and can be measured.
- An extended time, like a spatial area, may be conceived of as a bounded region.

This form of the mapping accounts for another family of expressions:

We are *getting closer* to the end of the summer; Fortunately, we *left* that horrible story *behind us*; I will *walk towards* the future with optimism. He *passed* the time happily. We are *approaching* the year 2000. We are *arriving* at the end of the millennium.

The two forms of the ego-based metaphor have a quite different inferential structure. In fact, as Lakoff points out (1993), they are sometimes inconsistent with one another: the same words used in both special forms have inconsistent readings. For instance, the approaching of 'The end of the world is approaching' (Form 1), and 'We are approaching the year 2000' (Form 2) take different arguments. Both refer to temporal events, but the former takes a moving time as a first argument and the latter takes a moving observer as a first argument. The same holds for arrive in 'the time has arrived', and 'We are arriving at the end of the millennium'. But despite these differences, there is an important entailment that is shared by both forms.

Entailment common to both Dynamic Ego-based forms:

- If Time *A* approaches the observer (Form 1) or if the observer approaches Time *A* (Form 2), the metaphorical distance between Time *A* and the observer:
  - gets shorter as the action 'approaching' takes place, and
  - will be shorter after the action 'approaching' is over.

It is very important to keep in mind that this entailment does hold for both forms of the general dynamic mapping, but *it is not* what is empirically observed. In everyday conversations, we don't normally use expressions such as 'Christmas and ourselves are approaching each other'. It is simply an empirical fact, that these kinds of expressions are not observed. Therefore, from a cognitive perspective, it would be a mistake to consider the two forms of the ego-based metaphor as 'models' of a unique abstract truth about the distance between the observer and a specific time. As a scientist, one should focus on what one does observe empirically, that is, expressions involving time which are accounted for either by Form 1 or by Form 2 of the conceptual metaphor. It is this observation that can tell us about the basis of how the mind works, not an *a posteriori* logical analysis of reason. This point is crucial. Ignoring it simply impedes us in understanding why and how reason is bodily grounded.

We, the experts

One of the most striking things about everyday conversations is that we seem to be absolute experts in mastering the subtleties involved in the network of conceptual mappings. For instance, in what concerns our example, we use the different forms of the general mapping Time Events Are Things In Space, unconsciously 'knowing' when and how to operate in the appropriate sub-mappings, and drawing the appropriate inferences. Not only do we often do it unconsciously, but also we do it effortlessly, and with an astonishing speed and accuracy. Consider, for instance, the following two expressions in Spanish (in which language the above mappings also occur):

1) El estudiará lo que aconteció con *posterior*idad a 1988.

2) El estudiará lo que aconteció desde 1988 en *adelante*.

Expression (1) means 'he will study what happened after 1988 ('*posterior* to')', and expression (2) means 'he will study what happened from 1988 on ('*front*ward')'. Both expressions *mean* the same,[5] that is, that 'he will study what happened in the years following 1988': they refer to the years in the *future* relative to 1988. But notice that expression (2) refers to *adelante* (front), and expression (1) refers to *posterior* (back, rear). So how come we metaphorically mean the same thing using expressions which are referring to opposite orientations such as front and back? This sounds quite paradoxical indeed. But in fact, if we look closely at the underlying conceptual mappings, it is not. Expression (1) is based on the time-based mapping described above, whereas expression (2) is based on the ego-based mapping. The *posterior* (back, rear) of expression (1) applies to *the back* of '1988' as a particular metaphorical object (a year-thing) characterized by the background conditions of the time-based mapping. And the *adelante* (front) of expression (2) applies to the front relative to the bodily orientation of the observer characterized by the background conditions of the ego-based mapping. As we saw, these two conceptual mappings are very different, but sometimes the extensionality of the cases holding their entailments may coincide, resulting in expressions which *mean* the same while making reference to opposite orientations! However, we seem to be experts at keeping our mappings straight, so we don't make mistakes, and we don't have troubles maintaining conversations. In fact, it is really amazing that in our everyday conversations, these things don't seem paradoxical at all. We are (unconsciously) simply wonderful experts at operating in different mappings at once, and at making inferences within the exact appropriate mapping. And we are experts at dealing with these sophisticated situations under extremely demanding real-time and real-world constraints. The naturalness and speed at which we master these cognitive mechanisms is often striking, and go beyond pure speech or written language. Consider the following example in French:

---

[5] In order to be precise, I should say that expression (2) may include the year 1988 in the period, whereas expression (1) leaves that year out. However, for the purpose of what I want to illustrate, that is not relevant. What matters here is that both expressions refer to years in the future relative to 1988.

Pierre: Notre grand-mère est née en 1930, n'est-ce pas?
           (Our grandmother was born in 1930, right?)

Natalie: Non! Elle est née bien *avant*! (No! She was born way be*fore*!).
           (And at the very same time, Natalie makes a gesture quickly moving her hand
           towards her back, with the palm facing backwards.)

Here Natalie says '*avant*' (meaning be*fore*, front), and her gesture indicates *backwards*. This may be seen as a contradiction between gesture and speech, a 'contradiction' which, as speakers, we are not aware of (unlike when somebody says to us 'go left', while indicating to the right, which usually does bother us and we immediately become aware of). But there is no such thing here. What happens is that the oral expression makes use of the time-based metaphor in which the time when grandmother was born has occurred much earlier (in front, according to that mapping) than the time Pierre had in mind. And the gesture makes use of the sequence defined by the time-based metaphor, but oriented in the precise direction required by the source domain of the ego-based metaphor, that is, a sequence with their metaphorical fronts facing the observer. The result is that the time when the grandmother was born is *in front* (earlier) of the time suggested by Pierre (*avant*, meaning be*fore*) in the time-based metaphor, but also that *that* time has passed already and it is *behind* the observer in the ego-based metaphor. As active sense-makers, we seem to use basic everyday cognitive mechanisms to sustain these very sophisticated inference patterns in a precise, fast, and unambiguous manner.

Learning From the Case Study: The Primacy of the Living Body

Now, let's step back for a moment, and reflect on what we have analysed so far. What we have done is to characterize just *one* among the thousands of conceptual mappings we use — often simultaneously, and combined in complex ways — in everyday conversations. We analysed a system of conceptual metaphors — the general mapping Time Events Are Things In Space. This mapping characterizes the conceptual structure that sustains the fundamental idea of time events and time flow, giving an account of hundreds of everyday linguistic expressions (like those listed above) and their semantic entailments. Because of its fundamental nature, this mapping provides some interesting insight into the nature of the human mind and human conceptual systems. Two aspects are especially relevant for this article: the universality of the use of unidimensional space as source domain of the mapping, and the primacy of the inherent bodily orientation.

Universality of space as a source domain of the mapping

What can we say about the universality of the basic mapping Time Events Are Things In Space? Do we find it in other languages and cultures as well? We now know that both conceptual metaphors for time and space — time-based and ego-based — have been observed not just in English, Spanish, or French, but also in many Indo-European languages, and even in non-Indo-European ones such as Chinese, Japanese, Korean, some African languages like Wolof in Senegal, and Hebrew, to name a few (Sweetser, 1999; Moore, 1999). These observations provide huge evidence in favour of the idea that the general mapping Time Events Are Things In Space is indeed a human universal. Of course, there are some details which may vary from culture to culture, and from language to language. For example, some languages may focus on dynamic aspects and others on static and positional aspects. Or some languages may differ in when they use Form 1 or Form 2 of the ego-based metaphor, and so on.

Sometimes the variation of the details of the mappings is in fact quite striking. For the last few years — in collaboration with some colleagues of Northern Chile — I have studied the structure of the general mapping Time Events Are Things In Space in native speakers of Aymara, an Amerindian language spoken in the highlands of the Andes mountains. We have found that besides a perfectly common time-based mapping, they use a form of the ego-based mapping which maps the front with the *past*, not with the future! We have observed this not only in purely linguistic expressions, but also in the manifestation of spontaneous gestures. For instance, when saying something like 'long time ago' they point towards the front of them. And when referring to some event that occurred even earlier than that, they point even further ahead (thus exhibiting the transitivity properties described earlier). The details of how this works in Aymara, how this situation is explained in bodily-grounded terms, and how all this relates to the Aymara culture are quite interesting, but go beyond the scope of this paper (see Núñez *et al.*, in preparation). But beyond these very striking differences, what matters for our argument here is that, as far as we know, it is very safe to say that the general mapping Time Events Are Things In Space is universal (or at least extremely predominant).

The issue of universality also raises another question: Is it possible to observe cultures in which time events are conceived in terms other than objects in space, say, in terms of sweet-and-sour tastes, chromatic experiences, or blood pressure sensations? The answer is quite simple: there is no evidence that such a case exists. In all the languages studied so far — oral and written — time events are in one way or another conceived in terms of *things* (entities or locations) *in space*. We simply don't observe the conceptual structure of time flow based on domains of human experience such as tastes, flavours, or colours. Given this, the future can't taste purple.

It is worth mentioning that in all cultures studied so far, the mapping Time Events Are Things In Space is not taught deliberately and systematically at school or through any form of specific instruction. These observations suggest that such mappings are not mere social agreements or conventions. If that were the case, one would expect as many experiential modalities generating these conceptual mappings, as social and cultural environments one encounters. The stability and universality of these conceptual mappings supports the idea that they are shaped by non-arbitrary species-specific peculiarities of our brains and bodies. These fundamental specificities allow individuals to use these mappings effortlessly, often unconsciously, and in an extremely fast, precise, unambiguous, and accurate manner. In sum, *human beings, no matter the culture, organize chronological experience and its conceptual structure in terms of a very specific family of experiences: the experience of things in space*.

The primacy of the inherent bodily orientation in the mapping

Now, let us analyse what we can learn from the fact that an inherent bodily orientation structures the very core of the mapping Time Events Are Things In Space. At first glance, this issue may seem a superficial one, but it isn't. In fact, it has deep theoretical and philosophical implications. Consider, for instance, the following question.

Why should an abstract inferential mechanism such as the one we use to make inferences about time events require an implicit, precise, and unambiguous bodily orientation? After all, if the cognitive mechanism were really inherently abstract (as traditional mainstream cognitive science has postulated for decades), it shouldn't need anything concrete such as a bodily orientation! When you do your empirical observations, however, you observe that the inherent bodily orientation is everywhere. How can that be?

The mapping Time Events Are Things In Space under the ego-based metaphor, does make explicit reference to a bodily orientation. To be more precise, both forms of that metaphor — Ego-Based Form 1: Time Passing Is Motion Of An Object and Ego-Based Form 2: Time Passing Is Motion Over A Landscape — explicitly state that:

- future times are in a specific orientation relative to the speaker's body, namely, in front of him/her (in our culture), and
- past times are in another specific orientation relative to the speaker's body, namely, behind him/her (in our culture).

Because of the way in which the source domain of the metaphor is structured (objects in space being in front/behind the observer, etc.), and because of the structure of the projections to the target domain (things in front of/behind the observer are future/past times, etc.), the mapping itself establishes a precise bodily orientation of the observer relative to the times.[6] If we want to understand the subtleties of this phenomenon and its theoretical implications, seeing human reason as a purely abstract logical phenomenon does not help. Such a disembodied view wouldn't help us to answer the following questions: First, why is there a bodily orientation at all? And second, why do we observe a specific bodily orientation? In fact, from a purely abstract point of view, you don't need a particular bodily orientation — say, with the future in front of us — in order to keep the inferential structure provided by the mapping. We could still keep the same rich inferential structure preserving, for example, order and transitivity if future times were, say, above one's head and past times under one's feet. In other words, if time–space cognition was a purely disembodied abstract logic phenomenon, one would expect different languages, cultures, or even individuals manifesting all kinds of different bodily orientations with respect to a one-dimensional landscape. Any kind of bodily orientation would do it, even an orientation having, say, the past in the upper right front of the body and the future in the lower left rear. But empirical data show that this is not the case. There is one bodily orientation that is predominant in a wide range of human cultures, namely, the future as being ahead of us and the past as being behind. If we are really serious about studying the mind, we can't ignore this simple but important fact. This means that in explaining this (and any) human conceptual apparatus, we must propose a research programme that considers, in an essential way, the primacy of the peculiarities of the

---

[6] One could argue that in our culture it is also possible to find situations in which the past is conceived as being on the left of the speaker, and the future as being on the right. Such is a specific case of the time-based metaphor, in which the times are ordered in a sequence before the eyes of the speaker (the speaker is outside the space of the sequence). In such a case, terms like 'before time *t*' apply to the *front* of time *t*, not to the front of the speaker. This specific case of the time-based mapping underlies more sophisticated conceptual domains such as graphics of functions in the Cartesian plane where time is the independent variable (Lakoff and Núñez, 2000).

human body, bodily experiences and actions which underlie basic forms of human sense-making. Regarding the bodily orientation involved in our time–space case study, this means proposing an explanation that, among others, takes into account fundamental features of human movement. For instance, normally when we walk,

- we do it in one direction (we don't do it in two directions at a time, or spreading out over a surface),

- we do it in the direction of the orientation of our visual field (not, say, towards the auditive field of our left ear),

- we do it keeping our heads stable relative to the ground (unlike our hands or elbows, which are not) so we experience vision as being stable,

- we do it faster when we move frontally rather than laterally or backwards,

- our movements frontwards are more precise than those done backwards or sidewards,

- our movement frontwards requires less attention and coordination efforts than, say, moving backwards,

- our head is the part of the body which is the most distant from the ground (unlike when we sleep),

- we spend energy, and eventually get tired,

- we are erect and at any moment we can run (unlike when we are in our knees),

- and so on.

The moral

So, what have we learned from our time–space case study? We have learned that the resulting explanatory proposals of these sort of phenomena should be made in terms of basic cognitive mechanisms emerging from fundamental human experiences in a real environment as they are shaped by bodily properties and biological constraints. When explaining human conceptual systems, purely abstract, logical, and *a priori* considerations about time and space are secondary to these most fundamental species-specific peculiarities of the human body and brain. As Esther Thelen puts it, 'Even when adult cognition looks highly logical and propositional, it is actually relying on resources (such as metaphors of force, action, and motion) developed in real-time activity and based on bodily experience' (Thelen, 1995, p. 323). It would be a big mistake to consider the predominant bodily orientation in the mapping Time Events Are Things In Space as a an accident, as anecdotal data, or as a mere physical instantiation secondary to a purely abstract form of reason. A theory of mind and cognition must consider the primacy of the specific constraints of our bodily grounded experience shaped by the peculiarities of our brains and bodies. In sum, what we have learned from our case study is that in order to understand cognition and the mind, one must conceive them as *fully embodied* phenomena.

But, what do we mean exactly by 'embodiment'? At this point we have to be careful, because the term 'embodiment' has become very fashionable (and polysemous) in contemporary cognitive science and philosophy of mind. Therefore, we need to be much more specific and avoid misunderstandings.

What Embodiment for Reclaiming Cognition?

In the last couple of decades, the study of the mind has experienced an interesting (and gradual) shift. There has been a tendency to move from a rational, abstract, culture-free, centralized, non-biological, ahistorical, unemotional, asocial, and disembodied view of the mind, towards a view which sees the mind as situated, decentralized, real-time constrained, everyday experience oriented, culture-dependent, contextualized, and closely related to biological principles — in one word, embodied (Núñez, 1995). This gradual shift has produced terms such as 'embodiment', 'embodied mind' or 'embodied cognition' (for details see Clark, 1997; Johnson, 1987; Lakoff, 1987; Varela *et al.*, 1991). These terms, however, have not been used in a monolithic and coherent manner in the various disciplines of cognitive science and its various theoretical approaches. Moreover, with rare exceptions they often have lacked a precise operational characterization. As a result, we find several notions of embodiment which, while sharing a common core, differ on important theoretical points and philosophical implications. Let us see them in more detail.

Levels of commitment: trivial, material, and full embodiment
From a very general perspective, we can at least distinguish three major levels of understanding the term 'embodiment' which are directly related with the levels of commitment involved. I will call them trivial, material, and full embodiment.

Trivial embodiment: It affirms what today is obvious for many, that is, that cognition and the mind are directly related to the biological structures and processes that sustain them. Nowadays, few scholars (perhaps with the exception of orthodox cognitivists and transcendentalists) would disagree with this idea. This view holds not only that in order to think, speak, perceive, and feel, we need a brain — a properly functioning brain in a body — but also that in order to genuinely understand cognition and the mind, one can't ignore how the nervous system works. Compelling evidence coming from contemporary neuroscience and neuropsychology has made this view quite popular. As a result, this level of commitment regarding embodiment is today quite uncontroversial.

Material embodiment: This view not only claims that cognition (and the mind) is made possible by the underlying neurobiological and bodily processes. It also explicitly develops a paradigm (and a methodology) that has two main features. First, it sees cognition as a decentralized phenomenon, and second, it takes into account the constraints imposed by the complexity of real-time bodily actions performed by an agent in a real environment. These features depart in an essential way from more classical approaches to cognitive science. Material embodiment has, in general, oriented itself towards low-level cognitive tasks. As a result, it does not have to confront certain basic issues of high level cognition such as the nature of conceptual systems as such. One can endorse material embodiment to study, say, visual scanning or locomotion, without being constrained to make any particular commitment about how these bodily actions may ground the very nature of human concepts and logic. In material embodiment, you may thus implicitly assume the existence of concepts in an *a priori* way (e.g., square-root-of-two, the-future- ahead-of-us) without being constrained to say much about their nature and inferential structure. The commitment does not apply to *all* of cognition.

Full embodiment: It shares the basic tenets of trivial and material embodiment, but it goes further. It has a commitment to all of cognition: from the most basic perceptive activity to the most sophisticated form of poetry and abstract thinking. Full embodiment explicitly develops a paradigm to explain the objects created by the human mind themselves (i.e., concepts, ideas, explanations, forms of logic, theories) in terms of the non-arbitrary bodily experiences sustained by the peculiarities of brains and bodies. An important feature of this view is that the very objects created by human conceptual structures and understanding (including scientific understanding) are not seen as existing in an absolute transcendental realm, but as being brought forth through specific human bodily grounded processes. Conceptual systems and forms of understanding are not considered *a priori*, but they become subject matters to be explained in real-time bodily grounded terms. From this perspective, not only are colour categories embodied and are not out there in the world, but so is the concept of democracy, the truth of Pythagoras' theorem, or the essence of any mathematical object (Lakoff and Núñez, 2000).

Needless to say, full embodiment is (still) way more controversial than trivial and material embodiment. However, our time–space case study — being about conceptual structure and high level cognition — clearly illustrates the necessity of endorsing full embodiment. If we need to understand the primacy of the inherent bodily orientation which is at the core of the conceptual mapping Time Events Are Things In Space, trivial and material embodiment do not suffice. We must understand the conceptual structure involving elements such as future-being-in-the-front, past-being-in-the-back, and so on, not as *a priori* transcendentally objective ideas, but through the peculiarities of our brains and bodies that make them possible. What is needed then is an embodied–oriented commitment to all of cognition. Full embodiment then becomes a must.

Choices of theoretical assumptions and methodologies

But beyond these three levels of commitment, there are more nuances we have to make regarding the term 'embodiment'. The gradual paradigm shift that I mentioned at the beginning of this section originated from a need to overcome the limitations of early cognitivism and the more recent connectionist approaches (for details, see Varela, 1989; Clark, 1997 and also in this volume; Freeman and Núñez, in this volume). Essential to this enterprise was to give form to approaches which would address a fundamental problem: the neglect of the body and the environment in which the cognizing organism exists. Although philosophers such as Edmund Husserl, and Maurice Merleau-Ponty, as well as ecological psychologists such as James Gibson, had seen already the centrality of body and environment for the understanding of the mind, cognitive science took a long time to make the first steps in this direction. Endorsing this new view meant to take seriously such domains as everyday life, the environment, bodily experiences, real-world and real-time action, and so on. As a result, an important group of new approaches, in various disciplines, proceeding with rather different methodologies, headed towards investigating cognition as a product of complex adaptive behaviour emerging from on-going action on the part of an agent which is always immersed in a real-world environment, and with physical and real-time constraints. This provided core properties for a new way of conceiving cognition.

The process, however, didn't produce a monolithic understanding of 'embodiment'. In fact, the heterogeneity of disciplines and methodologies generated a sort of polysemy of the term embodiment. Depending on what aspects of the core notion you privilege, you obtain a variety of different meanings with their own theoretical implications. For instance, taking the 'agent' to literally be any autonomous agent, natural or artificial, entails that fundamental properties of the organization of the living phenomenon are not taken as essential (e.g., morphogenesis, biochemistry of synapses, etc.). This gives you the notion of 'embodiment' used in modern cognitive robotics and autonomous agent theory (Brooks and Stein, 1993; Clark, 1997; Pfeifer and Scheier, 1999), which does not apply to the work by Maturana and Varela (1987) in theoretical biology, and Núñez (1995; 1997) in philosophy of mind. Similarly, taking as essential the explicit rejection of computationalism and representationalism in favour of the use of the tools of dynamical system theory, would give a notion of 'embodiment' that would apply to the work by Thelen (1995; see also Iverson and Thelen in this volume) in developmental psychology, and by Skarda and Freeman (1987; see also in this volume) in neuroscience, but not to the work by Mark Johnson (1987) in philosophy of mind, or to the one by Feldman and Regier in structured connectionism (Feldman *et al.*, 1996; Regier, 1996).[7] Or taking the human bodily grounded *experience* as essential would give a notion of 'embodiment' that would apply to the work of Lakoff and Núñez in mathematical cognition (1997; 2000), and to that of Csordas (1994) and Lock (1993) in anthropology, but it wouldn't apply to real-world robotics. In sum, among scholars directly involved with embodiment, there are deep theoretical, methodological, and philosophical differences. For some scholars, the embodied mind is literally computational, for others, it is metaphorically computational, and for others, it is not computational at all; for some, it is inherently an emerging phenomenon proper to certain forms of *living systems*, and for others, it is not; for others, its study requires the use of certain methodologies such as the tools of dynamical systems theory, and for others, it does not; for some, the living phenomenological bodily experience is essential, and for others, it is not; and so on. Basically, the situation is as if you pick your favourite feature of the core properties, bring them to the foreground, add your own theoretical flavour, stir with your own vintage, and you get your 'embodiment' *à la carte*. So, what is the view of embodiment I endorse here? In a nutshell, a non-computational view that emphasizes the primacy of the organization of the living and the resulting bodily experience it sustains.

Conclusion

The time–space case study provides a wonderful opportunity to learn several essential dimensions of the human mind. Conceptual structures are not purely abstract logical entities merely instantiated in our bodies, but are stable and precise forms of sense-making that come out of the peculiarities of brains and bodies, and the bodily grounded experience they sustain. The analysis of the conceptual mapping Time Events Are Things In Space has shown this for the particular concept of time flow. That analysis has allowed us to see how the primacy of the bodily orientation and

---

[7] Such a view wouldn't apply either to Andy Clark's idea of embodiment (1997). In fact, Clark calls such a view 'Radical Embodied Cognition' (p. 148).

real-time bodily action is at the very core of the cognitive mechanisms that make the concept of time flow possible. From that, we have learned that in order to study cognition and the mind in an adequate way, one needs to reclaim the traditional notions of mind, body and cognition. One needs to understand cognition and the mind as *fully embodied* phenomena. This is an interesting, though not simple, challenge. It implies endorsing a view of embodiment with the following specific features:

- A level of commitment seeking to understand the fundamental human bodily experience that grounds *all* forms of sense-making, from basic motion, to everyday common sense, to scientific theories and formal logic.

- A choice of theoretical and methodological orientations that emphasize the dynamic biological (structural) co-definition that exists between living organisms and the medium in which they exist, from which bodily grounded experience and cognition result as real-time enactive processes.

But there is more. Reclaiming cognition from this perspective (i.e., an embodiment-oriented commitment to all of cognition) implies also something much deeper: the rejection of at least five harmful dogmas that lie at the very foundations of much of current cognitive science and philosophy of mind (for details, see Núñez, 1997):

(1) the *a priori* assumption that there is a pre-given objective reality independent of any human understanding (including conceptual systems);

(2) the idea that epistemology is absolutely subordinated to ontology;

(3) the idea that there is a strict objective–subjective dichotomy;

(4) the assumption that the body must be excluded from the study of the mind; and

(5) the idea that the mind can be reduced to the study of neurophysiological processes of individual brains alone.

The analysis of our time–space case study implies that in order to adequately address the questions involved, one needs to reject these dogmas. The reasons are straightforward. Adopting a pre-existing God's-eye view (dogma 1) of what time and space objectively and transcendentally *are* (e.g., assuming that space *is* Euclidean and three-dimensional) hides the very phenomenon we want to study. That is, how humans through everyday cognitive mechanisms bring forth, create, and conceptualize the idea of time flow. Besides, assuming that 'merely' epistemic issues cannot address pre-existing ontologies (dogma 2), impedes us in asking the very question of the nature and origin of the pre-existing God's-eye view: What kind of brain sustains God's-eye view? As a result, this dogma doesn't allow us to understand how human beings enact systematic forms of everyday sense-making to create stable abstract concepts such as time and space (including the very idea of God's-eye time and space!). Similarly, a strict objective–subjective dichotomy (dogma 3) impedes us in understanding the rich spaces of commonalities and inter-subjectivity that exist in human conversations, preventing us from understanding the amazing inter-individual inferential stability based on shared species-specific bodily grounded experiences. Then, there is the absence of the body. Not considering the primacy of the body in the study of the mind (dogma 4) simply leaves us without tools to understand, for example, why there is a bodily orientation at the core of the concepts of time flow. And finally, reducing the biological processes to strictly individual brains (dogma 5), impedes us in understanding our brains as organs of social action, and therefore in comprehending the biological basis of inter-subjectivity. In sum, in order to reclaim mind, body, and cognition, we must free ourselves of these dogmas. The time to develop a richer and deeper science of the mind 'has come'.

Acknowledgements
I would like to thank Elizabeth Beringer, Jean-Louis Dessalles, George Lakoff, Kevin Moore, Vicente Neumann, Eve Sweetser, Francisco Varela, and two anonymous reviewers for comments and suggestions.

References

Brooks, R. and Stein, L. (1993), Building Brains for Bodies. Memo 1439, Artificial Intelligence Laboratory, Massachusetts Institute of Technology.

Clark, A. (1997), *Being There: Putting Brain, Body, and World Together Again* (Cambridge, MA: MIT Press).

Csordas, T.J. (1994), *Embodiment and Experience: The Existential Ground of Culture and Self* (Cambridge: University Press).

Dunkel, G.E. (1983), προσσω και οπισσω. *Zeitschrift für Vergleichende Sprachforschung*, **96** (1), pp. 66–87.

Fauconnier, G. (1997), *Mappings in Thought and Language* (Cambridge: University Press).

Fauconnier, G. and Turner, M. (1998), 'Conceptual integration networks', *Cognitive Science*, **22** (2), pp. 133–87.

Feldman, J., Lakoff, G., Bailey, D., Narayanan, S., Regier, T. and Stolcke, A. (1996), 'L0 — The first five years of an automated language acquisition project', *Artificial Intelligence Review*, **10** (1).

Gibbs, R. (1994), *The Poetics of Mind: Figurative Thought, Language, and Understanding* (Cambridge: University Press).

Iverson, J.M. and Thelen, E. (1999), 'Hand, mouth and brain: The dynamic emergence of speech and gesture', *Journal of Consciousness Studies*, **6** (11–12), pp. 19–40.

Johnson, M. (1987), *The Body in the Mind. The Bodily Basis of Meaning, Imagination, and Reason* (Chicago: University of Chicago Press).

Lakoff, G. (1987), *Women, Fire, and Dangerous Things: What Categories Reveal About the Mind* (Chicago: University of Chicago Press).

Lakoff, G. (1993), 'The contemporary theory of metaphor', in *Metaphor and Thought*, ed. A. Ortony (Cambridge: Cambridge University Press).

Lakoff, G. and Johnson, M. (1980), *Metaphors We Live By* (Chicago: University of Chicago Press).

Lakoff, G. and Núñez, R. (1997), 'The metaphorical structure of mathematics: Sketching out cognitive foundations for a mind-based mathematics', in *Mathematical Reasoning: Analogies, Metaphors, and Images*, ed. L. English (Hillsdale, NJ: Erlbaum).

Lakoff, G. and Núñez, R. (2000), *Where Mathematics Comes From: How the Embodied Mind Creates Mathematics* (New York: Basic Books).

Lock, M. (1993), 'Cultivating the body: Anthropology and epistemologies of bodily practice and knowledge', *Annual Review of Anthropology*, **22**, pp. 133–55.

Maturana, H. and Varela, F. (1987), *The Tree of Knowledge: The Biological Roots of Human Understanding* (Boston: New Science Library).

McNeill, D. (1992), *Hand and Mind: What Gestures Reveal about Thought* (Chicago: University of Chicago Press).

Moore, K. (1999), 'Language, point of view and metaphorical mappings from spatial to temporal concepts in Wolof', PhD Dissertation, Dept. of Linguistics, University of California at Berkeley.

Núñez, R. (1995), 'What brain for God's-eye? Biological naturalism, ontological objectivism, and Searle', *Journal of Consciousness Studies*, **2** (2), pp. 149–66.

Núñez, R. (1997), 'Eating soup with chopsticks: Dogmas, difficulties, and alternatives in the study of conscious experience', *Journal of Consciousness Studies*, **4** (2), pp. 143–66.

Núñez, R. and Lakoff, G. (1998), 'What did Weierstrass really define? The cognitive structure of natural and ε–δ continuity', *Mathematical Cognition*, **4** (2), pp. 85–101.

Núñez, R., Neumann, V., and Mamani, M. (in preparation), 'Time–space conceptual metaphors in the Aymara language: Evidence of an ego-based past-in-front mapping'.

Pfeifer, R. and Scheier, C. (1999), *Understanding Intelligence* (Cambridge, MA: MIT Press).

Regier, T. (1996), *The Human Semantic Potential* (Cambridge, MA: MIT Press).

Skarda, C. and Freeman, W. (1987), 'How brains make chaos in order to make sense of the world', *Behavioral and Brain Sciences*, **10**, pp. 161–95.

Sweetser, E. (1990), *From Etymology to Pragmatics: Metaphorical and Cultural Aspects of Semantic Structure* (Cambridge: Cambridge University Press).

Sweetser, E. (1999), Personal communication, September, University of California at Berkeley.

Thelen, E. (1995), 'Time-scale dynamics and the development of an embodied cognition', in *Mind as Motion: Explorations in the Dynamics of Cognition*, ed. R.F. Port and T. van Gelder (Cambridge, MA: MIT Press).

Turner, M. and Fauconnier, G. (1995), 'Conceptual integration and formal expression', *Journal of Metaphor and Symbolic Activity*, **10** (3), pp. 183–204.

Varela, F. (1989), *Connaître les Sciences Cognitives: Tendances et Perspectives* (Paris: Seuil).

Varela, F., Thompson, E. and Rosch, E. (1991), *The Embodied Mind: Cognitive Science and Human Experience* (Cambridge, MA: MIT Press).

Eleanor Rosch

Reclaiming Concepts

The story is told of a physicist who is invited by a dairy farmers' association to tell them how to get more milk from cows. The physicist begins: 'First we start with a spherical cow.' That is told as a joke! Yet far more strange is what cognitivism has done to what is supposed to be the study of human thought and human life. This chapter is about concepts, the central building blocks of cognitivist theory. I will first show how cognitivism necessarily cannot give an adequate treatment of concepts and will then, more importantly (who pays any attention to criticisms?), outline the foundations for a new nonrepresentational view of concepts which should place the study of concepts on a real (rather than a spherical cow) basis.

If ever there was a domain where you'd think cognitivism could get it right, it is concepts. One wouldn't necessarily expect cognitivism to illuminate biology or art or qualia or intuition or spirituality. But concepts are the very building blocks of cognitivist theories. Thus concepts are a particularly apt case study for this volume because to examine the case of concepts is to penetrate cognitivism on its own home ground.

Why should concepts be expected to be the *sine qua non* for cognitivism? To answer that requires some background: what do we mean by concepts and, for that matter, by cognitivism?

Concepts are one aspect of the study of categorization. One of the most basic functions of living creatures is to categorize, that is to treat distinguishable objects and events as equivalent. Humans live in a categorized world; from household items to emotions to gender to democracy, objects and events, although unique, are acted towards as members of classes. Some theories would say that without this ability it would be impossible to learn from experience and thus that categorization is one of the basic functions of life. Since at least the nineteenth century, it has been common to refer to the cognitive or mental aspect of categories as *concepts*.

What is it about concepts that is inaccessible to cognitivism? In the course of this article, I will argue that: Concepts are the natural bridge between mind and world to such an extent that they require us to change what we think of as mind and what we think of as world; concepts occur only in actual situations in which they function as participating parts of the situation rather than either as representations or as mechanisms for identifying objects; concepts are open systems by which creatures can learn new things and can invent; and concepts exist in a larger context — they are not the

Journal of Consciousness Studies, **6**, No. 11–12, 1999, pp. 61–77

only form in which living creatures know and act. In a cognitivist machine, on the other hand, concepts are inherently solipsistic; the nature of concepts can be nothing other than closed analytic definitions which admit of neither a participatory function nor of novelty; and there can be no context of other forms of knowing in which conceptual knowing can be situated. What then is cognitivism and where does it come from such that it has these limitations?

There is a long philosophical tradition treating concepts and categories, called 'the problem of universals', to which cognitive science is heir. At issue is how there can be a solid foundation for knowledge when what the senses provide seems so changeable and unreliable. But cognitivism is additionally heir to another set of constraints. As other papers in this volume have pointed out (see also Johnson and Erneling, 1997), there are actually two rather different endeavours called cognitive science. One is simply the attempt at interdisciplinary cooperation between artificial intelligence, psychology, philosophy, linguistics, neurophysiology and anthropology. The other form of cognitive science additionally assumes the particular philosophical stance now called cognitivism. The centrality of concepts in cognitivism has a somewhat different origin than it has in the philosophical and psychological traditions. Cognitivism treats the mind as a machine, more precisely as a computer program, more precisely still as the sort of program which functions as a series of computations (that is, rule-governed changes) on symbolic representations. The mind is considered to be a collection of mental representations precisely analogous to the computer's symbolic representations. The only question which we may ask of such a model or machine, the only appropriate test of it, is just the classical Turing test — that its output be indistinguishable from that of a human. Fodor (1998) calls this model 'The Representational Theory of Mind (RTM)' and indicates its importance for cognitivism: 'No cognition without representation' (p. 26).

Representation has a technical meaning here; it does not refer to a relationship between the mind and things of the world, but to symbols within the closed system of the machine. Cognitivists (Carnap, Chomsky, Fodor, etc.) have long maintained that while there may be semantic content (meaningful real-world reference) to language and thought, it is only the syntax (formal operations) which is accessible to cognitive science. Fodor (1982) calls this 'methodological solipsism'. Concepts are central to RTM because they are the units out of which rules are made and on which the rules operate. Thus, in a sense, concepts have been shifted from a philosophically to a technologically inspired role — a deep technology in which the requirements for setting up a machine's computations on its symbolic representations become our theory of mind.

The above model might be called *philosophical cognitivism* or *strict cognitivism*. Practician researchers are wont to ignore the methodological solipsism aspect of cognitivism and to treat cognitive representations as mirroring the world or, at least, as recovering information from the world. In other respects, such researchers may maintain the rest of the cognitivist model. We will call this *working cognitivism*.

Since concepts are central to the cognitivist representational model of mind, cognitivism is badly in need of a theory of concepts. Fodor (1998) again: 'RTM is simply *no good* without a viable theory of concepts' (p. 39). What then are the possible theories of concepts?

It is generally agreed that there are three major present approaches to concepts: (1) the so-called classical view (2) the graded structure and/or prototype view, and (3) various theories views. These all grew, like so many cow paths, out of the philosophical and psychological traditions. Although none originated in the cognitivist position *per se*, cognitivism has adopted or critiqued these views at length since it is in need of a theory of its central building blocks. One contention of this paper is that cognitivism has been forced by its own tenets to badly misunderstand and misrepresent each of these approaches to concepts and that its critiques of the traditions are, thus, actually biting critiques of its own architecture. In the following sections, I will present each view in terms of its own tradition, show how cognitivism is forced to misinterpret that tradition, indicate both the limitations and the (at times not obvious) wisdom of each view, and will conclude by presenting a non-cognitivist alternative to the present views.

I: Present Approaches to Concepts and Categories

A. *The classical view*

The classical view is the approach to concepts derived from the history of western philosophy. When humans begin to look at their experience by means of reason, questions about the reliability of the senses and the bases for knowledge naturally arise, as do more specific questions about how categories can have generality, how words can have meaning, and how concepts in the mind can relate to categories in the world. The Greeks, and most western philosophers since, have agreed that experience of particulars as it comes moment by moment through the senses is unreliable, therefore, only stable, abstract, logical, universal categories can function as objects of knowledge and objects of reference for the meaning of words. This is a significant choice historically. Greece was not the only culture that made this turn to essentialist thinking; it was shared by the Sanskritic tradition in India (Chakrabarti, 1975; Radhakrishnan and Moore, 1957). However, quite a different direction was taken by Chinese dialectical thinking (Nisbett *et al.*, under review; Peng and Nisbett, 1999).

To provide a proper basis for knowledge, Plato felt that he had to introduce metaphysical entities, the Forms, which underlay experience and which one came to know by recollection. For Aristotle, universals were *in* particular objects and were learned by experience with the objects. Universals were the common elements in particulars; that is, a universal X is whatever is common to, or shared by, all Xs. Individual objects were to be classified into kinds that shared the same properties, and kinds were to be subdivided into genus and species according to the properties that differed between them. One of the primary tasks of natural science was to divide and classify natural objects by genus and species into the real kinds to which, by nature, they belonged. The Aristotelian system should sound very familiar to us. Our dictionaries are organized that way, not to say much of our thinking. (By way of contrast, early Chinese dictionaries contained only synonyms and analytic dissection of Chinese ideograms — Baker, unpublished.)

The next major historical school with respect to categories is called Conceptualism and is attributed to the British Empiricists beginning with John Locke. Actually, Conceptualism is not a rival theory to Greek Realism but a matter of different emphases — how precisely does the mind form general concepts? Locke argued that this occurs through a process of abstraction from particular concepts; each time one encounters a member of a category, one subtracts from one's idea of the category features not common to all members of the category until one is left with only the features that are necessary and sufficient for membership in the category. Subsequent arguments between Locke, Berkeley and Hume concerned the nature of the resulting mental abstraction: Locke is credited (somewhat unfairly) with claiming it was a general, abstract idea or image; Berkeley with the argument that it was a particular idea which becomes general by coming to represent things of the same kind, and Hume with the (generally ignored) objection that it was not an idea at all but simply habit or custom. Note how similar much of Empiricist thought is to Aristotle in its general portrait of concepts and categories.

Psychology inherited a particular view of categories from the history of philosophy. To serve as a proper essentialist basis for knowledge, categories were required to: (1) Be exact, not vague — i.e. have clearly defined boundaries, and (2) Have attributes in common which were the necessary and sufficient conditions for membership in the category. From these it followed that (3) All members of a category must be equally good with regard to membership; either they have the necessary common features or they don't. Categories and concepts were thus seen as logical sets.

The philosopher's view of categories entered psychology explicitly in the form of concept learning research in the 1950s. Led by the work of Jerome Bruner and his associates (Bruner *et al.*, 1956), subjects were asked to learn categories which were logical sets defined by explicit attributes, such as *red* and *square*, combined by logical rules, such as *and*. Theoretical interest was focussed on how subjects learned which attributes were relevant and which rules combined them. In developmental psychology, the theories of Piaget and Vygotsky were combined with the concept learning paradigms to study how children's ill-structured, thematic, concepts develop into the logical adult mode. Artificial stimuli were typically used in experimental research at all levels, structured into micro-worlds in which the prevailing beliefs about the nature of categories were already built in.

The cognitivist version of the classical view, as stated by Fodor (1998), is that concepts are *definitions*, by which is meant equivalent and substitutable strings of symbols. 'A bachelor is an unmarried man' would be a prototypical example. Cognitivist machines do indeed work by means of substitutable strings of symbols, but was that really the point of concepts for any of the proponents of the classical view? When Aristotle defined objects in terms of genus and differentia, he was explicitly intending to describe empirical content. In the standard concept attainment paradigm in learning theory, psychologists sought to find out how creatures learn which attributes are relevant to a concept and which rules connect them; subjects were not being trained to substitute strings of symbols for each other. Categories may have been seen as logical sets, but they were sets designed to reflect a mapping between mind and world. Thus one problem with the cognitivist interpretation of the classical view of categories as definitions based on substitutable strings of symbols is that it does not in fact describe the classical view. But could the cognitivist argue that that is what the classical view *should* be?

When we say that someone has or understands a concept, do we mean that all he has is a substitutable string of symbols? This issue is analogous to the ever-popular debate about whether or not computers can think. Computers work by means of substitutable strings of symbols. But consider John Searle's now famous Chinese room example: a man who does not speak or understand a word of Chinese sits in an isolated room. Sentences in Chinese are passed in to him; he looks them up in a concordance; and passes out the replies that are listed. That is all that he does. His replies pass the Turing test, that is, are indistinguishable from those of a human who understands Chinese. Does he understand Chinese? Is he working with the concepts and categories of Chinese? NO! Yet what this man is doing is exactly substituting strings of symbols for one another.

The Chinese room example evokes our intuitions about concepts in several ways. One aspect that seems inherently wrong is the solipsism; another is the impoverished nature of the exchange. Concepts seem to require not only our dictionary but also the full range of the *experienced* human encyclopaedia of knowledge, not to mention interaction with the material and social world (for background on these issues see Dreyfus and Hall, 1982). The classical view of categories may not provide an account adequate to these requirements (as will be shown in succeeding sections), but it at least asks questions in the arena where the history of western thought has sought answers.

In fact, even cognitivists do not accept definition by means of substitutable strings of symbols, when it is stated baldly this way, as an adequate theory of concepts. This kind of account is actually a form of nominalism, the philosophical position which asserts that generalities reside in words alone. Nominalism has never been taken seriously by philosophy where it has been viewed as an incoherent position easily demolished (Woozley, 1967). Even Fodor, the arch-cognitivist, does not accept as adequate a definitional account of concepts; in fact he feels that all of the extant theories of concepts are 'radically and practically *demonstrably* untrue, and that something drastic needs to be done about it' (Fodor, 1998, p. viii).

How is it then that cognitivists, such as Fodor, can simply take for granted the definitional interpretation of the classical view? Perhaps this is because concepts for machines *are* substitutable strings of symbols; that is the only thing they *can* be. This is particularly blatant in the rule-based computational systems of the Representational Theory of Mind. In so far as Fodor's critique is successful, his criticism can be seen to bear more on cognitivism itself than on the classical view of concepts.

B. The graded structure/prototype view

Consider the colour red: is red hair as good an example of your idea or image of *red* as a red fire engine? Is a dentist's chair as good an example of *chair* as a dining-room chair? Such questions are nonsense within the classical view of categories where something either is a category member or it isn't, and all members are equivalent. Challenges to the classical view came from within psychology beginning with work on colour categories (Rosch, 1973). Colour categories do not have any obviously analysable criterial attributes, formal structure, or definite boundaries, and they have an internal structure graded in terms of how exemplary of its category people judge a colour to be. Furthermore, those areas of the colour space judged as best examples of colour categories seem to serve a special role; they are the most stable, cross-culturally agreed upon, and perhaps even the physiological origins of colour categories (Berlin and Kay, 1969; Hardin and Maffi, 1997; Rosch, 1977). Are other kinds of categories structured in a similar way?

An extensive research programme established a core of empirical findings (Rosch, 1978; 1994). In the first place, all categories show gradients of membership; that is, subjects easily, rapidly, and meaningfully rate how well a particular item fits their idea or image of the category to which the item belongs. Such judgments are the hall-mark of the graded structure/prototype view. Note that these are not probability judg-ments but judgments of degree of membership. Gradient of membership judgments apply to the most diverse kinds of categories: perceptual categories such as *red*, semantic categories such as *furniture*, biological categories such as *woman*, social categories such as *occupations*, political categories such as *democracy*, formal cate-gories that have classical definitions such as *odd number*, and *ad hoc*, goal-derived categories such as *things to take out of the house in a fire*.

Gradients of membership must be considered psychologically important because such measures have been shown to affect virtually every major method of study and measurement used in psychological research. I will belabour this point somewhat as a contrast to the way in which cognitivism treats the graded structure/prototype view which will be described shortly. (Unless otherwise indicated, the following studies are all reported or referenced in Markman, 1989; Mervis and Rosch, 1981; Rosch, 1973; 1978; 1987; Rosch and Lloyd, 1978; or Smith and Medin, 1981).

Learning: good examples of categories are learned by subjects in experiments and acquired naturalistically by children earlier than poor examples, and categories can be learned more easily when better examples are presented first — findings with implications for education. **Speed of processing**: the better an example is of its cate-gory, the more rapidly subjects can judge whether or not that item belongs to the cate-gory. This is important because reaction time is often considered the royal road to learning about mental processes. **Expectation**: when subjects are presented with a category name in advance of making some speeded judgments about the category, performance is helped for good and hindered for bad members of the category. Called priming or set in psychology, this finding has been used to argue (not indisputably) that the mental representation of the category is in some ways more like the better than the poorer exemplars. **Association**: when asked to list members of the category, subjects produce better examples earlier and more frequently than poorer examples. **Inference**: subjects infer from more to less representative members of categories more readily than the reverse, and the representativeness of items influences judg-ments in formal logic tasks, such as syllogisms. **Probability judgments**: representa-tiveness strongly influences probability judgments (Kahneman *et al.*, 1982) which is important because probability is thought to be the basis of inductive inference and, thus, of the way in which we learn about the world. **Natural language indicators of graded structure**: natural languages themselves contain various devices which acknowledge and point to graded structure such as hedge words like *technically* and *really* (Lakoff, 1987; Rosch, 1975). **Judgment of similarity**: less good examples of categories are judged more similar to good examples than vice versa. This violates the way similarity is treated in logic, where similarity relations are symmetrical and reversible.

Other research challenges directly the requirement of the classical view that cate-gories have defining features. Rosch and Mervis (1975) found that when subjects are asked to list attributes for category members, many categories show up with few or no attributes in common. Attributes appeared to have a family resemblance (Wittgenstein, 1953), rather than a necessary and sufficient structure.

The judged best examples of conceptual categories are called *prototypes*. While some may be based on statistical frequencies, such as the means or modes (or family resemblance structures) for various attributes, others appear to be ideals made salient by factors such as physiology (good colours, good forms), social structure (president, teacher), culture (saints), goals (ideal foods to eat on a diet), formal structure (multiples of ten in the decimal system), causal theories (sequences that 'look' random), and individual experiences (the first learned or most recently encountered items, or items made particularly salient because they are emotionally charged, vivid, concrete, meaningful, or interesting). One of the most philosophically cogent aspects of prototypes is that, far from being abstractions of a few defining attributes, they seem to be rich, imagistic, sensory, full-bodied mental events that serve as reference points in all of the kinds of research effects mentioned above.

A very important finding about prototypes and graded structure is how sensitive they are to context. For example, while dog or cat might be given as prototypical pet animals, lion or elephant are more likely to be given as prototypical circus animals. In a default context (no context specified), coffee or tea or coke might be listed as a typical beverage, but wine is more likely to be selected in the context of a dinner party. Furthermore, people show perfectly good category effects complete with graded structure for *ad hoc* goal-derived categories such as *good places to hide from the Mafia*. In fact, the effects of context on graded structure are ubiquitous (Barsalou, 1987). In the classical view, from the time of its origin in Greek thought, if an object of knowledge were to change with every whim of circumstance, it would not be an object of knowledge, and the meaning of a word must not change with conditions of its use. One of the great virtues of the criterial attribute assumption for its proponents had been that the hypothesized criterial attributes were just what didn't change with context. Barsalou has argued that context effects show that category prototypes and graded structure are not pre-stored as such in the mind, but rather are created anew each time 'on the fly' from more basic features or other mental structures. The extreme flexibility of categories to context effects may have even more fundamental implications.

Cognitivists normally define the graded structure/prototype view as a statistical theory. Smith and Medin (1981) call it the probabilistic view. Fodor (1998) defines it as a statistical theory based on 'how likely it is that something in the concept's extension has the property that [a given] feature expresses' (p. 92). But to define it thus is to miss the philosophically most prescient aspect of graded structure. All people for all conceptual categories judge members of categories to have different degrees of membership even when there is no question about, or statistical variation in, whether or not items are judged members of the category. (For example, people who judge 8421 as unequivocally an odd number also judge 7 to be a better example of their idea of the category odd number.) In fact, the phenomenon of judging different degrees of membership is so universal that, by a strange twist of logic, it has been used as a refutation of graded structure affects (Gleitman *et al.*, 1983). In addition, as was pointed out earlier, many prototypes are derived from sources other than statistical frequency. To define graded structure in terms of statistics alone is to shirk addressing one of the basic challenges that graded structure and prototype effects offer both to the classical view and to cognitivism. We will discuss later why cognitivism must define it that way.

Both strict and working cognitivists are highly critical of prototypes and graded structure as an account of concepts or categorization even when they accept the experimental findings. The only legitimate way in which prototypes can represent concepts in cognitivism is if the prototype can be substituted for the definition and manipulated accordingly. Can prototypes substitute for defining features in a formal semantic model that would account for logical and linguistic functions such as synonymy, contradiction, and conjunctive categories? This is known as the issue of componentiality. In an influential paper, Osherson and Smith (1981) modelled proto-type theory using Zadeh's (1965) fuzzy set logic, in which conjunctive categories are computed by a maximization rule, and showed that prototypes do not follow this rule; for example, *guppy*, which is neither a very good example of the category *pet* nor of the category *fish*, is an excellent example of the category *pet fish*. (This has become known as 'the pet fish problem'.) Of course, this is actually a critique of a particular formal model, Zadeh's fuzzy set logic, rather than of prototypes and graded structure. It is not clear which, if any, formal model is appropriate for prototypes and graded structure, or even if there is, in principle, any empirical evidence which could distin-guish between rival models (since each model of storage is always presented with complementary processing assumptions which allow it to match any kind of experi-mental data; Barsalou, 1990).

Why does cognitivism require that concepts be compositional within formal mod-els? In the symbol systems of strict cognitivism, it is necessary that one be able to get from *pet* and *fish* to *pet fish* by purely syntactic means — that is, by the manipulation of formal symbols by formal rules — since there can be no corresponding real-world semantics to rely on. But one cannot derive the prototype for pet fish by any manipu-lation of the symbols for pet and fish. Thus one cannot turn graded structure into a for-mal definition. Cognitivism takes this as damning evidence against prototypes and graded structure; but again, it could as well be taken as a critique of cognitivism.

One way in which working cognitivism has tried to come to terms with the empiri-cal evidence for prototypes and graded structure is to divide categorization into *core concepts* and *processing heuristics*. This is embodied in a class of models in which the actual meaning for category terms is a classical definition onto which is added a processing heuristic or identification procedure which accounts for graded structure aspects (Osherson and Smith, 1981; Smith *et al.*, 1974). In this way, *odd number* can 'have' both a classical definition and a prototype. This is a dangerous move for it decouples theory from any empirical referent. In this way, the 'actual meaning' for a category term can become a kind of metaphysical entity known by logic alone, a suit-able entity perhaps for a methodologically solipsistic, artificially intelligent, cognitivist system but not one that our empirical sciences would probably want to embrace.

In sum: the evidence for graded structure and prototypes violates the tenets both of the classical view and of strict and working cognitivism: (1) Graded structure catego-ries do not have clear-cut boundaries. This is not simply an issue of the probability that items will be classified as members of the category since for many categories, such as colour, subjects will assert that some items are genuinely between categories. (2) Many categories have no, and no category need have any, necessary and sufficient attributes which make an item a member of the category. On the contrary, category prototypes contain a great deal of rich information, such as full-blown imagery and much detail about particular situations, which is not common to all members of the category. In short, the idea of skeletal necessary and sufficient attributes appears to be irrelevant for concepts when they are used in performing most activities. (3) Items in a category are not equivalent with respect to membership but rather possess gradations of membership. Again this is not merely a matter of probability as people will assert directly that one member of a category is a better example of the category than another. (All curves are not necessarily probability distributions.) (4) Graded structures are not formal systems nor are any items in a graded structure necessarily implicatory or productive of any other items in the structure, nor need anything in a graded structure fill the role of substitutable strings of symbols. Graded structures and prototypes do not need to be modelled by a formal system, such as Zadeh's fuzzy logic, in order to exist. (5) Graded structures and prototypes, although they have default contexts, are otherwise flexible with respect to the ever-varying contexts of life situations. They cannot be modelled by a solipsism, methodological or otherwise, since they must take into account those ever-changing real world situations. For the same reason, it is unclear how they could be representational in the working cognitivist sense; what stable thing in the real world, either underlying or surface, could they represent?

C. Theories

The third present-day approach to concepts is in terms of theories. There are actually two groups of theory theorists, somewhat conflated by many people: cognitivist oriented cognitive psychologists, who primarily address categorization issues, and developmental psychologists of the 'theory theory' school, who address conceptual change. The first group (Medin, 1989; Medin and Wattenmaker, 1987; Murphy and Medin, 1985) have used the idea of theories primarily as criticism of previous categorization research. Typically they set up an opposition between alleged similarity based theories of categorization, in which they place the graded structure/prototype view (and possibly everything else), and their new theories view. Similarity views are then criticized on the basis that: (1) they rely on counting attributes, but cannot explain attributes, (2) similarity cannot be explained by existing formal models (namely Tversky, 1977), and (3) various experiments on prototypes and graded structure show context effects. They then assert that attributes, similarity, and all experimental findings on categorization are derived from and accounted for by 'theories'. It is true that none of our present accounts provide an adequate understanding of the nature of attributes, similarity, or context — all hoary old problems. But what is the theories view, and how does it account for them?

It is remarkable that the first group of theory theorists never define or describe what they mean by theory and never, in all their literature, offer a single example of an actual theory from which findings, even one finding, in categorization research could be derived. Nor is there any attempt to show how attributes, similarity, or context could be derived from theories, either in the abstract or from specific theories.

Is this theory of categories as theories a new claim of substance or only a battle cry? What is meant by a theory? Explicit statements that can be brought to consciousness? Any item of world knowledge? The complete dictionary and encyclopaedia?

Any expectation or habit? Anything in the mind or in a book? Any context? It is hard to escape the impression that for the cognitive psychology approach to categories as theories, presently absolutely anything can count as a theory and that the word *theory* can be invoked as an explanation of any finding about similarity or attributes or context. If perceptual constraints are in evidence, one talks of perceptual *theories* (and invokes evolution) — somewhat like the proliferation of *instincts* and *drives* in an earlier psychology. It is interesting that many of the arguments used to support a theory view (e.g. examples that require one to bring world knowledge into one's explanation) are the very kind of issue used in the Heideggerian phenomenological tradition to argue against theories and in support of the necessity of positing a non-theory based Background of habits and skills which underlie the categories and activities of human life.

How is it that a view that relies so much on hand waving has proven so seductive? I believe it is because the word *theory* manages to evoke and give the impression of satisfying two problematically contradictory understandings of the world. On the one hand, people (even psychologists) know very well that life activities take place in a much larger context than is offered by our laboratory experiments. Experience presents itself in interdependent meaningful wholes, not in isolated units. Meaningful wholes include: world knowledge, beliefs (which are generally not organized into anything like coherent theories), expectations, desires, habits, skills, intuitions, the body, the functioning of the senses, tacit knowledge, everything that is un- or non-conscious, customs, values, the environment, and so on. On the other hand, we feel that scientific knowledge must come in bounded, explicit (ideally formal) structures. The word *theories* is suggestive of both context and formalisms. But this is a mirage. The larger context of meaningful wholes is not composed of theories, and what we do call theories are themselves (as the debates among philosophers of science illustrate) far from unarguably specifiable. Were any attempt made to carry out the theories programme, it would run into the same limitations and be subject to the same critiques that it makes of other accounts. The theories view maintains the illusion of viability only because it has no content. It is thus seriously misleading. However, the popularity of theories is a signal that people have important intuitions about concepts and context which should be explored further, a task to be undertaken in Section II of this paper.

In contrast to the cognitive psychologist's lack of specification of what is meant by theory, the 'theory theory' school of developmental psychology (see Gopnik and Meltzoff, 1997) explicitly defines theory as analogous to scientific theories, much like Kuhnian paradigms, and argues that cognitive development should be viewed as the successive replacement of one paradigm theory held by the child by another. Interest in concepts tends to be from the point of view of change in the child's (rather than the researcher's) theory of what a concepts is. When specific concepts are studied (such as biological types — Carey, 1985; Keil, 1979), the thrust is to show them as parts of larger theoretical units.

Fodor (1998), representing strict cognitivism, defines the 'theory' theory as conceptual holism and dismisses it with the charge that if a concept is defined in terms of the theory of which it is a part, then conceptual change of the sort discussed by psychologists is impossible, since to introduce a new term into a system or change the meaning of an already incorporated term would change the meaning of all the other terms in the system. Indeed, a cognitivist machine programmed with such a version of conceptual holism would have this difficulty. In reality, theories for the developmentalists are treated in many different ways according to demands of experiments — for example, as loosely structured sets of implicatory belief statements where change in one belief, such as in what it means to be alive, results in changes in certain other beliefs, such as in what it means to be dead (Slaughter *et al.*, in press). Fodor would call this an incoherent presentation of the 'theory' theory. What Fodor has done in the last analysis, and what any strict cognitivist would have to do, is to reduce both graded structure/prototypes and theories to a definitional account of concepts in terms of substitutable strings of symbols. Cognitivism has to do this because that's all a cognitivist machine can work with. Yet, as we have seen previously, this is completely unsatisfactory, even to cognitivists. (Fodor, 1998, recognizes the problem and attempts to remedy it, without changing any of the tenets of the cognitivist world view, by . . . yes . . . a new set of definitions — for a critique see Rosch, 1999.)

It is going to take something far more radical than this to reclaim concepts, and indeed cognitive science as a whole, from cognitivism. It requires a genuine rethinking of mind, world, concepts and their relationship.

II: Foundations for a New View of Concepts

A. *Concepts are the bridge between mind and world*

We think of mind and world as separate things.[1] We also think of bodies (or organisms) and environments as separate things. That is because we look at them in a certain way. Looked at another way (Rosch, in press), this is obviously false.

The world as perceived or categorized is, to echo Skarda's (this volume) terminology, a seamless whole or seamless web, in which perceiver/categorizer and perceived/categorized are simply opposite poles of the same event. In consciousness, those poles appear to be divided into actual separate things. The first function of concepts is to reconnect those opposite poles into functioning, even if still apparently separate, units. Looked at this way, concepts are partially recovering a state which is more veridical, and thus potentially more scientific, than the way we normally look at the world.

The first cognitive scientist to articulate anything like this vision was J.J. Gibson.

> To perceive the world is to coperceive oneself. . . . The optical information to specify the self . . . accompanies the optical information to specify the environment . . . . The one could not exist without the other . . . . The supposedly separate realms of the subjective and the objective are actually only poles of attention. The dualism of observer and environment is unnecessary. The information for the perception of 'here' is of the same kind as the information for the perception of 'there', and a continuous layout of surfaces extends from the one to the other (Gibson, 1979, p. 116).

This kind of thing may be most immediately obvious in perceiving one's location. One knows where one is only in relation to where other things are and vice versa; that is, one's judgment about one's own location and one's judgment about the location of relevant environmental reference points are aspects of the same inclusive

---

[1] Note that I am starting with *mind* and *world* as those terms are generally used and attempting to guide the reader towards a different (arguably better) understanding of such matters. A reader who feels the need to begin instead with classical definitions of *mind*, *world*, *organism*, *environment*, *consciousness*, and so on, might pause here and contemplate the import of the first part of this paper.

informational field. The burgeoning discipline of ecological psychology contains many experimental demonstrations of how this fact affects behaviour (see Neisser, 1993). Likewise, interactions between people are codependently defined, experienced, and acted out; this has been called *intersubjectivity* and is also the impetus for much current research (Trevarthen, 1993 — see also Rosch, 1996). On the micro level, Skarda (this volume) has shown in detail and with great clarity how dualistic perception can arise from the unbroken field of a perceptual event. Jarvilehto (1998a, 1998b) has argued for the need to reconceptualize the relationship between organism and environment as a single system at both the micro (neural) and macro (behavioral) levels. Yet at the conscious level of perception, it is very obvious that the perceiver and the world perceived are experienced as different and separate. Hence concepts.

All of the approaches to concepts and categories discussed earlier struggled with the issue of how to put together the perceiver/cognizer and perceived/cognized world. Look at the claims: the world contains real universals which the knower recollects or discovers; or the world contains the objects categorized (*extension* of the category) while the knowing mind contains the concept; or world and knower together produce saliences which become prototypes around which categories form; or world and knower together produce situations within which flexible categorizations take place; or the world works in a certain way which the knower maps by theories; and so on. All of these accounts share the same underlying intuition that it is concepts that span the mind–world divide. Only strict cognitivism with its methodological solipsism denies this, or at least denies that science can be done on it.

Corollary A: Concepts are not representational
Since the subjective and objective aspects of concepts and categories arise together as different poles of the same act of cognition and are part of the same informational field, they are already joined at their inception. They do not need to be further joined by a representational theory of mind, such as that of working cognitivism, and they cannot be separated by the solipsistic representational theory of mind of strict cognitivism. Concepts and categories do not represent the world in the mind; they are a *participating part* of the mind–world whole of which the sense of mind (of having a mind that is seeing or thinking) is one pole, and the objects of mind (such as visible objects, sounds, thoughts, emotions, and so on) are the other pole. Concepts — red, chair, afraid, yummy, armadillo, and all the rest — inextricably bind, in many different *functioning ways*, that sense of being or having a mind to the sense of the objects of mind.

B. *Concepts and categorizations only exist in concrete complex situations*
No matter how abstract and universal a concept may appear to be (*square root*, for example), that concept actually occurs only in specific, concrete situations. Real situations are information-rich complete events. One does not stand in thin air gaping at a tree as one does in philosophical examples; there is always a rich context, so rich that it has been argued that it can never be fully specified (Searle, 1983). Situations/contexts are mind–world bonded parts of entire forms of life. Context effects tend to be studied in psychological research only as negative factors, artifacts that invalidate somebody's experiment or theory. But it may be that contexts or situations are the unit that categorization research really needs to study.

Corollary B-1: Concepts and categorizations never occur in isolation

Concepts only occur as part of a web of meaning provided both by other concepts and by interrelated life activities. Think of the concept *big*; it has meaning only in relation to *small*. Furthermore, we know that a big flea is smaller than a small elephant. People who build actual artificial intelligence systems and story understanders know that to teach a machine the meaning of one word means necessarily to teach it the meaning of many other words; likewise for those other words, an exponential explosion which can only be limited by artificial means. And words alone are not enough. Look up a foreign word in a foreign dictionary; the definition will do you no good if you don't already know the language. To teach the meaning of a word also means to teach the machine many facts, the encyclopaedia part of the problem. As pointed out earlier, it is the interrelated aspect of concepts and world understanding that gives the theories view its apparent plausibility. But the interrelated web of understanding is not delimited by specifiable theories. Facts, like words, require innumerable other facts. That they really are innumerable is pointed out in Searle's *ceteris paribus* argument: you do not stipulate when you order a hamburger that it not come encased in lucite, not be a mile long . . . , an infinite number of assumptions about meaning and context which can never be formally specified. Furthermore, for all the information fed them, machine story understanders still do a pretty lame job; they do not understand the point of stories. That is because concepts (and the rest of human mentation) are not *per se* abstractly informative; they are participatory.

Corollary B-2: Concepts have a participatory, not an identifying, function in situations

What a thing is is already pre-given as part of the total mind–world situation in which it occurs. The basic function of concepts is not to identify things (just as it isn't to represent); it is not the function of concepts that we say to ourselves continually, 'Lo, yonder object has four legs and barks; it must be a dog!' Rather concepts participate in situations in innumerable flexible ways. Much experimental research on concepts and categories, and certainly most models, assume that what is to be explained, disputed, or modelled is the identification function. To be sure, we can participate in specialized identification activities (such as taking a botany exam, playing twenty questions, or being a subject in a concept learning experiment) but these are better considered as particular language games (as in Wittgenstein, 1953) than as the prototypical conceptual activity. (Analogously there are also particular representational forms, like architectural drawings and the like. Even creating formal systems can be seen as an activity prized and rewarded in some situations by some cultures.)

Because concepts are situation based and participatory rather than identification functions, definitions can be viewed in a new light. One of the situations in which people in our culture find themselves is to be asked for the definition of a concept. In this situation, the entire background of practices, understandings and explicit teachings with which we have been raised come into play, and the correct answer is generally an Aristotelian classical category definition. This is another form of language game. 'Definition' by means of prototypes and graded structure may serve a different function. Prototypes with their rich non-criterial information and imagery can indicate, on many different levels, possible ways of situating oneself and navigating in complex situations.

Situations are important beyond categorization research. Because mind and world only occur as conjoined aspects of actual complex situations, situations are the units within which human interpretations, emotions, and motivations hold sway. (For example, puzzling cases of so-called weakness of the will are cases where motivations don't transfer across situations.) Situations are also the domain of actions (an old issue in personality psychology, see Buss and Cantor, 1989). So situations are surely the units that require study for psychology in general. How? What is a situation? What kinds of situations are there? How are situations related and classified? How are they structured into forms of life? — And how might concepts and categories relate to all that?

Because concepts and categorizations play a vital role in bridging the mind–world unit and in revealing situational contexts, they may be able to provide a point of entry for the study of situations. Instead of asking how categories can be universal or how concepts can represent an external world in an internal mind, we could ask where categories and category systems come from in the first place? (Why are chairs a different category from tables or sofas? Why does *chair* seem more like this object's real name than *piece of furniture*, *material object*, or *desk chair*? Why does the category *kangaroos weighing between 1.3 and 2.9 pounds* seem neither basic, coherent, nor likely?) In strict cognitivism and in the classical view, categories could just as well be arbitrary sets of attributes and, indeed, were just that in traditional laboratory concept learning tasks. In so far as working cognitivism is concerned only with issues of representation, it can provide no clue as to the ecological conditions of real world category formation (aside from the requisite hand-wave at evolutionary theory in general). One attempt to address these questions (Rosch *et al.*,1976) proposed that under natural conditions there is a great deal of relational structure between perceptions, actions, and life activities, and that categories form so as to maximally map that structure.[2] For taxonomies of common objects, they called this level the *basic level*, showed how it was the default level at which categories were interpreted, and offered evidence that it had perceptual, linguistic, and developmental priority. Since objects are 'props' in situations, can research such as this provide clues as to a basic level of categorization of situations? (See Cantor *et al.*, 1982; Rifkin, 1985: Rosch, 1978.) Might information-rich concepts act as situational *simulators* (Barsalou, in press; Kahneman *et al.*, 1982)? Are there other approaches to situations that can situate concepts in their natural environments? Cognitivism cannot deal with such questions; it cannot even ask them.

C. 'Causal' laws for mind–world nonrepresentational units

When mind and world are considered separate, causal or explanatory efficacy is attributed either to the mind or the world. Happenings in the world may be considered stimuli to which the organism or person responds, or the mind or person may be seen as the source of desires, intentions, theories, or actions on the things of the world. Polarization of these extremes leads to theories and research programmes which go in and out of fashion, but which do not progress. At best the mind and world will be said to interact. The new view requires rethinking how we want to model causality and prediction altogether. Mind and world occur together in a succession of situations which are somewhat lawful and predictable. We want to be able to find those laws and

[2]   To perceive something as a *chair* or as *red*, is equally an act of perception and of categorization. More generally, cognition and perception may be much closer than our current theories, particularly cognitivism, are willing to acknowledge (for example, see Barsalou, in press).

to find a level of description which neither turns human action into something mechanical like engineering nor something mental like fantasy.

Gibson's ecological psychology faced a similar problem. His ecological optics was accused of being a covert way of placing the environment inside the organism's head. Yet observe how he avoids falling off either side of the dualistic tightrope when discussing the action of the human hand:

> The movement of the hands do not consist of responses to stimuli . . . . Is the only alternative to think of the hands as instruments of the mind? Piaget, for example, sometimes seems to imply that the hands are tools of a child's intelligence. But this is like saying that the hand is a tool of an inner child . . . . This is surely an error. The alternative is not a return to mentalism. We should think of the hands as neither triggered nor commanded but *controlled* (Gibson, 1979, p. 235).

Gibson is trying to develop an analysis of perception and action which captures just that level of functional description between subject and object. He has not been followed in this. While researchers have flocked to develop his ideas on higher order invariances in perception, ideas which could be pursued without changing one's mind set, little has been done to implement a new form of description.

D. Novelty

Cognitivist machines are determinate. When a machine is simply manipulating strings of symbols, replacing one with another, anything that comes out has been put in, in some form. Ultimately it is put in by the human programmer. Descriptions of concepts and categories in the rule driven systems of strict cognitivism admit of no novelty. The actual situations in which concepts and categories operate are unique and infinitely different from each other. Working cognitivism can only picture concepts responding to a few commonalities across situations to produce the skeletal abstracted concepts of the classical view. In the new view, concepts are not representations but functioning, participating parts of total situations. Thus concepts allow for all the novelty of the situations themselves, and there can be genuine surprises, learning and invention.

E. Non-conceptual background

Concepts only exist against a non-conceptual background. We could not even think to talk about concepts and conceptualization without some contrast of what they are not. All systems, other than cognitivism, have some way of admitting and at least trying to approach the non-conceptual. Some examples: knowing how versus knowing that, Heidegger's Background of habits and practices, private experience with its so-called qualia, body-based knowing, Searle's *ceteris paribus*, intuitions, experiences of all the arts, and ineffable experiences in love, grief, doing mathematics, religions, and everything else. But in cognitivism, it's concepts all the way down! — there is no way that a cognitivist system can deal with the non-conceptual. Yet it is in just those experiences that people find meaning and integrity in their lives. In cognitivism such things must be relegated to a separate sphere where they are either denied to exist or put fundamentally beyond the reach of cognitive science. In the new view proposed here, the non-conceptual is inherently part of mind–world situations, perhaps of every situation. No science of human existence can afford to straight-facedly exclude what is most meaningful to people.

Conclusions

Cognitivist systems are made out of concepts. If cognitivism cannot give an adequate account of its basic building blocks, it is in trouble. I have tried to show how cognitivism must necessarily misunderstand previous theories and research on concepts and how it cannot cope with the basic facts about concepts. I've outlined the foundations for a new view of concepts and categories which should place their study on a real (not a fantasy spherical cow) basis. The logic of concepts and categories must be an open system, not a closed logic. Only then can the study of concepts be reclaimed from the sterile reaches of definitions and substitutable strings of symbols to become a genuine science of human knowing.

References

Baker, C. (unpublished), 'The Shuowen Jiezi: The organization of the first Chinese dictionary', Department of Linguistics, University of California at Berkeley.

Barsalou, L.W. (1987), 'The instability of graded structure: Implications for the nature of concepts', in *Concepts and Conceptual Development: Ecological and Intellectual Factors in Categorization*, ed. U. Neisser (Cambridge: Cambridge University Press).

Barsalou, L.W. (1990), 'On the indistinguishability of exemplar memory and abstraction in category representation', in *Advances in Social Cognition (Vol. III): Content and Process Specificity of Prior Experiences*, ed. T.K. Skull and R.S. Wyer Jr.(Hillsdale, NJ: Lawrence Erlbaum).

Barsalou, L.W. (in press), 'Perceptual symbol systems', *Behavioural and Brain Sciences*.

Berlin, B. and Kay, P. (1969), *Basic Color Terms: Their Universality and Evolution*. (Berkeley: University of California Press).

Bruner, J.S., Goodnow, J.J. and Austin, G.A. (1956), *A Study of Thinking* (New York: John Wiley).

Buss, D.M. and Cantor, N. (ed. 1989), *Personality Psychology: Recent Trends and Emerging Directions* (New York: Springer-Verlag).

Cantor, N., Mischel, W. and Schwartz, J.C. (1982), 'A prototype analysis of psychological situations', *Cognitive Psychology*, **14**, pp. 45–77

Carey, S. (1985), *Conceptual Change in Childhood* (Cambridge, MA: MIT Press).

Chakrabarti, K. (1975), 'The Nyaya-Vaisesika theory of universals', *Journal of Indian Philosophy*, **3**, pp. 363–82.

Dreyfus, H.L. and Hall, H. (1982) *Husserl, Intentionality, and Cognitive Science* (Cambridge, MA: MIT Press).

Fodor, J. (1982), 'Methodological solipsism considered as a research strategy in cognitive psychology', in Dreyfus and Hall (1982).

Fodor, J.A. (1998), *Concepts: Where Cognitive Science Went Wrong* (Oxford: Clarendon Press).

Gleitman, L.R., Armstrong, S.L. and Gleitman, H. (1983), 'On doubting the concept "concept"', in *New Trends in Conceptual Representation: Challenges to Piaget's Theory?*, ed. E.K. Scholnick (Hillsdale, NJ: Lawrence Erlbaum).

Gibson, J.J. (1979), *The Ecological Approach to Visual Perception* (Boston: Houghton-Mifflin).

Gopnik, A. and Meltzoff, A.N. (1997), *Words, Thoughts, and Theories* (Cambridge, MA: MIT Press).

Hardin, C.L. and Maffi, L. (ed. 1997), *Colour Categories in Thought and Language*, (Cambridge: Cambridge University Press).

Järvilehto, T. (1998a), 'The theory of the organism-environment system: I. Description of the theory', *Integrative Physiological and Behavioral Science*, **33**, pp. 321–34.

Järvilehto, T. (1998b), 'The theory of the organism-environment system: II. Significance of nervous activity in the organism-environment system', *Integrative Physiological and Behavioral Science*, **33**, pp. 335–42.

Johnson, D.M. and Erneling, C.E. (1997) *The Future of the Cognitive Revolution* (New York: Oxford University Press).

Kahneman, D., Slovic, P. and Tversky, A. (ed. 1982) *Judgment Under Uncertainty: Heuristics and Biases* (New York: Cambridge University press).

Keil, F.C. (1979), *Semantic and Conceptual Development: An Ontological Perspective* (Cambridge, MA: Harvard University Press).

Lakoff, G. (1987), *Women, Fire, and Dangerous Things: What Categories Reveal About the Mind* (Chicago: University of Chicago Press).

Markman, E.M. (1989), *Categorization and Naming in Children* (Cambridge, MA: MIT Press).

Medin, D.L. (1989), 'Concepts and conceptual structure', *American Psychologist*, **44**, pp.1469–81.

Medin, D.L. and Wattenmaker, W.D. (1987), 'Cognitive cohesiveness, theories, and cognitive archeology', in *Concepts and Conceptual Development: Ecological and Intellectual Factors in Categorization*, ed. U. Neisser (Cambridge: Cambridge University Press).

Mervis, C.B. and Rosch, E. (1981), 'Categorization of natural objects', *Annual Review of Psychology*, **32**, pp. 89–115.

Murphy, G.L. and Medin, D.L. (1985), 'The role of theories in conceptual coherence' *Psychological Review*, **92**, pp. 289–316.

Neisser, U. (ed. 1993), *The Perceived Self: Ecological and Interpersonal Sources of Self-knowledge* (Cambridge: Cambridge University Press).

Nisbett, R.E., Peng, K.,Choi, I. and Norenzayan, A. (under review), 'Culture and systems of thought: Holistic versus analytic cognition', *Psychological Review*.

Osherson, D.N. and Smith, E.E. (1981), 'On the adequacy of prototype theory as a theory of concepts', *Cognition,* **9**, pp. 35–58.

Peng, K. and Nisbett, R.E. (1999), 'Naive dialecticism and its effects on reasoning and judgment about contradiction', *American Psychologist*, **54**, pp. 741–54.

Radhakrishnan, S. and Moore, C.A. (1957), *A Sourcebook in Indian Philosophy* (Princeton, NJ: Princeton University Press).

Rifkin, A. (1985), 'Evidence for a basic level in event taxonomies', *Memory and Cognition*, **13**, pp. 538–56.

Rosch, E. (1973), 'Natural categories', *Cognitive Psychology*, **4**, pp. 328–50.

Rosch, E. (1975), 'Cognitive reference points', *Cognitive Psychology*, **7**, pp. 532–47.

Rosch, E. (1977), 'Human categorization', in *Studies in Cross-cultural Psychology (Vol. 1)*, ed. N. Warren (London: Academic Press).

Rosch, E. (1978), 'Principles of categorization', in *Cognition and Categorization*, ed. E. Rosch and B.B. Lloyd (Hillsdale, NJ: Lawrence Erlbaum).

Rosch, E. (1987), 'Wittgenstein and categorization research in cognitive psychology', in *Meaning and Growth of Understanding: Wittgenstein's Significance for Developmental Psychology*, ed. M. Chapman and R.A. Dixon (Berlin: Springer-Verlag).

Rosch, E. (1994), 'Categorization', in *The Encyclopedia of Human Behavior* (San Diego, CA: Academic Press).

Rosch, E. (1996), 'The environment of minds: Towards a noetic and hedonic ecology', in *Cognitive Ecology*, ed. M.P. Friedman and E.C. Carterette (San Diego: Academic Press).

Rosch, E. (1999), 'What are concepts?' Review of Fodor (1998), *Contemporary Psychology*, **44**, pp. 416–7.

Rosch, E. (in press),. '"Spit straight up — learn something": Can Tibetan Buddhism inform the cognitive sciences?', in *Meeting at the Roots: Essays on Tibetan Buddhism and the Natural Sciences*, ed. B.A. Wallace (Berkeley, CA: University of California Press).

Rosch, E. and Mervis, C.B. (1975), 'Family resemblances: Studies in the internal structure of categories', *Cognitive psychology*, **7**, pp. 573–605.

Rosch, E. and Lloyd, B.B. (ed. 1978), *Cognition and Categorization*. (Hillsdale, NJ: Lawrence Erlbaum).

Rosch, E., Mervis, C.B., Gray, W.D., Johnson, D.M., and Boyes-Braem, P. (1976), 'Basic objects in natural categories', *Cognitive Psychology*, **8**, pp. 382–439.

Searle, J.R. (1983), *Intentionality* (Cambridge: Cambridge University Press).

Slaughter, V. Jaakkola, R. and Carey, S. (in press), 'Constructing a coherent theory: Children's biological understanding of life and death', in *Children's Understanding of Biology and Health*, ed. M. Siegal and C. Peterson (Cambridge: Cambridge University Press).

Smith, E.E. and Medin, D.L. (1981), *Categories and Concepts*. (Cambridge, MA: Harvard University Press).

Smith, E.E., Shoben. E.J. and Rips, L.J. (1974), 'Structure and process in semantic memory: a featural model for semantic decisions', *Psychological Review*, **81**, pp. 214–41.

Trevarthen, C. (1993), 'The self born in intersubjectivity: The psychology of an infant communicating', in *The Perceived Self: Ecological and Interpersonal Sources of Self-knowledge*, ed. U. Neisser (Cambridge: Cambridge University press).

Tversky, A. (1977), 'Features of similarity', *Psychological Review*, **84**, pp. 327–52.

Wittgenstein, L. (1953), *Philosophical Investigations* (New York: The Macmillan Company).

Woozley, A.D. (1967), 'Universals', *The Encyclopaedia of Philosophy* (New York: Macmillan Publishing Co.).

Zadeh, L. (1965), 'Fuzzy sets', *Information control*, **8**, pp. 338–53.

Christine A. Skarda

The Perceptual Form of Life

To view organismic functioning in terms of integration is a mistake, although the concept has dominated scientific thinking this century. The operative concept for interpreting the organism proposed here is that of 'articulation' or decomposition rather than that of composition from segregated parts. It is asserted that holism is the fundamental state of all phenomena, including organisms. The impact of this changed perspective on perceptual theorizing is profound. Rather than viewing it as a process resulting from internal integration of isolated features detected by receptor neurons into a perceptual whole, the new theory suggests that the task of perceptual processing is to break up what initially exists holistically in sense organs into features and eventually perceived objects. Similarly, the goal of perceptual activity is not Sherrington's, that of integrating essentially unrelated organisms with their environmental surround, but rather to generate percepts in which the environment appears as a field of objects and events independent of the perceiver which are available for manipulation. Perception is a process by which organisms use their embeddedness in physical reality as if they were independent of it. There are a number of interesting results of this conceptual reorientation. The binding problem is eliminated because the percept's holistic character is the precondition for neural activity, not its product. The concept of representation can be dispensed with since the fundamental conceptual motivation for its introduction — the assumed need to produce an internal copy of what was assumed to exist independently outside the organism in order to integrate organismic behaviour with its environmental causes — is rejected outright. And finally, the issue of perceptual consciousness is addressed: how does the percept acquire its objective status vis-à-vis a perceiver, and what is the basis of the experiential character of perception?

The real task today for those critical of cognitivism is to develop viable theoretical alternatives to such models. Perceptual theorists, for example, ought to ask what form a truly nonrepresentation-based perceptual theory might take and how might it reinterpret the data gathered by perceptual system researchers over the past half-century? I have developed one such theory whose basic propositions are summarized in the present paper.

By way of introduction, the new theory represents a profound rethinking of the perceptual problematic. Of course, the proposed theory makes no use of the concept of representation, hence it is non-cognitivist in spirit. It also rejects the naive realism of

Journal of Consciousness Studies, **6**, No. 11–12, 1999, pp. 79–93

representational theories because it does not equate what is perceived by the organism with the physical causes or stimuli that give rise to perception. This means that the new model does not 'perceptualize' the physical stimuli such as light or sound waves transforming them into colour or sound that is assumed to exist before perception takes places. All *perceived* features and objects are the *products* not the causes of perceptual processing. The nature of physical reality prior to perception is not that of the perceptual unities (objects and their features) of the percept and cannot merely be copied by perceptual systems. But rejection of naive realism does not imply that the current theory is a form of subjectivism, for it does not claim that the percept is merely subjective. Subjectivism violates the principle, fundamental to the new model, that all phenomena (whether physical or mental) are inexorably embedded in a causal network of reality. The notion of a cut off or private (acausal) interiority is essentially meaningless. The new model attempts to walk a middle road between the extremes of naive realism which underpins cognitivism's representation-based theory of perception and the dead end of a meaningless subjectivism which too often is assumed to be the only theoretical alternative when representational thinking is abandoned.

Two premises form the basis of the new model. They are set in opposition to two of the fundamental assumptions of cognitivist theories of perception: that the aim of perceptual functioning is to produce a copy (internal representation or correlate) of what is thought to exist independently outside the perceiver, and thus to co-ordinate or integrate what goes on within the perceiving organism with its external environment. The two counterpremises proposed by the new model are that (1) what is perceived, the percept[1], cannot be copied from the so-called stimulus array because the contents of the percept (perceived objects and features) are not 'out there' simply waiting to be detected and copied. The percept is uniquely perceptual in nature, it is the product (not the physical cause) of perceptual system functioning. And what is more, (2) the new model proposes to view integration of the organism and its environment as a *given*. It is not the task of perception to integrate the organism with the rest of the physical world. The view of the organism as an isolated or essentially unrelated physical system is a fiction, a theoretical construct that ought to be rejected if we are to achieve a more satisfactory account of organismic functioning.

Regarding the issue of integration, if one views the relation between what occurs within the organism and the environmental surround as essentially integrated from the start, this rules out the existence independent or isolated internal sphere of reality that is presupposed by subjectivist accounts. Nothing takes place within the organism that is not always already related to what goes on outside of its skin. Nor, it should be added, are internal organismic processes correctly viewed as inherently independent. The holism of the new model applies across the board to all aspects of organismic functioning, internal and external. Yet, historically the view of integration as the goal of

---

[1] By 'percept', I mean the object, state of affairs or event that is perceived. So all directly experienced objects are percepts, i.e., objects that are not merely imagined, remembered, etc. In the language of philosophy, a percept is the intentional object or objective correlate that stands in relation to a perceiver in any act of perception. Percepts can be conscious or unconscious. Awareness or the direct experience of them does not create the objective status of their content, i.e., awareness is not itself objectifying. In the present paper, however, I restrict the discussion to conscious acts of perception. My interest is not in what makes a percept conscious, but rather in what it is about organismic functioning that contributes to the creation of the percept.

organismic functioning, as a product of compilation, is the operative concept in terms of which organisms have been understood at least since the time of Sherrington (1906). It is traditional to view the organism as a collection of parts, each of which serve individual functions that are welded together by nervous system functioning to produce the global behaviour of the organism. Perception's function, given this interpretation, is to integrate what goes on inside the organism with what goes on outside, as if the two spheres were fundamentally unrelated in spite of their obvious causal interaction.[2] Moreover, mind, according to this conceptual scheme, is just another isolated component standing in need of integration, this time with the physical realm. Thus, it is not surprising that Sherrington's ultimate goal was to discover the final integration which unites the 'two great, and in some respects counterpart, systems of the organism' — the physical body and the mind. The new model proposes instead that we think of holism as the initial condition of all phenomena and of parts as produced. This is the conceptual inversion that is applied here to interpret organismic functioning.

Yet if holism is the initial condition of everything that exists, then the entire landscape for perceptual theorizing changes drastically. The new model proposes that the real problem of perception is that of *articulation*.[3] The goal of perceptual functioning is to break up the holistic fabric of reality into perceived objects and states of affairs and perceivers of them. Thus, the perceptual problematic is inverted from that of integration to a situation in which perceivers generate percepts in which the world appears to the perceiver as a collection of objects and events independent of the observer which are available for manipulation. Perception is a process in which perceivers use the seamless state of their embeddedness in physical reality as if they were independent of it by creating percepts.

The new model's view on the question of integration also has implications for the first of cognitivism's fundamental assumptions, the idea that perceptual systems internalize copies of the extra-organismic environment. Cognitivist theories require a copy of the external stimulus because they assume what goes on inside the organism is essentially cut off from what occurs outside its skin. Internal representations form the 'glue' to integrate the two spheres. However, having rejected all talk about genuine independence, indeed of gaps in physical reality, the new model has no need for internal copies, images, or maps. While the new model accepts that the extra-organismic environment is indeed 'structured' and that it therefore shapes and constrains organismic functioning, it claims that the effects of such processes are accomplished without creating internal representations. Perception is a complex set of processes interpreted as 'uses' that progressively articulate the interactive

---

[2] The conceptual conflict inherent in viewing organismic functioning as both the product of causal interaction with external stimuli and in need of reintegration with the external world pervades contemporary thinking in science, philosophy, cognitive science and psychology, producing many of the problems that bedevil theorists today as in the past. Logically speaking, there are only two possibilities. Either something is related or it is not. If it is related, it does not stand in need of relating and hence the question of integration is moot. If its nature is to be unrelated, then unrelated it will remain. According to the present model, what exists (that which is real, reality) is by nature related. To be subject to causation, I hasten to add, is to be related from the start. Causation does not create relations among unrelated phenomena, it is a particular kind of relationship of influence holding between already related phenomena.

[3] 'Articulate' here is used in its Latin sense of *articulatis*, i.e., something divided into components, or jointed as a limb is jointed. Thus, I describe a process of articulation as that of 'shattering', a process of deconstruction.

organismic events taking place at the interactive interface between the organism and its environment into percepts.[4] The organism and its world are always already integrated and the articulation of them into correlated spheres of functioning does not and cannot remove the relatedness. The prior integration of percepts with their physical causes is implicit in every facet of their articulation.

There is a conceptual distinction essential to the proposed new model's rejection of representation-based theory that cognitivism fails to recognize. The distinction is that between *physical reality* and *perceived reality*. By physical reality, I mean the pre-perceptual, physical bases of perception, its raw material. Perceptual reality, on the other hand, refers exclusively to the percept, to that which the perceiver perceives. So, with respect to perceptual system functioning, 'physical reality' refers to things like light or sound waves and neural activity, while 'perceptual reality' refers to the product of perceptual functioning, i.e., the percept.

There is a substantial and convincing body of scientific research to support this distinction. It has been demonstrated, for example, that perceived features are unique to the percept. In the field of visual research, for example, Helson (1938) and Land (1977) demonstrated that perceived colour has no counterpart in the physical phenomena with which the eyes interact (see also Gibson, 1979). Light and its physical properties are not colours. Colour is generated first by so-called opponent mechanisms that are part of the visual system and which contrast different wavelengths of light to produce perceived colour. Moreover, there is not even a strict correspondence between a given wavelength of light and the colour perceived, as one might expect given representation-based thinking. For example, humans perceive the colour green when more middle-wave light is reflected from the stimulus array, but an object can be perceived as green even when the array, in fact, reflects more short- or long-wave light. And it has also been shown that the very same wavelength of light can be perceived as different colours depending on the context in which it occurs (see also Gurwitsch, 1964). Finally, there are findings that indicate that objects themselves are uniquely perceptual phenomena. Gouras and Zenner (1981) concluded from their research that it is impossible to separate the object perceived from colour because colour contrast itself forms the object. Thus, even objects are essentially perceptual phenomena since colour is a property of percepts, not of the physical stimulus array. Perceived features, not the external stimuli, help to create visually perceived objects.

All of which means that somehow a shift is made from the pre-perceptual, physical reality to the percept, but it is wrong to further equate this transformation with an abrupt break that isolates the percept from the pre-perceptual bases of perceptual functioning. The percept is inextricably embedded in the causal web of physical reality, which means that constraints always operate on the percept from its causal antecedents. The physical stimuli operate as a set of boundary conditions upon the percept. Somehow the percept and its contents are created with an apparent (perceived) independence from the observer while the percept itself and the perceiving

[4] Strictly speaking, an environment is the product of perceptual system functioning, not its precursor. That is to say, prior to perceptual functioning that defines for the organism what counts as 'self' and what is 'not self' or extra-organismic, what occurs 'within' is just as 'external' as any other physical process occurring in the physical world. It is organismic functioning itself that establishes the distinction. In other words, the organism–environment contrast is not a fact about the world, it is a fact created by organismic functioning itself.

organism remain embedded in physical reality. Explain this, and we will have gone a long way towards understanding perceptual functioning.

Proposition One *The first product of perceptual system functioning is the 'phenomenal fabric' of the percept that is created by the sense organs.*

The new model's first proposal, then, is that the sense organs make *direct contributions to the percept* that are not, and need not be, filtered through neurons. Thus the new model rejects neurocentrism, the view that neural activity *alone* contributes to the percept, a view that has already come in for criticism (Damasio, 1994). But the real problem with neurocentrism in relation to perceptual theory, as I see it, is not just its preoccupation with neurons, but the fact that it leads to an oversight that is responsible for one of the chief mysteries of perception: why perceivers perceive percepts and not neural activity or physical reality. Photons may stream through the lens of the eye and neurons become active, but neither photons nor neural activity are perceived. Why? There is a deafening silence today regarding this issue that can no longer be brushed aside. It is, after all, the defining property of perception that it is the percept that is perceived. It is also the basic stumbling block to the acceptance of materialist perceptual theories that they have been unable to explain how neural activity relates to perceived qualities.

The solution may be found by reconsidering the available data. The new model proposes that there is a solution to the mystery, and that it is to be found by looking at what is created by the sense organs and not within the neural system. Consider visual perception. All visual percepts are basically complex configurations of light. Neural activity is not light, nor does such activity produce light whatever promises may have been made by representation-based cognitivist theories in the past. However, the sense organ, the lens in this case, does create a complex light phenomenon that is projected onto the back of the eye. This phenomenon is the actual stuff or raw material of visual perception and ought to have interested theorists enormously, but it hasn't. The conceptual bias of neurocentrism did not allow theorists to admit that the pre-neural phenomena in sense organs could be directly perceptually accessible without neural intervention. Yet the sense organs do create within the perceptual system the complex light phenomena that are eventually articulated into perceived features and objects. Recognizing this the new model proposes that neural activity need not be 'transformed' into a light phenomenon in the brain, nor is there any need to represent the light neurally. One does not need to copy or create a neural copy of a phenomenon that is already present in and created by the perceptual system and hence available to the organism.

However, this does not mean that the percept already exists in the sense organ. Sense organs alone do not create the percept. The percept is the articulated phenomenal event as that articulation is used by the whole organism, and articulation is the result of neural activity according to the new model. The 'phenomenal fabric' is but the cloth out of which the percept is articulated. The phenomenal basis of the percept is created within sense organs. Its phenomenality consists in the fact that it is a complex, but as yet undifferentiated, sensory content *directly apprehended by* the organism without the intervention of neural activity. It is not articulated into features, nor is it as yet 'objectified' by the sense organ, and hence is not to be identified with the percept, but it is the phenomenal basis for all subsequent levels of perceptual system functioning.

Proposition Two *Level One neurons in perceptual systems have two functions: they 'shatter', with their selective response abilities, the holistic phenomenal fabric created by the sense organs, and secondly, they create features by means of contrast usage.*

Today it is generally accepted by neuroscientists that neural activity in perceptual systems can be divided into two distinct levels of functioning (see Engel *et al.*, 1992).[5] The new model accepts the division, but redefines both levels in keeping with its nonrepresentational interpretation. Level One neurons include receptor and post-receptor neurons. These neurons function singly or as members of neural networks (groups of neurons that are linked together through feedback connections and that become active as a group rather than singly). Level One neurons are organized in a hierarchical fashion with succeeding levels of activity dependent upon the results of earlier levels of functioning. Neural activity at this first level is highly parallel in nature.

According to current thinking, Level One neurons first represent isolated, and then sequentially more complex, features of the stimulus array. They are thought to internalize, represent, and begin to assemble an internal correlate of the stimulus. Hence receptor neurons have been dubbed 'feature detectors' (Barlow, 1953), although this representational reading of the data was based solely on the discovery that individual receptors are *selectively sensitive* to particular stimuli. For example, the selective sensitivity of a ganglion cell in the eye of a frog to the presentation of a moving black spot led to the claim that this neuron was a 'bug detector' whose activity represented the presence of a fly. The conceptual leap from selectivity of response to internal representation passed without comment in the literature. Later research shifted attention to neural activity located deeper within the perceptual system (Hubel and Wiesel, 1962). In the visual cortex, for example, neurons were discovered that responded selectively to the movement of a light stimulus and to more complex kinds of stimuli. These post-receptor neurons received input simultaneously from different classes of receptor neurons selective for different stimuli and were therefore thought to represent compound features of the external stimulus. It was a small step to conclude from these findings that the combinatorial process continued within the brain to produce eventually a complete internal neural correlate of the external stimulus array.

But small though it may have seemed, it was a decisive and eventually fatal step for perceptual neuroscience which became the experimental quest in search of the complete neural correlate of the percept. Nearly half a century later the goal of finding this reconstructed correlate still eludes researchers. The problem today is referred to as the 'binding problem'. In response to this, the new model makes the following proposal. First, jettison the neurocentric premise regarding perceptual system functioning and adopt the hypothesis that sense organs (such as the lens of the eye) make direct contributions to the percept without neural intervention. The new model suggests that the binding problem remains unsolved, not because the neural processes responsible for binding have eluded researchers, but because the neurocentric conceptual framework with which neuroscientists have approached their data forced neuroscientists to look for a neural process that does not exist.

---

[5]   The data to support the distinction between two levels of neural functioning are based on early work by researchers such as Barlow (1953) and by Hubel and Wiesel (1959). Barlow's seminal paper of 1972 heavily influenced subsequent research in the field of perception. What follows is a brief summary of the findings on the subject collected during the past half century.

The new model proposes that the percept's holistic nature is the second major contribution to the percept made by sense organ functioning, the first being its creation of the percept's phenomenal fabric. Recall that I said earlier that the complexity of the phenomenal fabric, although present in the sense organ, remains unarticulated until neural activity takes place. This means that the phenomenal fabric exists as a unified whole at the outset. Its holism or undifferentiated character is also perceptually available without neural intervention according to the new model, just as the phenomenality of that undifferentiated whole is available to the organism. Thus I suggest that creation of the percept as a holistic phenomenon is accomplished by the sense organs *at the very outset of the perceptual process* and that it is not the task of neurons to compile such wholes. If this interpretation is correct, then it is understandable why a complete neural correlate of the percept has never been discovered by researchers: there isn't one. Thus, the binding problem is eliminated.

But if neurons are not in the combinatorial business of building up perceptual wholes out of features and, furthermore, if they do not represent features as cognitivism maintains, then what do neurons contribute to perception? Proposition Two characterizes the achievements of the two forms of neural activity taking place at Level One without appeal to representations or the concept of binding. The first form of functioning involved at Level One is, as we have seen, the selective functioning of receptor neurons that are stimulus-specific in terms of their response characteristics. Each receptor neuron responds maximally (i.e., with a burst of intense electrical activity) to a specific type or class of stimuli. Moreover, receptor neurons do not interact among themselves. The activity of each receptor neuron occurs independently of the activity of other receptors because they share no connections and therefore have no feedback connections. This means that receptor neurons are not only narrowly selective in relation to the phenomenal event created by the sense organs, but that receptors also effectively hold the stimuli that they react to in isolation because they do not interact among themselves.

Based on these data concerning neural functioning, the new model proposes that receptor neurons are the means by which the holistic phenomenal events created by sense organs are *shattered*, broken up, or articulated into parts. Because there are many different types of receptors selective for many different kinds of physical stimuli, they effectively create complexity *vis-à-vis* the previously unarticulated phenomenal event in the sense organ by means of their finely tuned response characteristics. Shattering into isolated parts is by no means a trivial accomplishment. Remember that holism (integration, connection) is the fundamental condition of reality and that the phenomenal event in the sense organ shares this universal characteristic. But because receptors operate in isolation from one another, they effectively isolate their triggering stimuli for the organism, even though the stimuli actually exist in an undifferentiated state in the sense organ. In this way aspects of the phenomenal event to which individual receptor neurons react *inherit discreteness* from the way in which they are 'used' by the organism. The phenomenal event in the sense organ acquires its articulation because perceptual systems interact with it in an analytical fashion.

Receptor functioning is followed by post-receptor activity in perceptual systems, a second form of Level One neural activity. Post-receptors are differently connected than receptor neurons and hence function differently with respect to the sense organ event. Because they function in a new way, they constitute a different form of use.

This form of use results in the generation of perceptual features. Post-receptor functioning includes the activity of neurons acting singly and within neural networks of all sizes, but its defining characteristic is that post-receptor neurons interact via feedback relations with other neurons. Post-receptors deal with the phenomenal event occurring within the sense organ at one remove, as it were — they react to the phenomenal fabric as shattered by receptor neurons.

Post-receptors engage in different forms of contrast functioning. Contrast is a form of neural activity that involves so called 'opponent mechanisms'. A simple example of contrast functioning is provided by the activity of a type of post-receptor neuron in the eye, the ganglion cell. Ganglion neurons receive input from two different types of receptor neurons sensitive to the same local region of the retina (called the ganglion's receptive field). One of these receptor neurons inhibits (stops or slows) the bursting activity of the ganglion and the other excites (starts or speeds up) the activity of the ganglion. The input from the two types of receptors can be organized in various ways at the ganglion. For example, a ganglion neuron may have what is called an on-centre, off-surround arrangement. This means that when the ganglion neuron receives input from a receptor sensitive to the central area of its receptive field, the ganglion neuron becomes active. However, if receptor neurons sensitive to the outer portion of its receptive field (the surround) should become active, they will inhibit the activity of the ganglion neuron. In essence, therefore, the ganglion post-receptor neuron effectively contrasts one receptor neuron's input with another receptor's input. All post-receptor neurons act in this way, as contrast mechanisms, although not all contrast is achieved by on-centre, off-surround mechanisms. The contrasted input may be from different locations in the receptive field of the post-receptor neuron, or it may involve input from receptors sensitive to different kinds of triggering stimuli or input from different post-receptor neurons (either individual post-receptors or from networks of post-receptor neurons). But in the end it comes down to this: post-receptor activity is the way that perceptual systems have evolved to be able to contrast what was isolated by receptor neurons from the holistic phenomenal fabric created by sense organs.

Features are dependent phenomena. Every feature acquires its unique nature as a feature in a process of contrast. Contrast, in turn, requires the prior isolation of the phenomena that are brought into the relation of contrast. The colour red, for example, acquires its character because (1) a particular wavelength of light is isolated and (2) is then contrasted within the visual system with the wavelength of light that humans perceive as green (rather than yellow or blue). Features require contrast mechanisms and this is why, in the absence of light opponency mechanisms like those that occur in the visual system, there are no colours. Light remains, but not colour. Thus, an argument can be made that it is fundamentally incorrect to refer to receptor neurons as 'feature detectors', although this is the term that is generally used today, because contrast does not occur at the level of receptor activity (it first occurs at the level of *post*-receptor activity).

But how can we explain the genesis of features by post-receptors without appeal to the notion of representation? Here the new model introduces the concept of 'use': it views receptor and post-receptor activity as *ways in which the organism does something with the phenomenal event*, rather than as neural representations of that event. This is significant. Forms of use are inherently 'creative', and their creativity creates the meaning of the phenomena that are used. One example of this creativity has already been mentioned, that of stimulus specificity in relation to receptor activity. The discreteness of components of the phenomenal event created by sense organs, I claimed, is not copied from that event, it is bestowed on it by virtue of the isolated way in which it is used by receptor neurons.

The concept of use can also be applied to feature generation by post-receptor neurons. The colour feature red, for example, is not just a particular frequency of light wave considered in isolation. Different frequencies of light acquire their unique colour characters from being contrasted with one another. Light is not a perceived feature, colour is; and the physical phenomenon, light, acquires its unique colour character when contrasted by post-receptor neurons in the human visual system with light isolated by so-called green receptor neurons. Moreover, red would not be the phenomenal colour that it is if the same light frequency were contrasted with some other light frequency rather than that which green receptor neurons isolate. For example, in avian visual systems, although the same light frequency that generates the colour red in the human visual system is isolated by receptor neurons, it is contrasted with infra-red and ultra-violet spectra of light. Hence the self-same frequency does not have the same phenomenal character as red does for human beings. What has changed is not the type of light wave isolated by receptor neurons in each case, but the contrast relations into which it has been brought by post-receptors. The result is a new perceived colour, a colour feature incommensurate with our own (Varela *et al.*, 1991).

Use makes it so. Use endows what is used with its character, its functional meaning for the organism. It is not the physical nature of what is used that determines its functional meaning and character. Of course, the physical substrate must be able to support the kind of use to which it is put; it must be the sort of phenomenon that can be used as colour (*viz.*, light). But it is the use of that physical substrate by way of contrast that endows it with its quality as a feature of a particular sort. Thus, representations are replaced with the organism's use of the results of its own interactive functioning in the world.

> **Proposition Three** *Level Two neural activity is the activity of globally distributed neural masses. Its primary function is to create 'system behaviour' and thereby to articulate the organism whole into distinctly functioning subsystems of use.*

Level Two neural activity is sometimes identified with the activity of very large neural networks, so-called supernetworks. The data, however, indicate that neurons within supernetworks exhibit synchronized activity when presented with a stimulus such as a moving bar of light with a particular orientation (Gray and Singer, 1987). Desynchronizing connections are thought to exist between neurons with widely divergent orientation preferences. Does such activity represent a distinct form of functioning, a form of activity that could be categorized as belonging to a different functional level than that which is defined as Level One? I think not.

According to researchers, neurons in supernetworks are narrowly tuned and stimulus specific. Moreover, it seems clear that synchronization and desynchronization are essentially contrast mechanisms that pit one stimulus against another. These data suggest, therefore, that supernetwork activity is actually a form of Level One neural activity. The scale of neural involvement differs — e.g., from single neurons and small neural networks to supernetworks — but the functional attributes of that activity are the same as that occurring in Level One. For this reason, the present model classifies supernetwork activity as a form of Level One neural activity.

According to the new model, true Level Two neural activity can be defined as neural behaviour collectively generated by tens of thousands of interconnected neurons. This form of activity is referred to as 'neural mass action' (Freeman, 1975). Unlike network activity, neural mass action is true system behaviour, like that of a wave in the ocean. Just as there are innumerable water molecules in a wave, there are tens of thousands of neurons involved in neural mass action within the brain. Like the wave, neural mass action occurs only when masses of neurons function as a unit to create a coherent form of system behaviour whose properties and effects are irreducible to the properties and effects capable of being produced by its components.

Perceptual functioning, not coincidentally, also happens to be a form of system activity. This raises an important issue that is not ordinarily addressed. How is such system activity created so that it is available to the organism as an identifiable system? Level One neural activity is not system activity; and even supernetwork activity is not system-wide activity. For example, in the olfactory bulb, research has shown that neural network activity involves perhaps 1–5 per cent of bulbar neurons (Skarda and Freeman, 1987); hardly the whole system. So how is the system qua system generated?

The new model suggests that a neural system is created when neural mass activity is generated within the brain. Systems, thus defined, are dynamic realities, not statically defined anatomical regions. The anatomical definition of the system is actually a third person or observational reality. Dynamically created system activity is the way in which the system qua system is available to the organism itself as part of its functioning. Therefore, one essential task for brains must be to create systems of unified functioning for the organism as a whole. This is the function of neural mass activity according to the present model.

Neural mass action is a form of neural activity involving mechanisms not utilized in the formation and activity of neural networks and other Level One processes. In the brain, whole *populations* of neurons are interconnected by excitatory and inhibitory connections, forming pairs of excitatory and inhibitory neural populations. Negative feedback takes place between excitatory and inhibitory populations, and unfolds in a number of stages (Freeman, 1991). First, the excitatory population becomes active and activates the inhibitory population. The activity of the inhibitory population is then fed back onto the excitatory population, thus inhibiting its activity. This is negative feedback. Negative feedback occurs throughout the brain. When the excitatory activity is inhibited, however, the excitatory population's input to the inhibitory population decreases. This allows for a new burst of activity on the part of the excitatory population, and the whole cycle repeats itself. The result is a pattern of on-off activity that is referred to as 'oscillatory behaviour'. Oscillatory behaviour is the behaviour characteristic of neural masses.

Returning to the new model's claim that neural mass action represents a distinct level of neural functioning in the brain, the reasoning proceeds as follows. When an entire neural mass becomes active, a mass that involves *every* neuron in the region, contrast functioning among neurons in the system is ruled out within that population. There is nothing else going on in the region with which such global activity can be contrasted. Thus, mass action is not a form of contrast functioning. And since contrast functioning is defined by the present model as a form of functioning responsible for creating perceptual features, neural mass action is not in the business of feature generation. So what is it doing and how can it be understood if we reject the representational reading of cognitivist neuroscientific accounts?

Before answering this question, we need to review a few more findings about neural mass activity in perceptual systems (Freeman and Schneider, 1982; Freeman and Skarda, 1985). In the olfactory system, the correlation between the stimulus and neural mass activity was anything but the tight fit predicted by the representational hypothesis. For example, when an identical odourant stimulus was presented to different animals, animals trained to give identical responses to it, different patterns of neural activity were recorded from each animal. The patterns, thus, correlated with the animals, not the stimulus. Additionally, the patterned activity that resulted from the presentation of the same odourant to the same animal was never twice the same. Instead, the patterns possessed a kind of family resemblance. And when animals trained to respond to odourants were taught to respond to a new odourant, or when the response contingency for a previously learned odourant was changed, all of the recorded patterns associated with previously learned odourants changed. Finally, and very significantly for the new model, when feedback connections between the perceptual subsystem and the rest of the brain were severed, neural mass action failed to occur. What can we conclude from these data?

First, the data lend support to the interpretation of neural activity in terms of the concept of use rather than that of representation. Forms of use are context dependent, they acquire their distinctive character in relation to other forms of use (Wittgenstein, 1953). The context dependency of use is significant because it would offer a simple and rather elegant explanation as to why the introduction of a new odourant stimulus led to changes in all of the recorded EEG patterns associated with previously learned odourants. If the recorded patterns of neural mass activity correlate with perceptual forms of use, then one would predict that the introduction of new forms of use would change the meaning of the entire context of related uses. A form of use cannot remain the same when the context that defines it changes, and the data reflect this context dependency.

Moreover, it is also characteristic of forms of use that they are never strictly identical over time even though they may serve as functional equivalents. We know this from ordinary everyday experience of forms of use. Take bicycle riding as one example. As a form of behaviour, bicycle riding is not definable in terms of a set of features that apply to every instance. What counts as bicycle riding changes depending on the context. It is one set of bodily movements, postures, balances and counterbalances when riding up a steep hill, but can be quite different behaviour when riding downhill or over bumpy terrain. It can be done without using one's hands, without putting both feet (or any foot) on the pedals, while standing or sitting down. The behavioural variations are endless. My point is that there is no one form of behaviour that is bicycle riding in every case, although there can be said to be a family resemblance among all instances of the form of use that is referred to generically as bicycle riding. A form of use is a generic (never identical) form of context dependent interaction engaged in by organisms. The family resemblance of the recorded EEG patterns of neural mass activity generated in response to an identical stimulus support the interpretation of them in terms of the concept of use. If the patterned activity were an internal representative (correlate, stand in for, image, copy) of the external stimulus (the organism's only clue to the external state of affairs), then one would expect that external identity would map with identity in its internal correlate. But this is not what we see.

The finding, however, that is of central importance for determining the essential role of neural mass action (Level Two activity) in perception, is that such activity does not occur if connections between the olfactory bulb (sensory subsystem) and the cortex are severed. Computer simulation of the olfactory system, including the prepyriform cortex, indicates that neural mass action taking place in the cortex is required to push the olfactory bulb away from its rest state. This, in turn, creates the state of instability required for neural mass action to be generated in the bulb. Once destabilized, Freeman's simulation showed, the neural mass in the bulb is able to respond to local network activity (Level One neural activity) with global, system-wide oscillatory behaviour. Thus, neural systems as dynamic entities are created by means of feedback with other systems. They are co-defined and co-created by way of interaction taking place among neural masses. Extrapolating from these data, the new model proposes that neural mass action effectively carves up (articulates) organismic functioning by dynamically creating subsystems of functioning. This is the chief contribution of Level Two neural functioning according to the new model.

Before continuing to the final proposition concerning perceptual system functioning, however, one more feature of neural mass activity warrants comment. Neural mass activity is self-organized. Briefly, self-organized behaviour is behaviour generated by enormous groups of component elements that share dense feedback connections. The collective form of behaviour that arises is not imposed from outside the system, rather it is spontaneously generated by the collective itself under certain conditions. Much has been made, in recent years, of the fact that neural activity is self-organized but in relation to the proposed new model it would seem that self-organization relates in an interesting way to the claim that Level Two neural activity is responsible for articulating the organism into behavioural subsystems. Organismic articulation cannot be borrowed or copied from the extra-organismic world. It must be internally generated by the organism itself if it is to exist at all. Which means that organisms must themselves generate a form of dynamics that can provide such articulation. Self-organized neural activity fits the bill. The new model, then, proposes that Level Two neural activity is essential for perception because without it there would be no *perceptual system event*. Although its physical nature is holistic (inescapable connectivity is the fundamental nature of reality), organismic functioning is a tale of articulation.

Proposition Four *Whole-organism functioning, which contrasts two results of Level Two neural activity, creates the subject–object form of use. This form of use is what is referred to as perceptual consciousness.*

Proposition Four extrapolates speculatively from the data that form the bases for the first three propositions of the new model. Speculative though it may be, it addresses an issue that is of pivotal importance for any theory of perception. For what is perceived appears as something other than the perceiver in the sense of being an object of perceptual activity; and just as all of the other features of the percept, its objectivity, too, must be perceptually generated. It is an issue that cannot be ignored, not even by those wedded to materialistic interpretations of perception, since it is undeniable from the way that percepts function in behaviour that they enjoy an objective status from the perspective of the behaving organism. Moreover, since what is objective acquires its objective status only in relation to what is subjective, the implication is that eventually perceptual theorists must address the issue of perceptual experience as well. Proposition Four is offered as a possible way forward on this front.

The view of perceptual consciousness adopted by the new model is novel. Consciousness is defined first as a form of whole organism *use*. Unlike materialism, the present model does not equate consciousness with any part or component of the organism, such as neurons, networks or neural masses. Neurons are not conscious; perceiving organisms are. Nor is perceptual consciousness something purely subjective as opposed to physical. Such a view would make the present model into a form of idealism, an extreme of thinking that it disavows along with materialism.

The new model proposes that perceptual consciousness is *based on* the results of neural activity, but is not itself a form of neural activity. Thus, the new model draws a distinction between the *bases used* (results of Level Two neural activity) and the *use* made of them. Consciousness is here defined as a 'structure of behaviour', a field of action defined by a specific form of use engaged in by organisms. Moreover, consciousness is a form of use that itself generates both the perceiver as subject and the percept as object. Therefore, consciousness is the precondition for subjectivity and objectivity and thus should not be identified with the subjective sphere, as idealism claims, or with the objective sphere (which is ultimately defined as material reality), as materialism and scientific realism claim. It is a form of use engaged in by the whole organism in which a gap is introduced that generates the perceiver as subject and the percept's objective status.

Perceived objectivity results, according to the new model, when the organism makes use of the results of the perceptual system event as 'other'. This presupposes that three interrelated process occur.

(1) That there be a system event that defines the perceptual system qua system providing the basis for objectivity.

(2) That the result of the perceptual system event be distinguished by the organism from the rest of its own functioning occurring simultaneously, a process that generates the 'self' by way of contrast.

(3) That the entire process be 'undergone' by the organism, be part of its own functioning, thus generating its 'experiential' character.

Two aspects of global mass action within the perceptual system itself create the basis for the form of use by which the whole organism generates an objectified percept. When a perceptual system event is created by neural mass action, that event adopts a unique form of patterned activity. Neural mass action exhibits a specific pattern in each case because it is shaped by the forms of Level One feature functioning taking place within the perceptual subsystem. According to Freeman's model, this is how it happens. The neural mass is destabilized from its ground state just prior to becoming active as a whole. At this point Level One activity, taking place in localized neural networks responsible for feature generation, biases the neural mass in such a way that the resulting mass action reflects the results of Level One feature activity by adopting a unique form or pattern of mass activity. Here the suggestion is made that this process effectively reconfigures the results of Level One functioning in relation to the phenomenal fabric so that they are made available as the output of the system to the rest of the organism. The reconfiguration is necessary, not to bind together features, but (1) to provide the organism as a whole with access to the results of the complete articulation of the sense organ event and (2) to create a system event which it can then use as a unified entity. This is the first step required for the generation of perceptual consciousness. There must be some unified basis available to the organism for use as objective.

The second feature of mass action as a system event that contributes to perceptual objectivity is that the perceptual system event is generated only when there is feedback from the rest of the brain. The correlation that is here observed reflects the interdependency that always exists between subject and object. An object can only exist in relation to a subject, and, correlatively, nothing can be a subject except in relation to something objective. The two concepts acquire their meaning only in relation to one another. This is the sort of relational process, as Freeman's model indicates, that occurs within the brain. Perceptual system activity is created and defined as a subsystem event by way of feedback occurring between that system event and the rest of the brain. This implies that global mass action taking place within the perceptual system not only generates the unified basis for what gets used as perceived object, but that it also effectively generates what becomes defined as the 'rest of the organism' by way of feedback. That remainder, which effectively comprises the organism whole, becomes the basis for the form of use that pits the subject against the object. The result of this process of contrast between the subsystem and the rest of the organism provides the bases for two segregated forms of use. This form of use is that which creates the subject/object poles, the gap between the self and its objects. For 'in order to become aware of appearances we must first be free to establish a certain distance between ourselves and the object' (Arendt, 1961). Thus, the new model suggests that the organism uses its own dynamics in a way that allows it to carve up its own holistic nature (its physical reality) into what it then uses as 'self' and 'other'.

The final piece to the puzzle of perceptual consciousness relates to its experiential character. This feature, I believe, is contributed by the fact that the entire process which generates the subject/object relation as well as the holism and articulation of the phenomenal event taking place in the sense organs is *undergone by the organism* itself. That is, the process is something that the organism *lives* in its own functioning. This is the basis of the experiential character of perceptual consciousness. It is the living of the process that gives rise to the percept that explains the experiential character of perception. And the percept's perceptual reality is crucial here. For it is the fact that percepts are facts about a form of life that lends to the perceived object and its features its character of being something experienced.

Thus, perceptual objectivity and its correlate, the perceiving subject, have their bases in organismic functioning. Perceived objects are not located outside of the perceiver any more than subjectivity is localized within the organism. Both are the results of a uniquely structured form of use generated by life forms that have the ability to articulate their own dynamical functioning in a particular way. One form of use cannot exist without the other. And as the result of a form of use, percepts cannot be copied from the extra-organismic environment. Indeed, it is organismic functioning itself that defines what counts as extra-organismic in the first place.

But why is it that perceivers can behave successfully as if they were independent of the world that they perceive? Perception works, I believe, essentially because nothing exists in the way that perceivers perceive it to exist. The holism, the inescapable relationships remain in place. Percepts are used as if there were independent states of affairs, but they are actually an articulation of the organism's causal embeddedness in physical reality. All organismic processes necessarily reflect the causal history of embeddedness of the organism. So although perception essentially obscures its own origins by projecting discreteness upon what exists holistically, it can do so and still facilitate a form of behaviour because those articulations remain part of the fundamental holism of physical reality. This is the remarkable achievement of the perceptual form of life.

References

Arendt, H. (1961), *Between Past and Future* (New York: Viking Press).

Barlow, H.B. (1953), 'Summation and inhibition in the frog's retina', *Journal of Physiology*, **199**, pp. 69–88.

Barlow, H.B. (1972), 'Single units and sensation: A neuron doctrine for perceptual psychology', *Perception*, **I**, pp. 371–94.

Damasio, A.R. (1994), *Descartes' Error* (New York: Putnam).

Engel, A., Koenig, P., Kreiter, A., Schillen, T. and Singer, W. (1992), 'Temporal coding in the visual cortex: New vistas on integration in the nervous system', *Trends in Neuroscience*, **15/16**, pp. 218–22.

Freeman, W. (1975), *Mass Action in the Nervous System* (New York: Academic Press).

Freeman, W. and Schneider, W. (1982), 'Changes in spatial patterns of rabbit olfactory EEG with conditioning to odours', *Psychophysiology*, **19**, pp. 44–56.

Freeman, W. and Skarda, C. (1985), 'Spatial EEG patterns, non-linear dynamics and perception: The neo-Sherringtonian view', *Brain Research Reviews*, **10**, pp. 147–75.

Freeman, W. (1991), 'Insights into processes of visual perception from studies in the olfactory system', in *Memory: Organization and Locus of Change*, ed. L. Squire (New York: Oxford University Press).

Gibson, J. (1979), *The Ecological Approach to Visual Perception* (Boston: Houghton Mifflin).

Gouras, P. and Zenner, E. (1981), 'Colour vision: A review from a neurophysiological perspective', *Progress in Sensory Physiology*, **1**, pp. 139–79.

Gray, C. and Singer, W. (1987), *Society for Neuroscience Abstracts*, **13**, p. 1449.

Gurwitsch, A. (1964), *The Field of Consciousness* (Pittsburgh: Dusquesne University Press).

Helson, H. (1938), 'Fundamental problems of colour vision. I: The principles governing changes in hue, saturation, and lightness of nonselective samples in chromatic illumination', *Journal of Experimental Psychology*, **23**, pp. 439–76.

Hubel, D. and Wiesel, T. (1959), 'Receptive fields of single neurones in the cat's striate cortex', *Journal of Physiology*, **148**, pp. 574–91.

Hubel, D. and Wiesel, T. (1962), 'Receptive fields, binocular interaction, and functional architecture in the cat's visual cortex', *Journal of Physiology*, **160**, pp. 106–54.

Land, E. (1977), 'The retinex theory of colour vision', *Scientific American*, **237**, pp. 108–28.

Sherrington, C. (1906), *The Integrative Action of the Nervous System* (New Haven: Yale University Press).

Skarda C. and Freeman, W. (1987), 'How brains make chaos in order to make sense of the world', *Behavioral and Brain Sciences*, **10** (2), pp. 161–73.

Varela, F., Thompson, E. and Rosch, E. (1991), *The Embodied Mind* (Cambridge, MA: MIT Press).

Wittgenstein, L. (1953), *Philosophical Investigations* (Oxford: Blackwell).

M.T. Turvey and Robert E. Shaw

Ecological Foundations of Cognition

I. Symmetry and Specificity of Animal–Environment Systems

Ontological and methodological constraints on a theory of cognition that would gen-eralize across species are identified. Within these constraints, ecological arguments for (a) animal–environment mutuality and reciprocity and (b) the necessary specific-ity of structured energy distributions to environmental facts are developed as coun-terpoints to the classical doctrines of animal–environment dualism and intractable nonspecificity. Implications of (a) and (b) for a cognitive theory consistent with Gib-son's programme of ecological psychology are identified and contrasted with con-temporary cognitivism.

I: Hume's Touchstone

Prior to studying cognition, it is essential that we ask what there is to be known. How we pose our question, and what we take as an appropriate answer, must be true to *Hume's touchstone* (Hume, 1983; Massey, 1993). That is, the question and answer cannot be expressed in a way that renders the knowables uniquely human. Rather, they must be expressed in a manner so general as to be inclusive of all species. In so doing, we keep at bay the temptation to propose hypotheses and explanations that derive from, and apply selectively to, the human condition. David Hume's touchstone rejects as bogus any account of human cognition that does not apply evenhandedly to all other animals (except for accounts of those phenomena that can be shown unequivocally to be particular to humans).

II: The Furniture of the Earth

Eschewing exactness in the use of the word *know*, all animals can be said to know about their niches in the sense of being able to cope with them, more or less ade-quately. Intuitively, there are several modes of *knowing about* with perceiving–acting foremost among them. Following Gibson (1979a), what animals perceive, and act in respect to, are the substances, surfaces, places, objects and events of the environment. The perceptibles are not the bodies in space referred to in textbooks on physics and promoted by traditional analyses of perception and action. Further, still following

Journal of Consciousness Studies, **6**, No. 11–12, 1999, pp. 95–110

Gibson, what animals discriminate, and act toward differentially, are the meaningful properties of substances, surfaces, places, objects and events. The discriminables are not the primary and secondary qualities of things (as identified by John Locke: form, size, position, solidity, duration and motion, on the one hand, and colour on the other).

The *substances* of the environment (e.g., rock, soil, wood, plant tissue) are, to varying degrees, rigid, nondeformable, impenetrable, and unyielding in shape (Gibson, 1979a). They differ in hardness, viscosity, density, cohesiveness, elasticity and plasticity. They also differ in solubility, volatility and stability. These differences, which follow from differences in chemical composition, have implications for the biochemistry, physiology and behaviour of animals. The ecological notion of substance is not to be confused with the physical notion of matter.

The *surfaces* of the environment are the interfaces between substances and the medium (air or water). They are nondenumerable by virtue of the fact that within the upper and lower limits of the earth's scale, surfaces are indefinitely nested within other surfaces. At all nestings, surfaces have characteristic properties that can persist or change, such as layout, texture, degree of reflectance, and state of illumination (shaded or unshaded). Most occurrences of significance for animals are at surfaces. Thus, light is structured, chemicals are transformed, vibrations in substances are transmitted to the medium, and contacts are made with limbs and bodies. The ecological notion of surface is not to be confused with the geometrical notion of plane.

The *places* of the environment are more or less extended surfaces. They are not located by coordinates but by the relation of inclusion. Places are nested in other places and, in the general case, places lack boundaries. The ecological notion of place is not to be confused with the geometrical notion of point.

The *objects* of the environment are substances that are either completely or incompletely surrounded by the medium, either detached from or attached to the substratum (Gibson, 1979a). In other words, whereas the surface layouts of detached objects are topologically closed, those of attached objects are not and their substances are continuous with the substances of other surfaces. The ecological notion of object is not to be confused with the physical notions of particle and body nor is it to be confused with the object in the philosophical notion of subject–object dichotomy.

The *events* of the environment are changes in the environment of three major varieties, namely, of surface layout (e.g., translation, deformation), of surface colour and/or texture (e.g., ripening fruit) and of surface existence (e.g., melting). Events, like other constituents of the ecology, are nested, and their localization is not by clock time but by inclusion. They are not limited, therefore, to the translational and rotational motions of classical mechanics (Shaw *et al.*, 1996). Further, events are reversible in some cases and irreversible in other cases. The ecological notion of event, therefore, is not approachable through the physical–mathematical notion of time. It cannot be related to the time reversal premise of classical mechanics nor to the irreversible arrow of time defined by the second law of thermodynamics.

The substances, surfaces, places, objects and events of the environment are opportunities or possibilities for behaviour. These action possibilities are defined in the complementation of the properties of substances, surfaces, etc., and the properties of the animal. When juxtaposed with the properties of a given animal, a substance such as water affords drinking and pouring (for some mammals) and walking upon (for some insects). Similarly, a surface that is more nearly horizontal than vertical and rigid to a particular degree (that is, it can provide a maximal reactive force of so many Newtons) affords standing upon and 'footing' for a terrestrial mammal of a particular size. These opportunities or possibilities for action, referred to as *affordances* (Gibson, 1966; 1979a), are not phenomenal or subjective. They are, to repeat, defined by the complementation of objective, real and physical properties. At the same time they are not simply physical (in the sense currently conceived by physical science). They are *ecological*. They are properties of the environment *relative to* an animal.

The preceding ecological realities, more so than the realities of physics and the ontology of Locke, are the starting point for the study of what animals perceive, what they learn and know, and how they behave (Turvey, 1992a). It is valuable to highlight a few of their implications. As Gibson (1979b) notes, the ecological fact that surfaces, places and events are nested means that these ecological realities are neither discrete nor (perforce) countable. As such, any account of the cognition of these realities based on the theory of sets (with its assumption of denumerable things) must necessarily fail. That is, surfaces, places and events cannot be classified or categorized. Similarly, they cannot be grouped. The so-called organizational principles of grouping advanced by Gestalt Psychology and promoted in various forms by modern psychologists assume discrete things.

Another noteworthy implication is that the Newtonian notions of space and time are inappropriate for the study of cognition. In broaching questions of the knowings of animals, space and time must be understood not in the abstract but as relations among the ecological realities. Ecologically, *space* is objects, surfaces and places together with their mutual embeddings and mutual separations (Gibson, 1979a; Turvey, 1992a). Ecologically, *time* is events together with their mutual embeddings and mutual sequences (Gibson, 1979a; Turvey, 1992a). Historical issues of how an animal perceives the third dimension and how it perceives time are, from a consideration of nature's ecological scale, misguided.

Even more noteworthy, perhaps, is the implication of affordances: what a substance, surface, etc., *is* and what a substance, surface, etc. *means* are not separate (Reed and Jones, 1986). Meanings and values do not have to be imposed, by mental processes, upon a meaningless niche. A surrounding environment's meanings await discovery not invention. Implicit in the preceding is the rejection of the traditional and sacrosanct distinction between percepts and concepts and, perhaps, the rejection of the very terms themselves. The constraining of behaviour by detected affordances seems to include, in one unitary activity, what are usually referred to as perceiving and conceiving (Gibson, 1975).

III: Symmetry of Animal–Environment Systems

If it is essential to face the issue of what there is to be known, it is also essential to face squarely the question of the relation between any animal, as a knowing agent, and the environment in which it functions. The answer given to this question, perhaps more than the answer to any other question, dictates the form and content of a theory of cognition.

1. Doctrine of animal–environment dualism

It has long been argued that a major impediment to a successful account of cognition is the inability to reconcile *mentalese*, the language of mind in which cognitive processes are couched, with *physicalese*, the language of matter used to describe features and processes of body and brain. This, of course, is the heralded mind–body dualism. The modern computational–representational view of cognition divests itself of this dualism through the twin assertions of *Turing reductionism* and *token physicalism* (or *multiple realizability*). Namely, that mental states are the computational states of a Turing machine and that such a machine, while necessarily physical in its composition, does not require that the physical composition be of a particular type (e.g., Fodor, 1981; Haugeland, 1982). For example, it need not be neurophysiological (specifically), it need not be carbon based (generally). Any material instantiation will suffice as long as it can support the formal relations of mental processes. The promise of Turing reductionism–token physicalism is adherence to both mentalism and materialism (at the same time) without contradiction. It goes without saying that such thinking has not been without critics. They can be found within the ranks of philosophers of mind (e.g., Churchland, 1989; Searle, 1992), neuroscientists (e.g., Edelman, 1992) and physicists (e.g., Penrose, 1989). Among other things, the critics deny that intentionality is satisfiable within strictly syntactic processes and they are sceptical about mental states being caused by anything other than brain states.

From an ecological perspective, the debates that consume proponents of functionalism (that a mental state is defined by its causal relation to other mental states) and their critics are at some remove from the major issue. This is because the dualism of mind and body is subordinate to a dualism that is more far reaching and more redoubtable — the dualism of animal and environment (Turvey and Shaw, 1995). Scientific inquiry into knowing tends to proceed according to the doctrine that the two primary points of reference in the theory of cognition are logically independent. This doctrine of *animal–environment dualism* sanctifies explanations of cognition wholly in terms of the causal powers of the animal's mind or brain or in terms of the causal interconnections between an animal and its environment treated as rigidly separable things.

Dewey and Bentley (1949), philosophers in the American pragmatist tradition, referred to these latter styles of explanation as *self-actional* and *interactional*, respectively. Dewey and Bentley contrasted these latter explanatory styles with the style of non-mechanical explanation introduced by Einstein (see Einstein and Infeld, 1938), a style they described as *transactional*.

2. Transactional explanation

Consider the question of why an apple falls down. Zee (1989) suggests that Aristotle would give the self-actional answer 'because it wants to get to the earth, where it belongs', Newton would give the interactional answer 'because the apple and the earth exert a force on each other', and Einstein would give the transactional answer 'because the earth warps the space and time around the tree, and the apple merely follows the natural curved path in that space'. For Einstein, Newton's forces of interaction arise from space–time curvature. That curvature is induced by and, in turn, affects the distribution of matter and energy. The influence of objects (such as an apple and the planet earth) on space–time and the influence of space–time on objects is mutual (without necessarily being equal). In the practice of law, mutual concessions characterize a transaction. Hence, Dewey and Bentley's (1949) choice of terminology.

Not surprisingly, a good way to lay hold of transactional explanation is to analyse a typical human transaction. There are three components in the transaction of borrowing money — a lender, a borrower and a loan. The reality of these components is dependent upon the field defined by the complete cluster of components and their interactions. Studying any one component in isolation is impossible given that the components necessarily co-implicate their complementary aspects. A borrower cannot borrow without a lender to lend (borrower and lender are not definable independently of one another) and it is impossible to isolate borrowing from the components (Handy, 1973). The field satisfies a governing principle or rule that is non-local. Although money passes from person A to person B in borrowing, stealing and gift-giving, the interaction is commutative in the first two cases and non-commutative in the third case. The differences are due to the culturally instantiated rules or contexts that specify (among other things) that B must repay A eventually in the case of borrowing, immediately in the case of stealing, and never in the case of gift-giving.

With respect to cognition, transactional and similar forms of so-called naturalistic inquiry and explanation (e.g., Kantor, 1920) emphasize that any instance of knowing about is inseparable from the field of which it is a part and that, with respect to strategy, the full complement of field factors constituting knowing about should be sought rather than a thing or a locus.

3. The symmetry operation of duality

Transactional explanation must be elaborated in ways that ensure against regression to ever more embracing contexts. So elaborated, it becomes *coalitional* explanation (Shaw and Turvey, 1981). We have argued that coalitional explanation goes hand-in-hand with the ecological doctrine of animal–environment mutuality and reciprocity (Gibson, 1979a; Kugler *et al.*, 1990; Michaels and Carello, 1981; Shaw and Alley, 1985; Turvey, 1992a; Turvey and Shaw, 1979; 1995; Turvey *et al.*, 1981).

Whereas mutuality implies sameness, reciprocity implies differences that are complementary. Consider numbers that add up to one. The numbers are mutual in that they may be added to one another (expressing a symmetry) but reciprocal in that they may have different values (expressing an asymmetry). The symmetry property can be reconciled with the asymmetry property by elevating analysis to a level at which the numbers are complementary in the sense that they complete one other. Likewise, taken together, an animal and its environment may be mutual and reciprocal, symmetric and asymmetric, but at a higher level they are complementary duals. They combine to make a whole, namely, an (epistemic) ecosystem. For a synergy of animal and environment, the asymmetry of dualism (where animals and environments are merely incommensurate kinds) must give way to the symmetry notion of *duality* (where the two are commensurate kinds) (Turvey and Shaw, 1995). The ecological counter to the traditional doctrine of animal environment dualism is, therefore, animal–environment duality.

Mathematically, a duality refers to a specific relational property or correspondence that holds between a pair of otherwise different systems, S1 and S2. The formal requirement is that some property of S1 plays the same role among its total property set as that played by a property of S2 among S2's total property set. That is, there is a symmetry T such that $T(S1) \rightarrow S2$ and $T(S2) \rightarrow S1$. Specifically, for any relation r_1 in S1 there exists a relation r_2 in S2 such that $T(r_1) \rightarrow r_2$ and $T(r_2) \rightarrow r_1$.

Identifying the duality symmetry that relates animal and environment will be a major scientific challenge. Affordances are a reflection of this symmetry. It is nonetheless the case that the consequences of the doctrine can be expressed straightforwardly. Namely, given the doctrine of animal–environment duality, cognition is not understandable in terms of processes within the animal and it is not understandable in terms of interactions between an animal and its environment. The knower (animal) and the known (environment) are not rigidly separable components, they are not definable independently of each other, and knowing cannot be isolated from these components.

4. Behavioural regulation and animal–environment spectroscopy

A classical concern of the experimental investigation of learning, the nature and mechanism of reinforcement, highlights the possibility that major aspects of cognition may not be approachable outside of animal–environment duality.

It is self-evident that a behaviour can be facilitated, improved and rendered more likely to occur by particular consequences historically referred to as *reinforcements*. Efforts to achieve a viable theory of the self-evident fact of reinforcement have been thwarted, however, by the difficulties of identifying reinforcers and the mechanism by which they work. The questions to be answered are: 'What determines whether something can function as a reinforcer or, synonymously, how can reinforcers be predicted?' and 'How does a reinforcer produce its effects, that is, how does it bring about an increase in the probability of a behaviour?'

In earlier conceptions, reinforcers were identified with a special class of stimuli. These special stimuli were viewed as categorically distinct from the behaviours that they reinforced and their effect was interpreted as one of strengthening the instrumental response. The preceding ideas were repudiated by the observation that reinforcers were oftentimes behaviours rather than stimuli of a particular kind (Premack, 1965). Response R_n can act as a reinforcer of response R_m but not because of anything intrinsic to R_n. Rather it is the probability with which R_n occurs relative to R_m. Premack's differential probability principle says that the reinforcing response R_n is simply a response that is more likely to occur than the instrumental response R_m (Domjan, 1998). If conditions rendered R_n less probable than R_m, then the roles would be reversed; R_m would be the reinforcer for R_n. Reinforcers do not exist in any absolute sense.

Research inspired by Premack's principle has led to an even more radical understanding. Far from being absolute, reinforcers are created online by the contingencies of instrumental learning. Any behaviour that functions as a reinforcer is a restricted behaviour in the sense that the opportunity to engage in the reinforcing behaviour is contingent upon engaging in the instrumental behaviour. A low-probability behaviour, so restricted, will act as an effective reinforcer of a high-probability behaviour. It is not the relative response probabilities that matter, as Premack had presumed, but rather the degree to which the instrumental response–reinforcer contingency disrupts behavioural stability. Apparently, the success of the contingency in increasing the instrumental response rests on its forcing the animal away from a preferred or optimal distribution of activities and on the animal's inclination to so behave as to maintain that distribution (e.g., Allison, 1989; Domjan, 1998; Timberlake, 1980). In sum, reinforcers are predictable and the resultant increases in behaviour are understandable in the light of a broad principle of behavioural regulation.

Given the preceding, it is evident that the theory of reinforcement will demand much of science. The notion of a preferred distribution of behaviours invokes principles of *homeokinetic physics* (e.g., Iberall and Soodak, 1987; Soodak and Iberall, 1978). The identification of these principles in relation to behavioural regulation begins with the observation that an animal is marginally unstable within its environment (when at rest, it cannot stay at rest; when active, it cannot stay active). This inherent instability is a patterning of a limited (but reasonably large) number of behavioural modes (Iberall and McCulloch, 1969). The patterning is in the form of cycling among the modes — a pattern of recurrences. There is nothing fixed about the patterning except that all modes are visited a requisite average number of times (e.g., eat 2–4 times a day, groom about every 10 minutes). An animal threads its way among the modes ergodically, meaning that there is global stability of behaviour in the statistical mechanical sense (Bloch *et al.*, 1971). For adaptive success, however, the animal must, in addition, thread its way *comfortably*. The pattern of behavioural recurrences should fit, rather than oppose or be indifferent to, the environmental recurrences. That is, the behavioural recurrences should be entrainable and preferably in small numbers. Humans eat, void, sleep, just a few times a day. Such recurrences can fit the circadian cycle comfortably.

Homeokinetics is a term so used because regularities in inanimate and animate systems are the result of thermodynamic engine cycles of energy dissipation and replenishment. Form and function are sustained as limit cycle processes (e.g., Iberall, 1977; 1978; Iberall and Soodak, 1987; Yates *et al.*, 1972). The constants that inspired the notion of physiological homeostasis are the mean states of oscillatory mechanisms.

In the homeokinetic perspective, an animal is a multitude of biochemical chains with characteristic rate-governing steps that foster, in each case, a limit-cycle process. The fact that their time scales are often lengthy (minutes, days) suggest the prominent roles of endocrinological mechanisms and the chemical messengers, the hormones. The variable operating characteristics of this battery of biochemical oscillators results in the aforementioned behavioural patterning. More specifically, the cyclic processes are motor-sensory-internal organ motions (Bloch *et al.*, 1971). A biospectroscopy is implied at the macroscopic scale of the animal–environment system with individual animal–environment systems distinguished, in principle, by the particulars of their spectra. Iberall (1972, p. 170) summarizes the condition as follows: 'One is hard put not to conclude that in order to attain global stability, the large number of only loosely coupled chains must represent an extensive collection of atomistic elements which have to be canonically constrained in accordance with an ergodic hypothesis.'

The key idea of biospectroscopy, namely, homeokinetics, addresses why there are multiple behavioural modes, why the system must ring through these modes, why the modes are coupled largely as a Markov chain, and why switching among the modes must be simple (e.g., Iberall and Soodak, 1987). It does not (as yet) address directly, however, the question as to why the baseline frequencies of the different behavioural modes differ as they do. This limitation aside, the ultimate significance of a recurring

ring of behavioural modes is that it allows the animal to balance the entropic degrada-
tions associated with the chains of processes thereby ensuring persistence of its characteristic forms and functions (Iberall and Soodak, 1987).

Returning to the concept of reinforcement, it can be seen, therefore, that to restrict
any particular activity in which an animal engages is to perturb the distribution of
all of the animal's activities, that is, to bring about a global disequilibrium. In this
light, increases in instrumental behaviours are by-products of a comprehensive
reorganization of behaviour due to homeokinetic processes attempting to return the
animal–environment system to global equilibrium. Despite its profoundly personal
nature, reinforcement may be understandable only in terms that address the
animal–environment system in full.

IV: Specificity of Animal–Environment Systems

The keystone of the contemporary theory of cognition is *representation*. Mind is said
'to represent' — both the facts and principles of the world (acquired or given
innately) and the facts and principles of the body (again, acquired or given innately).
Accordingly, a general theory of representation is high on the agenda of the science of
cognition as currently conceived. From an ecological perspective, however, representation is (minimally) a subordinate notion. Of considerably larger significance is
specification (Shaw *et al.*, 1982; Shaw *et al.*, 1996; Turvey and Shaw, 1995). What
makes it possible for one thing to specify another and what form does this specificity
take? If there is to be a general theory of representation, it will, in our view, be
founded on a thoroughgoing theory of specification.

1. *Specificity of information to environment*

Perceptual contact with the substances, surfaces, places, objects and events of the
environment is supported by light in the case of seeing, sound in the case of hearing,
chemicals in air or water in the cases of smelling and tasting, and mechanical forces in
the case of touching. How should these supporting energy distributions be characterized? With the study of vision leading the way, the well-entrenched historical answer
is that they are inadequate — incomplete at best, ambiguous at worst. Thus, the established starting point for accounts of visual perception is the inability of light to specify the facts of the environment. This lack of specificity means that the light available
to eyes, whatever their design (chambered or ommatidia), is intractable or refractory
(e.g., Epstein, 1977). That is, from the perspective of visual nervous systems, and the
requirement of maintaining visual contact with the environment, working with light
is a challenge.

Identifying the particulars of how this challenge is met defines the traditional task
of visual perceptual theories. The mechanisms typically proposed of associative
memory, inference engines, concepts, and organizational dispositions are *epistemic
mediators* (Shaw and Bransford, 1977; Turvey, 1977). Their role is to provide the
epistemic link between the animal and its environment, the link that is missing, in
large part, because of the assumed non-specificity of light distributions (e.g., Epstein,
1977; 1982). In order to fulfill this role the proposed mechanisms and their intellectual products — the final and intermediate mental representations of the environment
— must be connected to the facts of the environment. That is, they must be *grounded*.

There is, however, no account of this grounding that does not presuppose, to greater or lesser extent, ungrounded (*a priori*) knowledge of environmental facts (see overview in von Eckardt, 1993). The doctrine of intractable nonspecificity has spawned perceptual theories for which scientific closure is unattainable.

The ecological approach to cognition adopts a different doctrine, one that is in keeping with the overarching doctrine of animal–environment duality. It holds that ambient energy distributions are *necessarily specific* to the facts of the environment and of an animal's movements relative to the environment (Shaw and Turvey, 1981; Shaw, *et al.*, 1982; Turvey and Shaw, 1979; 1995). By this doctrine, the optical support for vision, for example, is not ambiguous, is not impoverished and, therefore, not intractable or refractory.

This ecological doctrine of necessary specificity places demands upon experimentation and theory that are radically different from those imposed by the conventional doctrine of intractable non-specificity. In a nutshell, any conclusion that a patterned energy distribution underspecifies the environment must be viewed with circumspection. In particular, failure to find specificity does not give the scientist licence to hypothesize specialized brain and/or cognitive mechanisms *sui generis* that resolve the ambiguity. On the contrary, such failure identifies an inadequacy or incompleteness in the physical and mathematical characterization of the patterned energy distribution. The failure implies the need for a deeper mathematical–physical analysis coordinate with a sharpening and refinement of the descriptors of the animal–environment system. The failure does not imply the need for more speculation about brain and allied cognitive mechanisms — speculation that is likely to violate Hume's touchstone. Repeated failure to find specificity is compelling reason for entertaining the possibility that the current physics and mathematics may not be up to the task and further development is in order. Minimally, the ecological doctrine of necessary specificity is a call for scientific patience and continued commitment to the standard practices of the natural sciences. The search is for the laws of the ecological scale — the scale at which animals and environments are defined — that make cognition possible (e.g., Swenson and Turvey, 1991; Turvey *et al.*, 1981).

2. *Specificity of perception to information*

A prominent influence on cognitive theory is a doctrine inspired by the perceptual constancies and illusions. Namely, that perception is not necessarily constrained by the conditions of stimulation (e.g., Epstein, 1977). In the visual constancies, perception remains the same despite marked variations in the retinal image. In the visual illusions, perception not only differs markedly from the retinal image but it can vacillate despite a retinal image that is unchanging. If perception truly displays a notable degree of independence from the facts of ambient energy distributions structured by the environment and by the perceiver's movements, then perception would be a questionable basis for achieving and maintaining epistemic contact between animal and environment (Shaw *et al.*, 1982; Turvey and Shaw, 1979; Turvey *et al.*, 1981). Gibson (1950; 1966; 1979a) suspected that the perceptual constancies and illusory percepts were being misinterpreted. In his view, the constancies were indicative of (a) a specificity of perception to invariants of stimulation and (b) the inadequacy of traditional characterizations of the forms of stimulation (e.g., the retinal image in the dominant case of vision). The doctrine of the independence of perception from stimulation should be replaced by the doctrine of the dependence of perception on stimulation. The latter expresses Gibson's (1959) generalized psychophysical hypothesis or, in modern terms, the hypothesis of perception as specific to information (Gibson, 1979a; Turvey, 1992b).

(i) The inertia tensor and perception by dynamic touch

Research on dynamic touch, a subsystem of haptic perception, provides ample support for the preceding hypothesis. Dynamic touching occurs when a utensil (e.g., a fork), a tool (e.g., a hammer), or any medium sized object (e.g., a book), is grasped and lifted, turned, carried, and so on. Muscle and tendon deformations (induced by the forces and torques engendered by holding and manipulating) characterize dynamic touch more so than deformations of the skin and changes in joint angles (Gibson, 1966; Turvey and Carello, 1995).

Wielding an object grasped firmly in the hand entails time-variations in torque, angular velocity and angular acceleration together with variations, relative to gravity, in the planes of wielding motions. Experiments suggest that the basis for non-visual spatial perception connected with wielded hand-held objects is the time-independent and gravity-independent inertia tensor I_{ij}. This tensor quantifies an object's invariant resistance to variable 3-dimensional rotational accelerations about a fixed point O. When I_{ij} is rendered in diagonal form, the three components are the principal moments or eigenvalues I_k (where $k = 1, 2, 3$, identifying the major, intermediate, and minor eigenvalues, respectively). Diagonalization refers the rotational inertia relative to O to the principal axes or eigenvectors $\mathbf{e}_k$ through O about which the off-diagonal terms disappear (e.g., Goldstein, 1980). The eigenvectors $\mathbf{e}_k$ constitute the only non-arbitrary coordinate system at O, one that is physically determined by the object itself. Any hand-held object possesses an intrinsic reference frame.

Research investigating dynamic touch's dependency on I_{ij} has involved the wielding of occluded hand-held objects in simple variants of the magnitude estimation paradigm developed by Stevens (1961; 1962). It reveals that the perceptions of spatial properties such as length, width, and shape, are single-valued functions of the eigenvalues I_k, and the perceptions of object-to-hand relations (e.g., 'how is the object oriented in the hand?') and hand-to-object relations (e.g., 'where is the grasp relative to the object?') are single-valued functions of the directions of $\mathbf{e}_k$ (see reviews by Turvey, 1996; Turvey and Carello, 1995). The single valuedness indicates specificity.

In the context of the doctrine of specificity to information, it is especially useful to consider how the prominent concerns of perceptual constancy and illusion are resolved within dynamic touch.

Constancy. In everyday dynamic touch, a hand-held object is displaced by rotations that take place about a number of joints (wrist, elbow, shoulder, hip, ankle). As the distance from the rotation point increases, an object's inertia for rotation increases as the square of the distance. Consequently, a hand-held object's rotational inertia is larger for wielding about the elbow than about the wrist, and larger for wielding about the shoulder than about the elbow, implying that the perceptions should be correspondingly larger. That is, there should be no constancy of size perception. A further problem arises because an ordinary act of wielding using the whole arm leads to distances of the object from the shoulder and elbow that change in time, meaning that the object's I_{ij} defined at each of these joints is time varying. It is the case, however, that the distance of a grasped object from the rotation point in the wrist is constant and so is the object's I_{ij} defined at the wrist. Neither vary with the number of joints used to wield the object or with time. Experiments have confirmed that in free wielding, and in wielding singly about shoulder and elbow, the perception of a wielded object's length is precisely the same as in wielding only about the wrist (Pagano et al., 1993). In all cases, the object's I_{ij} about the wrist, the only invariant tensor of inertia, fully constrains the object's perceived length (Pagano et al., 1993).

The time-varying forces on the body's tissues during free wielding are very complex. So are the accompanying tissue distortions and patterns of neural activity. In the face of all this complexity, the sensitivity of dynamic touch to the constant I_{ij} at the wrist indicates an ability to tune into invariants in the mechanical and neural fluxes. The constancy of length perception by dynamic touch is in agreement with Gibson's claim that the constancies are specific to invariants.

Haptic size–weight illusion. In the size–weight illusion, bigger versions of an object of fixed mass are perceived as lighter (e.g., Charpentier, 1891; Stevens and Rubin, 1970). Historically, the illusion has been interpreted to mean that one's perception of the heaviness of an object does not refer to the object's weight (Jones, 1986). A traditional substitute for the failed hypothesis that weight is perceived is the hypothesis that there could well be a mental state corresponding to the object's weight but it is unconsciously modified by the mental state corresponding to the object's size. The two mental states or percepts are coupled and 'heaviness' refers, therefore, not to a physical property of the object as such but to the mental state formed by the coupling.

The percept–percept coupling hypothesis of perceived heaviness is unsatisfactory for two major conceptual reasons and one major experimental finding (Turvey et al., 1999). First, there is no rationale for why weight and size should be linked. The lack of clear purpose suggests that the proposed mechanism is not a proper function (Millikan, 1984; 1993). Second, no principled reason is advanced for the particular directional influence of the size percept on the weight percept (why, precisely, should largeness imply lightness?). The experimental finding is that the application of the methods and analyses developed by Ashby and Townsend (1986) for addressing the relations between perceptions have revealed that haptically perceived heaviness is independent of haptically perceived size (Amazeen, 1999).

The set of objects that typically give rise to the size–weight illusion (e.g., Charpentier, 1891; Stevens and Rubin, 1970) prove to be characterized by a particular pattern of I_k (Amazeen and Turvey, 1996). As the objects in the set become larger with no change in mass, I_3 increases and I_1 and I_2 (where $I_1 \approx I_2$) are relatively constant. In contrast, the set of objects showing heaviness perception increasing with weight (namely, objects that increase in mass) are characterized by an increase in all three eigenvalues. These inertial patterns can be simulated with objects that are otherwise identical in linear dimensions and mass; so-called *tensor objects* (Amazeen and Turvey, 1996). Experiments on the wielding of such objects without benefit of vision showed (a) a decrease in perceived heaviness when I_3 increased relative to constant I_1 and I_2 (the classic size–weight illusion) and (b) an increase in perceived heaviness when all eigenvalues increased (a new size–weight illusion).

Outcome (a) suggests that for objects of the same mass, as their corresponding eigenvalues approach equality ($I_1 = I_2 = I_3$), they feel lighter. That is, an object that is dynamically centrosymmetric (Hestenes, 1986) about the rotation point will feel lighter than an object of the same mass that is dynamically asymmetrical ($I_1 > I_2 > I_3$) or possesses dynamical axial symmetry (e.g., $I_1 = I_2 > I_3$). In other words, an object that offers (to muscular exertion) equal resistances to rotational acceleration in all directions feels lighter than an object of the same mass that offers unequal resistances. Outcome (b) then suggests that, when the resistances are disparate, the larger the values of the resistances for fixed object mass, the heavier is one's perception of the object.

The research of Amazeen and Turvey (1996) and Turvey *et al.* (1999) moves the 'natural kind' for heaviness (a property or kind that could support a generalization of facts about heaviness) in the direction of dynamical symmetry taken in relation to the biological movement system that patterns the forces applied to handheld objects. The 'perception of heaviness' is, more appropriately, 'perception of wieldability (or steerability)'. That is, it is the perception of the opportunities a hand-held object affords for varying the patterning and level of muscular forces required to move the object in a controlled fashion. A hand-held object is unwieldy (more difficult to wield) to the extent that there are relatively few ways in which forces can be applied to the object in order to bring about a desired trajectory. The objective, real and physical basis for the wielding an object affords is the handheld object's mass distribution — specifically the symmetry of the inertia tensor — taken with reference to the force-producing neuromuscular system.

In sum, the haptic size–weight illusion is *not* an illusion. It is simply one of several natural consequences of the differential sensitivity of the haptic system to the affordance *wieldable* rather than the physical measure *weight*. In our penultimate section, we suggest that the notions of affordances and invariants provide the appropriate tools for dismantling a major historical barrier to a realist account of cognition, namely, that perceptions are mediated by concepts.

V: Revisiting the Conceptual *versus* Perceptual Distinction — Toward a New Theory of Cognition

Characteristic of most attempts to explain the ability of animals to know about their surroundings is the dogma that perception is unreliable in the absence of concepts. Knowledge is said to come primarily through reasoning, not sensing. Concepts and the capacity for inference are presumed to have primacy over sensitivity. Although cognitivism is founded upon the preceding expressions of rationalist dogma (e.g., Fodor, 1975), most explanations of the adaptiveness of animal behaviour similarly emphasize conceptualization over perception.

From an ecological perspective, the classical distinction between *perceptual* and *conceptual* is misapplied, outworn and in need of radical revision (see, additionally, Brooks, 1991, and comments by Kirsch, 1991a,b). As noted in Section II, affordances and their specification dispense with the classical usages of the terms percept (an internal semantic-free representation) and concept (that which endows the perceptual representation with meaning). Affordances and their specification are counterpoints to the view that the meanings constraining animal behaviour reside in the brain (or in some nether world as intimated by Fregean semantics). More generally, defining a niche as an affordance space, a space of real opportunities for action, has potentially profound implications for the study of cognition. A prevalent presumption of cognitivism is that knowledge in the form of concepts of denumerable static objects and relations is needed for an animal to behave felicitously (e.g., Nilsson, 1991). Affordances contradict this presumption. As an animal moves with respect to surrounding substances, surfaces, places, etc., some opportunities for action persist, some newly arise, and some dissolve, even though the surroundings analysed formally as conceptualized objects and relations remain the same. A change of pace or a change of location can mean that a brink in the ground now affords leaping over whereas at an earlier pace or location it did not (Turvey and Shaw, 1995). Further, subtle changes of action can give rise to multiple and marked variations in the opportunities for subsequent actions. The environment-for-the-animal is dynamic and action oriented while the environment-in-itself, that which has been the target of cognitivist modelling by means of concepts, is fixed and neutral with respect to an animal and its actions (Kirsch, 1991a).

A rethinking of the conceptual versus perceptual distinction in terms of *invariance detection* is encouraged by arguments for the symmetry and specificity of animal–environment systems. Non-visual perception of a handheld object was shown above to be constrained by an invariant (e.g., the eigenvalues of the inertia tensor) detected over time — more particularly, over time-varying torques and time-varying angular accelerations. In the above examples, the perception of what the hand-held object was doing (how it was moving) would be constrained by the detection of what is variant over time. A similar concern for invariant and variant (persistence and change, same and different) marks the study of concepts and their formation. In this latter kind of knowing, the invariants and variants in question are defined for a set of substances, or surfaces, or places, or objects, or events. At issue is what properties are the same and what properties are different over the given entities that allow the entities to fall — where possible, given the nondenumerable factor — into classes or categories.

Detecting invariants (and the transformations over which they are realized) is, therefore, at the heart of what an animal knows about its environment. An animal detects invariants over time that specify continuing entities and it detects invariants over entities that specify ecologically reliable partitionings. The two kinds of invariance detection are different but complementary. The kind of detection which underlies the perception of persisting substances, surfaces, places, etc., is more fundamental than the kind of detection which underlies the perception of similarities (and differences) among persisting substances, surfaces, places, etc. Whereas Gibson suggested the term *resonance* (of perceptual systems) for the more fundamental kind, he suggested the term *abstraction* (by perceptual systems) for the kind of invariance detection that relates to classes and categories (e.g., Gibson, 1976; 1979a).

The less fundamental kind of invariance detection accommodates functions typically ascribed to concepts. It unifies perceptions into equivalence classes and it provides the basis for inferences. Further, given an invariance detected across entities and the ability to reify it, combinations with other invariances, and considerations of relations to other invariances, become possible.

VI: Hume's Touchstone: A Postscript

It is, perhaps, at the level of a capacity to predicate, and more certainly for the levels
of intellectual skills that depend on predication, that we are no longer held accountable to Hume's touchstone. Nonetheless, a developing, ecologically-grounded theory
of human cognition is well served by the hypothesis that peculiarly human skills (e.g.,
explicit inferring and language, and the shared use of depictions, indications, signals
and symbols) derive from, and consolidate the gains of, direct perception (Barwise
and Perry, 1983; Gibson, 1966; 1979a; Reed, 1988; Turvey and Carello, 1985). Our
suspicion, however, is that real progress in explaining peculiarly human instances of
knowing about will have to await an understanding of the symmetry and specificity
that fix the directness of perceiving as a property of all animal–environment systems.

Acknowledgements

The preparation of this manuscript was supported, in part, by NSF Grants SBR 97–09678,
SBR 97–28970

References

Allison, J. (1989), 'The nature of reinforcement', in *Contemporary Learning Theories: Instrumental Conditioning and the Impact of Biological Constraints on Learning*, ed. S.B. Klein and R.R.
Mowrer (Hillsdale, NJ: Erlbaum).

Ashby, G. and Townsend, J.T. (1986), 'Varieties of perceptual independence', *Psychological
Review*, **93**, pp. 154–79.

Amazeen, E. (1999), 'Perceptual independence of size and weight by dynamic touch', *Journal of
Experimental Psychology: Human Perception and Performance*, **25**, pp. 102–19.

Amazeen, E., and Turvey, M.T. (1996), 'Weight perception and the haptic size–weight illusion are
functions of the inertia tensor', *Journal of Experimental Psychology: Human Perception and
Performance*, **22**, pp. 213–32.

Barwise, J. and Perry, J. (1983), *Situations and Attitudes* (Cambridge, MA: MIT Press).

Bloch, E., Cardon, S., Iberall, A., Jacobowitz, D., Kornacker, K., Lipetz, L., McCulloch, W.,
Urquhart, J., Weinberg, M., and Yates, F. (1971), *Introduction to a Biological Systems Science*
NASA Contractor Report CR–1720, Washington.

Brooks, R. (1991), 'Intelligence without representation', *Artificial Intelligence*, **47**, pp. 139–59.

Charpentier, A. (1891), 'Analyse experimentale de quelques elements de la sensation de poids'
[Experimental study of some aspects of weight perception] *Archives de Physiologie Normales et
Pathologiques*, **3**, pp. 122–35.

Churchland, P.S. (1989), *Neurophilosophy* (Cambridge, MA: MIT Press).

Dewey, J., and Bentley, A.F. (1949), *Knowing and the Known* (Boston: Beacon).

Domjan, M. (1998), *The Principles of Learning and Behaviour* (Pacific Grove, CA: Brooks/Cole).

Eckardt, von B.(1993), *What is Cognitive Science?* (Cambridge, MA: MIT Press).

Edelman, G.M. (1992), *Bright Air, Brilliant Fire* (New York: Basic Books).

Einstein, A., and Infeld, L. (1938), *The Evolution of Physics* (New York: Simon and Schuster).

Epstein, W. (1977), 'Historical introduction to the constancies', in *Stability and Constancy in
Visual Perception: Mechanisms and Processes*, ed. W. Epstein (New York: Wiley).

Epstein, W. (1982), 'Percept-percept coupling', *Perception*, **11**, pp. 75–83.

Fodor, J. (1975), *The Language of Thought* (Cambridge, MA: Harvard University Press).

Fodor, J. (1981), 'The mind–body problem', *Scientific American*, **244**, pp. 114–123.

Gibson, J.J. (1950), *The Perception of the Visual World* (Boston: Houghton Mifflin).

Gibson, J.J. (1959), 'Perception as a function of stimulation', in *Psychology: A Study of a Science,
Volume I*, ed. S. Koch (New York: McGraw-Hill).

Gibson, J.J. (1966), *The Senses Considered as Perceptual Systems* (Boston: Houghton Mifflin).

Gibson, J.J. (1975), 'What is the relation of concepts to percepts?', Unpublished manuscript,
Department of Psychology, Cornell University, Ithaca, NY.

Gibson, J.J. (1976), 'Memo on the process of perception: Invariance detection', unpublished manuscript, Department of Psychology, Cornell University, Ithaca, NY.

Gibson, J.J. (1979a), *The Ecological Approach to Visual Perception* (Boston: Houghton Mifflin).

Gibson, J.J. (1979b), 'A note on substances, surfaces, places, objects, events', unpublished manuscript, Department of Psychology, Cornell University, Ithaca, NY.

Goldstein, H. (1980), *Classical Mechanics* (Reading, MA: Addison-Wesley).

Handy, R. (1973), 'The Dewey-Bentley transactional procedures of inquiry', *Psychological Record*, 23, pp. 305–17.

Haugeland, J. (1981), 'Semantic engines', in *Mind Design*, ed. J. Haugeland (Cambridge: MIT Press).

Hestenes, D. (1986), *New Foundations for Classical Mechanics* (Dordrecht: Kluwer Academic Publishers).

Hume, D. (1983), *A Treatise of Human Nature* (Oxford: Clarendon Press).

Iberall, A. (1972), *Toward a General Science of Viable Systems* (New York: McGraw Hill).

Iberall, A. (1977), 'A field and circuit thermodynamics for an integrative physiology, I', *American Journal of Physiology: Regulatory, Integrative and Comparative Physiology*, 2, R171–R180.

Iberall, A. (1978), 'A field and circuit thermodynamics for an integrative physiology, II', *American Journal of Physiology: Regulatory, Integrative and Comparative Physiology*, 3, R3–R19.

Iberall, A., and McCulloch, W. (1969), 'The organizing principle of complex living systems', *Journal of Basic Engineering*, 91, pp. 290–4.

Iberall, A., and Soodak, H. (1987), 'A physics for complex systems', in *Self-organizing systems: The emergence of order*, ed. F. E. Yates (New York: Plenum Press).

Jones, L.A. (1986), 'Perception of force and weight: Theory and research', *Psychological Bulletin*, 100, pp. 29–42.

Kantor, J.R. (1920), 'Suggestions toward a scientific interpretation of perception', *Psychological Review*, 27, pp. 191–216.

Kirsch, D. (1991a), 'Foundations of AI: the big issues', *Artificial Intelligence*, 47, pp. 3–30.

Kirsch, D. (1991b), 'Today the earwig, tomorrow man?', *Artificial Intelligence*, 47, pp. 161–84.

Kugler, P.N., Shaw, R.E., Vicente, K.J., and Kinsella-Shaw, J. (1990), 'Inquiry into intentional systems 1: Issues in ecological physics', *Psychological Research*, 52, pp. 98–121.

Massey, G. (1993), 'Mind–body problems', *Journal of Sport and Exercise Psychology*, 15, S597–S115.

Michaels, C., and Carello, C. (1981), *Direct Perception* (New York: Appleton-Century-Crofts).

Millikan, R. (1984), *Language, Thought and Other Biological Categories* (Cambridge, MA: MIT Press).

Millikan, R. (1993), *White Queen Psychology and Other Essays for Alice* (Cambridge, MA: MIT Press).

Nilsson, N.J. (1991), 'Logic and artificial intelligence', *Artificial Intelligence*, 47, pp. 31–56.

Pagano, C.C., Fitzpatrick, P., and Turvey, M.T. (1993), 'Tensorial basis to the constancy of perceived extent over variations of dynamic touch', *Perception and Psychophysics*, 54, pp. 43–54.

Penrose, R. (1989), *The Emperor's New Mind* (Oxford: Oxford University Press).

Premack, D. (1965), 'Reinforcement theory', Nebraska Symposium on Motivation (Vol. 13, pp. 123–180). Lincoln; University of Nebraska Press.

Reed, E.S. (1988), *James J. Gibson and the Psychology of Perception* (New Haven, CT: Yale University Press).

Reed, E.S, and Jones, R. (1982), *Reasons for Realism: Selected Essays of James J. Gibson* (Hillsdale, NJ: Lawrence Erlbaum and Associates).

Searle, J. (1992), *The Rediscovery of the Mind* (Cambridge, MA: MIT Press).

Shaw, R.E. And Alley, T. (1985), 'How to draw learning curves: Their use and justification', in *Issues in the Ecological Study of Learning*, ed. T.D. Johnston and A.T. Pietrwicz (Hillsdale, NJ: Lawrence Erlbaum and Associates).

Shaw, R.E. and Bransford, J. (1977), 'Introduction: Psychological approaches to the problem of knowledge', in *Perceiving, Acting, and Knowing*, ed. R.E. Shaw and J. Bransford (Hillsdale, NJ: Lawrence Erlbaum and Associates).

Shaw, R.E. and Turvey, M.T. (1981), 'Coalitions as models for ecosystems: A realist perspective on perceptual organization', in *Perceptual Organization*, ed. M. Kubovy and J. Pomerantz, (Hillsdale, NJ: Lawrence Erlbaum and Associates).

Shaw, R.E., Flascher, O., and Mace, W. (1996), 'Dimensions of event perception', in *Handbook of Perception and Action, Volume I*, ed. W. Prinz and B. Bridgemean (San Diego: Academic Press).

Shaw, R.E., Turvey, M.T. and Mace, W. (1982), 'Ecological psychology: The consequence of a commitment to realism', in *Cognition and the Symbolic Processes*, ed. W. Weimer and D. Palermo (Hillsdale, NJ: Lawrence Erlbaum and Associates).

Soodak, H., and Iberall, A. (1978), 'Homeokinetics: A physical science for complex systems', *Science*, **201**, pp. 579–82.

Stevens, J.C., and Rubin, L.L. (1970), 'Psychophysical scales of apparent heaviness and the size–weight illusion', *Perception and Psychophysics*, **8**, pp. 225–30.

Stevens, S.S. (1961), 'The psychophysics of sensory function', *American Scientist*, **48**, pp. 226–52.

Stevens, S.S. (1962), 'The surprising simplicity of sensory metrics', *American Psychologist*, **17**, pp. 29–39.

Swenson, R., and Turvey, M.T. (1991), 'Thermodynamic reasons for perception-action cycles', *Ecological Psychology*, **3**, pp. 317–48.

Timberlake, W. (1980), 'A molar equilibrium theory of learned performance', in *The Psychology of Learning and Motivation, Vol. 14*, ed. G. H. Bower (New York: Academic Press).

Turvey, M.T. (1977), 'Contrasting orientations to the theory of visual-information processing', *Psychological Review*, **84**, pp. 67–88.

Turvey, M.T. (1992a), 'Affordances and prospective control: An outline of the ontology', *Ecological Psychology*, **4**, pp. 173–87.

Turvey, M.T. (1992b), 'Ecological approach to cognition: Invariants of perception and action', in *Cognitive psychology: Conceptual and Methodological Issues*, ed. H. Pick, P. Van den Broek, D.C. Knill (Washington: APA Books).

Turvey, M.T. (1996), 'Dynamic touch', *American Psychologist*, **51**, pp. 1134–52.

Turvey, M.T., and Carello, C. (1985), 'The equation of information and meaning from the perspective of situation semantics and Gibson's ecological realism', *Linguistics and Philosophy*, **8**, pp. 81–90.

Turvey, M.T., and Carello, C. (1995), 'Dynamic touch', in *Handbook of Perception and Cognition: V. Perception of Space and Motion*, ed. W. Epstein and S. Rogers (San Diego, CA: Academic Press).

Turvey, M.T., and Shaw, R.E. (1979), 'The primacy of perceiving: An ecological reformulation of perception for understanding memory', in *Perspectives on Memory Research*, ed. L-G. Nilsson (Hillsdale, NJ: Lawrence Erlbaum and Associates).

Turvey, M.T., and Shaw, R.E. (1995), 'Toward an ecological physics and a physical psychology', in *The Science of the Mind: 2001 and Beyond*, ed. R. Solso and D. Massaro (Oxford: Oxford University Press).

Turvey, M.T., Shaw, R.E., Reed, E., and Mace, W. (1981), 'Ecological laws of perceiving and acting: In reply to Fodor and Pylyshyn (1981)', *Cognition*, **9**, pp. 237–304.

Turvey, M.T., Shockley, K. and Carello, C. (1999), 'Affordance, proper function and the physical basis of perceived heaviness', *Cognition*, **73**, B17–B26.

Yates, F.E., Marsh, D.J., and Iberall, A.S. (1972), 'Integration of the whole organism — a foundation for a theoretical biology', in *Challenging Biological Problems: Directions Toward their Solution*, ed. J.A. Behake (Oxford: Oxford University Press).

Zee, A. (1989), *An Old Man's Toy* (New York: Macmillan).

Robert E. Shaw and M.T. Turvey

Ecological Foundations of Cognition

II. Degrees of Freedom and Conserved Quantities in Animal–Environment Systems

Cognition means different things to different psychologists depending on the position held on the mind–matter problem. Ecological psychologists reject the implied mind–matter dualism as an ill-posed theoretic problem because the assumed mind–matter incommensurability precludes a solution to the degrees of freedom problem. This fundamental problem was posed by both Nicolai Bernstein and James J. Gibson independently. It replaces mind–matter dualism with animal–environment duality (isomorphism) — a better posed scientific problem because commensurability is assured. Furthermore, when properly posed this way, a conservation law is suggested that encompasses a psychology of transactional systems, a biology of self-actional systems, and a physics of interactional systems. For such a solution, a theory of cognition for goal-directed behaviour (e.g., choosing goals, authoring intentions, using information, and controlling actions) is needed. A sketch is supplied for how such a theory might be pursued in the spirit of the new physics of evolving complex systems.

I: How Might Other Disciplines Inform the Theory of Cognition?

The investigation of cognition has flirted seriously with a variety of physical and mathematical tools in an attempt to become a mature science in the spirit of physics, while hopefully avoiding simplistic reductionism. The problems are difficult and, to a great extent, interdisciplinary. Here is a summary declaration of what we students of cognition typically seek from collateral fields:

- From philosophy we seek certain guidelines that might make incorrigible problems more corrigible, render incommensurate kinds more commensurate, or, more often, help transform mysteries into puzzles, hopefully, putting them on the road to becoming scientific problems. Failing this, we might at least learn how to ignore them.

- From physics we hope to find analogues to psychological phenomena so the unlawful might be rendered lawful.

Journal of Consciousness Studies, **6**, No. 11–12, 1999, pp. 111–23

- From mathematics we seek ways to make ill-posed problems well-posed, to find connections among variables that appear unrelated, and a variety of generally consistent schemes of description from which we might select one appropriate to our problem.

Hence philosophy may help us frame our problems, physics may predispose us to certain forms of principled explanations, and mathematics may provide us with schemes for expressing the principles and problems that, when judiciously used and empirically interpreted, lend consistency and validity to our efforts.

As a field, cognition faces an additional problem. Sufficient agreement exists in experimental design and statistical methods to have similar standards and criteria for peer review across its best journals and major funding agencies. In contrast, no consensus exists concerning which problems are most fundamental or which theoretic principles are most useful. Because there are no clearly defensible laws, the job of students of cognition is made more difficult.

In the present article we identify and explore the *generalized degrees-of-freedom problem*. This problem is a major bulwark to a successful theory of how animals can know about their environments and behave adaptively with respect to them. Moreover, it is a problem that seems sufficiently fundamental and general to warrant concern by all scientists. In preview, a solution to this problem seems to entail the discovery of a new conservation law, one defined at the level of animal–environment systems.

II: The Generalized Problem of Degrees of Freedom

Bernstein (1935/1967) was the first to draw attention to the fundamental importance of this problem for explaining the control of behaviour. He posed the following difficult but incisive question 'How can a neuromuscular system with an exorbitant number of degrees of freedom be made to act as if it had but one degree of freedom?' (see Turvey, 1990). Gibson (1966; 1979) promoted a related idea. He argued that, in contrast to its use in communication theory, *information* should be construed as specificity of the useful rather than as the uncertainty of the specific. The latter sense of information was, for Gibson, the best way to understand how perception could control behaviour (see Turvey and Shaw, 1999). Perception controls behaviour by detecting informational constraints specific to goal-paths (Gibson, 1979; Shaw and Kinsella-Shaw, 1988). Goal constraints, as compared to physical constraints, can be considered extraordinary (Kugler and Turvey, 1987; Turvey, 1986), for they take the form of a rule that prescribes how one should act if some outcome is intended. More to the point, the prescriptive rule asserts that one should act so as to change the current information, which is less specific to an intended outcome, into information that is more specific. With these intentional rules for action, Gibson opened wide the door to an ecological approach to cognition.

Such 'rules', he assures us, however, are not computational; they are more in the nature of laws at the ecological scale. When presented in dimensionless form, they may apply across species. Their effects may be observed whether or not an animal or human is aware of them. Perhaps they conform to Wittgenstein's notion of a rule entailed by the behaviour and its context rather than to a rule procedurally involved, like a recipe, of which the agent must be aware.

It is worth noting that this 'rule' is more like a law, but a law of quantum physics rather than classical physics. Quantum laws apply at the subtle scale of weak potentials (usually but not necessarily identified with smallness); they relate expectations across measurement situations, and in this sense are informationally based. They are prima facie about what we can know, not about what *is* (except under certain realist philosophical assumptions). Classical laws relate facts about energy, forces, etc. across physical situations. Hence a Gibson rule for the perceptual control of action is a law in the quantum sense, but not in the classical sense. Because this is so, we might tend to find help in characterizing them from a study of the mathematics used in quantum physics. Keeping this in mind will help in understanding the main points that follow.

The answer to Bernstein's question, couched in terms of a Gibson rule, is that a system with many degrees of freedom can act as if it were a simpler system only if sufficient constraints, or linkages, are established among its components by coupling them into a synergy. What are the possible sources of freedom (the removing of constraints) and constraint (the curtailing of degrees of freedom)? Three sources might be identified: external force fields of physical origin, internal force fields of biological origin, and information fields of psychological origin — in the ecological sense.

The coupling of external, environment-based force fields with internal, organism-based force fields by means of forceless information fields (Kugler and Turvey, 1987; 1988; Kugler *et al.*, 1985) poses as yet unsolved problems for cognitive theory. It is possible, however, to identify some of the most important characteristics that these solutions should have. The starting point is information's nature. How can it couple two systems, such as an animal to its environment?

III: Systems Informed by Interaction, Self-action, and Transaction

In physics, the freezing out of degrees of freedom is achieved automatically by laws that entail Hamilton's least action principle (where action is mass x distance x speed). Such physical degrees of freedom are a function of whether the particle is located in a field-less region of space (free fall), or whether in a region of space dominated by a field of external forces (e.g., gravitational or mechanical). Following Rosen (1978), we might say that a system moved from a force-free region of space–time to a force-dominated one becomes dynamically informed (literally, 'takes on form'). Under this view, information is simply that which imparts 'form' where the latter is simply just another word for 'constraint'. Just as 'free-form' means an unconstrained, or random, shaping of material, so 'free-fall' means an unconstrained shaping of its trajectory, a random path. Information in the preceding sense can be called *dynamical*.

To obtain the sense of information that will do most work for a theory of how animals know about their environments, we need to add 'specificity' to the connotation of the term (Gibson, 1966; 1979; see Turvey and Shaw, 1999). Information in the specificational sense picks out a few things from a background of many things. It is selective or choice-like. Preserving both connotations, we then have a concept of total information that is the sum of two complementary functions: the dynamically informing and the intentionally specifying — where being *intentional* is to be *about something*, to refer beyond itself, in the philosophical sense.

It is, we think, an egregious and dangerous error for a theory of cognition to take either meaning alone, say, to take selectivity alone, as done by Shannon (Shannon and Weaver, 1949), for this leaves information perspective-free, unscaled, and unconstrained by dynamical law. Better to keep information bound to physical parameters (e.g., Boltzman's constant in its negentropic interpretation), and start from there to determine its additional uses in biology and psychology (Kugler and Turvey, 1987; Shaw, 1985). Right or wrong, this is our attitude and plays a major role in how ecological psychologists do their work as scientists.

It is important to admit right off that informable systems are the general case. Systems in free-form or free-fall are the exception since they only exist temporally and locally, and, strictly speaking, only at limit — in the mathematical sense. Informable systems may be of two kinds: being externally informed or self informed. Ordinary physical systems are externally informed systems. These are the special case since they do not have all the properties of living systems, while living systems have all the properties of physical systems plus additional ones. Perhaps the chief difference between physics and the life sciences is that the former studies simple systems relative to the complex systems studied by psychology and biology. This claim was heresy among nineteenth-century physicists but today it is a growing consensus among the 'new' physicists, among whom we must number ecological physicists. Let us consider these differences.

Traditionally defined physical systems are solely *interactional*; when acted upon by a force, an immediate reactive force tempers their response (in accordance with Newton's law of action and reaction). However, there are some systems that are *self-informable*, in the sense of being self-motivating and self-controlling. All life-forms fall into this class, at least as far as we know. In addition to being interactional, as all systems with mass must be, these are also *self-actional*; they are capable of a delayed reaction in addition to the immediate reaction which they may modulate by the addition of self-generated counter-forces. To do so, however, they must have complex interiors, an on-board (metabolic) potential capable of biogenic forces that may be used to cancel, modulate, or delay their immediate reaction to an external force (Kugler and Turvey, 1987). Biogenic force modulation requires scaling information if the potentially deleterious effect of an external force is to be controlled. Such systems are informable by evolutionary or self-styled edicts as a function of, or sometimes independent of, local environmental force fields.

Systems that are more dependent on information fields, with information defined in the specificational sense, may be driven by intentions that are specific to a non-local, essentially forceless field — a goal. The field of goal-specific information is force-free when, at the scale of the system's mass, no significant momentum transfer takes place from the external field (Kugler and Turvey, 1987; Kugler *et al.*, 1985; Turvey and Shaw, 1995). When so, this leaves the system's internal force field in sole charge of its control while the environmental information field makes goal-specific demands on that control. Systems that conduct their business with the environment through information will be said to be *transactional* rather than simply self-actional (Dewey and Bentley, 1949; Shaw and Turvey, 1981; Turvey and Shaw, 1999).

Transactional systems do not merely modulate their reactive forces haphazardly, or, at least, natural selection weeds out those that do. Fit transactional systems are adaptive because they are not merely informed but are specifically informed about life-sustaining resources whose procurement is possible and intended. Across all species this is true. It is not (yet) true of artefacts (e.g., computers or robots). Why this difference? We think it exists because the natural species share a secret design feature that the artificial species do not. Organisms deal with their environments in a lawful manner that has not yet been built into machines. We will suggest that they conserve *total generalized action*.

Being information-driven does not preclude a transactional system from being force-driven. Transactions subsume self-actions and interactions. Biology makes possible extra degrees of freedom that may be, but need not be, used to modulate external forces by internal forces. Ecological psychology, among other things, is the study of how these extra degrees of freedom may be used to orchestrate intentional acts.

IV: The Plenitude Hypothesis and the Sculpting Metaphor

The most straightforward strategy for solving the degrees of freedom problem is to search for sources of constraint. The sources of constraint must be one less than the degrees of freedom. Why one less? A degree of freedom must exist if the system is to follow the path that the laws of nature dictate. In nature, there truly are no dynamically stultified systems, they are all in process and continue to change over time. The apparent persistence of certain phenomena, like the oceans and the mountains — even the moon orbiting the earth, or the sun rising and setting, is not permanent but only temporary.

Indeed, the dean of contemporary cosmologists, John Archibald Wheeler, asserts that the only fundamental law is mutability itself — a law without laws (Wheeler, 1982). No laws existed prior to the 'big bang'. Rather, the laws developed as the universe evolved. No laws of chemistry until particles coalesced into molecules, no laws of geology until the earth formed, no laws of biology until there was life, and so forth. Laws express what is structurally stable in nature so predictions are possible. Under the law of universal mutability, structures are but slow functions; hence laws must evolve. In the beginning, there was only a singular compacted action (a dimensionless form of Planck's constant) — energy to be distributed over time — so everything must come from it (unless there are also local sources of continual renewal, see below). The initial condition was a singular action integral to be differentiated over emergent space–time into energy, forces, momenta and velocities.

On earth, the environment is also characterized by change rather than absolute persistence, or as Gibson (1979) put it, persistence amidst change and persistence of change. We live in a plenum of possibility where all perception is event perception, the perception of changing things (Turvey, 1992), and is so whether events transpire slowly or swiftly (Shaw and Pittenger, 1978; Shaw *et al.*, 1996; Warren and Shaw, 1985). This event perception thesis follows directly from the new physics emerging at the end of the millennium. The fundamental reality consists of events and not of things. Recent theories of the cerebellum suggest a useful metaphor. The cerebellum is that part of the brain most centrally involved in the dynamical control of movement, and seems to be continually firing (Braitenberg *et al.*, 1997). Timing signals arise from systematic suppression of unwanted firings. They are, so to speak, 'sculpted out of this background of excitation' as miniature tidal waves that flow as coherent paths of excitation (Shaw *et al.*, 1997; Kadar *et al.*, 1997). This metaphor of 'sculpting from a background of excitation' captures the fundamental mechanism of the new physics.

This image of reality arising out of a background of raw dynamical potentiality, by processes that carve relative persistence from chaotic excitation, also fits the quantum theorists' notions about the creation–annihilation dynamics of the (false) vacuum. Under this modern view, the vacuum is not an empty, inert void but a cauldron of subtle energy from which particles emerge and into which they disappear (Finkelstein, 1996). Postulating the vacuum as the engine of continual renewal, endorses a truly Heraclitean view of nature. To avoid *creatio ex nihilo* — creating the world from nothing — the vacuum is no longer a dead-end void but a creative conduit to a plenum (a Greek word meaning 'bountiful' but used philosophically to denote a cornucopia of pure potentiality).

The plenitude hypothesis asserts that nature is a plenum, or superabundance, of real possibilities — a view consistent with ecological psychology (Turvey, 1992; Turvey and Shaw, 1995). The historical development of the plenitude hypothesis was traced earlier this century by Lovejoy (1936) in his widely acclaimed and prophetic book, *The Great Chain of Being*. This idea is of a plenum that sits behind observed reality is also fundamental to modern physics, and sums up its weakened view of determinism: If laws of nature do not disallow something, then it exists.

Modern physics is not about finding the causes that make something's existence lawfully necessary. It is about exclusion or censorship of those things not allowed. As in the case of change, no positive cause for it is required or even possible. All entities or processes are suspected of existing until cleared of the charge. All phenomena, no matter how weird, have equal ontological status. For one thing to exist rather than another, it needs to be most compatible with the other potential existents. It then moves across the ontological threshold to become actualized.

This is analogous to an argument by Leibnitz, one rooted in his notion of compossibility, or mutual compatibility. He held this argument in abeyance as a possible replacement for the more simplistic, but more Church approved, postulate of pre-established harmony. His follower Herbart (1824) framed a theory of consciousness from this notion of compossibles residing in a psychical plenum. Ideas remain below the threshold of consciousness until they reach a critical apperceptive mass, then they come into awareness. The most buoyant ideas are those that are most compatible with the most other ideas. If we forget the mental aspects, and focus only on the mutual compatibility argument, then we have the equivalent of the dynamical phase correlations that underlie the sculpting mechanism used in the new physics (Feynman and Hibbs, 1965).

Similar arguments have been applied in neuroscience by quantum brain dynamicists (Jibu and Yasue, 1995). They argue that coherent brain processes, called Bose condensates, arise from a creation–annihilation dynamic between photons, electrons, and water molecules in the brain. Such arguments reflect a deep belief in the plenitude hypothesis and the sculpting mechanism. A cerebral mechanism sculpts from a background of excitation — a plenum of possible excitonic waves — whatever cerebral events are needed to underwrite thoughts, feelings, memories, perceptions, or other experiences. In this way they hope to offer a credible scientific account that resolves Chalmers' easy problem.

V: Chalmers' Problems

Chalmers (1996) has identified two types of problem faced by neurocognitive science if either of these strategies were adopted — the 'easy' problems, that presumes a necessary alignment between experiences and events in a physical (e.g. neurological) substratum, and the 'hard' problem that we will define in a moment. The so-called 'easy' problems (which of course may be extremely difficult scientifically) are faced by the theory of mind which seeks to determine what events in the physical substratum (the brain?) take place concomitantly with the associated phenomenological events (experiences).

The Fechner–Spinoza principle of psychophysical correspondence is an example of an attempt to solve the 'easy' problems by postulating the solution: For every physical event there is a concomitant mental event, and for every mental event there is a concomitant physical event (Bain, 1873). What is the evidence for such a strong claim? There is none, since the induction is hardly fulfillable but, more fairly, none is intended. It is rather a strategy, a kind of licence for a neuropsychology fishing expedition. In any case, such views leave aside the issue of how experience arises, where its content comes from, and only addresses those aspects of the physiological processes to which an experience corresponds.

By contrast, the 'hard' problem requires more than mere concomitance but must explain how the character of experience (content) necessarily derives from, rather than merely corresponds to, the character of physiological events. It must address the sufficiency problem as well: How can the psychological content of experience be restricted to the variables describing the physical content? Consequently, it is difficult to see how a prima facie case can be made that either of these strategies gives us a theoretical handle on how to address the 'hard' problem scientifically. The problem remains a mystery — a puzzle with an ontological enigma at its core.

VI: Peirce's Chance Discovery of Order in Entropy

From the discussions above, the metaphor of 'sculpting from a background of excitation' may be too exemplary of nature to be treated as mere metaphor. Perhaps, it models the mechanism of the millennium, having power to heal the schisms that separate psychologists from other scientists and from each other. What might be its physical origins?

The brilliant friend of William James and co-founder of American pragmatism, Charles Sanders Peirce, presented an elegant argument for why physical reality might naturally contain a sculpting mechanism (Peirce, 1892). After describing the inevitable heat death of the universe as entropy accrues according to the second law of thermodynamics, Peirce observes: being part of the entropy producing process, no force left to its own designs can counteract the lawful degradation of order. However, opposing this lawful disordering process is 'chance' — the opposite of law. Chance offers an extra degree of freedom to physics by which life-forms can emerge.

Chance, by opposing the second law, allows a kind Prigoginean order (Prigogine, 1980; 1997) to be brought into the world in spite of its entropic tendency. Put into current idiom: Chance is just a looseness in the determinism of physical laws which, by the plenitude hypothesis, gets automatically filled up with excitation from the vacuum — a kind of space–time foam (Wheeler, 1982).

The sculpting of intentional paths from this background of excitation cannot come from opposing, head-on, the laws of nature as conceived by nineteenth-century science. The dissipation of energy by the regular laws of nature allows, by their inherent laxity, a home for another process, a counter-process whose tendency is a filling in of the entropic seams with order. This counter-process is not really chance, or randomness, but merely seems so from the perspective of the ordinary laws. From their perspective, the counter-process appears both non-holonomic and non-linear, as the ordinary laws do from its perspective.

The aforementioned dual, dynamical, antisymmetric processes are equally real and equally physical; but being contrary in direction, they just are not physical in the same sense. Nature no more favours one than the other. Neither holds hegemony over the other. There must be a point however, as Peirce assures us, at which the two tendencies achieve balance. The catabolic Second Law and its anabolic Prigoginean dual may hunt for a metabolic equilibrium without reaching ultra-stability (Swenson and Turvey, 1991). Their balance will be at best approximate, and therefore tolerant of life for as long as this balance holds.

The master equation for describing this hunting for balance is not a genetic algorithm with an *a priori* God-given harmony to achieve, but more like a dynamical ecosystem equation of the Lotke–Volterra variety (such as might be formulated for species competing over limited resources). Balance must be earned and sustained by continual work. This dynamical balance law foreshadows the existence of a deeper conservation than energy or momentum conservation, one that subsumes the other conservations. This new, putatively conserved quantity, is the generalized total action referred to above. It is defined as the time integral of energy used to inform and control a relatively closed system so that it behaves in a manner that satisfies certain boundary conditions (one of which, for certain systems, might be a goal). Application of this universal 'sculpting mechanism' to ecosystems naturally follows (e.g., Shaw and Kinsella–Shaw, 1988; Shaw *et al.*, 1992).

These counter processes are dual in the sense of being mutual and reciprocal (Turvey and Shaw, 1995; see Turvey and Shaw, 1999), that is, where one is free, the other is constrained. We might say, one is the environment of constraint to the other's freedom as process — a kind of cosmic ecosystem. The physics appropriate to studying this fundamental engine for life must be defined at this proto-ecological scale. Such an ecological physics must countenance both processes, and aims at laws that allow one process to be harnessed by the other. Its task is to explain how their loose coupling, or *graded determinism*, supports life and the informing of the controlling forces that sustain it.

VII: Ontological Descent from Possibility to Actuality

In an attempt to catalogue the extra degrees of freedom that accrue from these dual processes, and for which ecological physics is ultimately accountable, we offer the following bookkeeping scheme.

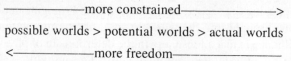

————————more constrained————————>

possible worlds > potential worlds > actual worlds

<————————more freedom————————

Actual worlds obey a cascade of constraints from the left to the right. The cascading constraints imposed squeeze out the available degrees of freedom. Existence occurs when sufficient constraints confine variables to single values, and they become constants, or the 'value of a bound variable' (as Quine suggests). Variables are bound to the degree they satisfy the contexts under which they are nested. Hence for a system to exist in an observed state, it must have 'constraint satisfaction'; its degrees of freedom problem must be implicitly solved, otherwise its behaviour would be indeterminate — an impossibility in nature. The constraints to be satisfied are denoted below.

generalized action conservation > other conservations > boundary conditions
> dynamical laws > perspectives > scales > action rules > values

Considerable discussion is needed to clarify this ontological descent scheme. We can venture here only a brief sketch of its levels.

At the possible worlds level, the only constraint is a weakened version of the classical logic law of non-contradiction. Under the strict determinism of the old physics, this law had to take the form of an exclusive disjunctive proposition (not both p and not-p), where nothing can be both true (existent) and false (non-existent) under exactly the same circumstances. By contrast, under the current graded determinism of the new physics, with its endorsement of the plenitude hypothesis, the law of non-contradiction is weakened. It now becomes an inclusive disjunctive proposition (either p or not-p or both p and not-p). Consequently, states are sometimes indefinite superpositions of all possibilities that laws allow.

Moreover, increased freedom allows for the dual dynamical processes discussed earlier and gives legitimacy to quantum physics' view of reality as a plenum of possibilities. Specifically, it allows for the superposition of quantum states, as exemplified in the famous Schrödinger cat problem. By Heisenberg's uncertainty principle, if a cat can be potentially killed by a random quantum event, the outcome remains in limbo, with the cat being neither living nor dead, or both, if you prefer, until observed or measured. Observing the situation is an occasion on which the wave function of the cat-and-apparatus collapses into one of the two possible, superposed states. Observation is supposed to cascade sufficient constraints to make the cat's indefinite possible state (both alive and dead), into a potential state (either alive or dead), to being a unique value of a bound variable — a constant (say, alive).

Does observation really cause the collapse of the indefinite into the definite, implying Cartesian interaction between mind and matter? Wigner (1970) has suggested it does, because consciousness plays the role of a state reduction operator. Or, does it simply occasion it, implying Leibnizean parallelism? The jury is still out on this issue (see Penrose, 1989). As ecological realists, we consider the question to be ill-posed. We agree that it is a degrees-of-freedom problem and, perhaps, the most fundamental example. But we disagree that mind–matter dualism is a useful way to approach its solution. The ecological approach would substitute the animal–environment duality in the place of the mind–matter dualism (Michaels and Carello, 1981; Shaw and Turvey, 1981; Turvey and Shaw, 1995; Turvey *et al.*, 1981; see Turvey and Shaw, 1999). How might this orthogonal strategy work in this case?

The issue is not whether we observe the cat and find it either living or dead, but whether the cat observes us as well. If it does, then neither it nor we can be dead, indefinite, or non-existent, in the usual meaning of these words. If the observer and the cat can have mutual and reciprocal perspectives, then they are dual observers, a social dyad, sharing an environment (laboratory). Hence an ecosystem exists. Two beings that share the same environment must logically be recipients of the same cascade of constraints. As partners, they make the ontological descent together, coupled by information, satisfying the same conservations, governed by the same laws, under reciprocal perspectives, at a mutual scale, until each becomes the value of the other's bound variable.

Hence, at the macroscale where such mutual information exists, all the variables of the situation are mutually constrained. Their wave functions are perfectly in phase. When this happens, harmony is achieved *a posteriori*. It is established dynamically. There need be no assumption of it's *a priori* pre-establishment. In physics this is known as solving an eigenvalue problem. The best formulation of this actualizing process, in our estimation, takes the form of a Feynman path integral. In ecological terms, it represents how a process, termed an *effectivity*, squeezes out the degrees of freedom from an *affordance*, to yield a specific, actual action (Shaw *et al.*, 1995; Shaw *et al.*, 1997; Kadar *et al.*, 1997). It works this way.

VIII: Feynman Path Integrals and the Generalized Action Principle

Let's not argue about whether the ontological descent scheme is true, or even if it could be true. Not being philosophers, we are only interested if it might be a useful idea. Does it have any legitimate scientific interpretation? More exactly, does it suggest a way that a particle (or an actor) might get from one point to another in the environment, say, by following a Gibson-like action rule that solves its degrees-of-freedom problem?

In the old physics, Hamilton's principle of least action somehow picks out the path actually followed from all possible, less stationary (under variation) paths. It is not at all clear how it does so, that is, how the particle is constrained to a unique path. Poincare (1905/1952) said the claim that Hamilton's principle provided an explanation was an offence to reason. He argued that finding the least action path implied that the particle first had to try out all the other paths to see which was least effortful. How could it do so? Does it explore all the paths 'off-line' from ordinary reality, while time is mysteriously suspended? Such an exploratory activity could not be dynamical since this would violate the conservation principles. The fact that we do see paths of objects does not help since perception is imprecise, being limited to a grain coarser than that of quantum uncertainty. As discussed earlier, conventional formulations of quantum mechanics provide no help since they have no state reduction operator. Is the path concept therefore useless?

Many physicists resigned themselves to the indefiniteness of paths, while others continued to believe that definite classical paths really do exist even if Hamilton's principle does not explain how. An impasse was reached and Poincare's objection continued to stand unchallenged for a half century or more, mostly because it was ignored (Yourgrau and Mandelstam, 1979). In this context of extreme pessimism and the misplaced optimism regarding the reality of paths, Feynman's eventual insight was quite unexpected.

Working from a suggestion by Dirac, Feynman ignored the pessimism and took a positive attitude toward this problem. He reasoned that if a unique path is not possible, then *all possible paths are allowed*, a clear application of the plenitude hypothesis. Furthermore, he showed how the classical path could be recaptured: if each path could be weighted at each step by an action factor, then so can each space–time point the path crosses. Next, remove the paths from consideration, what is left is a space–time array of point–action weights.

Finally, let each of these action values be represented by a vector perpendicular to each point. In this way, they describe an action topography with peaks and valleys. The minimum path is then found by the particle making the steepest descent through the hills and dales. This descent avoids being trapped in local minima because a non-local correlation is introduced (analogous to 'cooling' in simulated annealing techniques). The amplitude of these action peaks are cross-correlated by a field process in the manner of Huygens' principle of constructive and destructive wave interference.

Under this principle those paths that are most highly correlated are those that are most stable, least busy, and these are exactly those with minimal values. Feynman proved this with his path integral approach. His approach provides a different take on physics, and, although difficult to solve in closed form if more than two degrees of freedom are involved, can be readily simulated 'on-line' by Monte Carlo techniques (Landau and Mejia, 1997). The Feynman path integral provides one way to represent mathematically an effectivity, or sculpting mechanism. It carves out an actual tolerance region around the classical path. The former is psychologically more realistic than the latter because perception (measurement) has finite resolution.

In sum, Feynman showed that each path is more or less in dynamical phase with the other possible paths. Thus they each contribute to the sum of amplitudes which is greatest in the vicinity where the classical path is to be found by standard variational techniques. The path integral approach essentially gives a 'global' formulation to classical field theory and, for our purposes, to intentional dynamics.

Feynman's elegant generalization of Huygens' action principle achieves an explanation for the appearance of tolerance corridors around the minimum path. Application of Hamilton's action principle requires that paths have stationary end-points, that there be no background of excitation. By contrast, Feynman's action principle uses this non-stationary property to allow phase-correlated paths to emerge from that noisy background. In this way the excitation acts as a sculpting tool *sui generis* for carving out the most tolerant path.

The final trick is to make each level in the ontological descent a sum of Feynman path integrals. With such a fundamental principle available, perhaps, the science of cognition and the other sciences can work from the same page in nature's book on mathematical strategies.

IX: The Usefulness of an Action Conservation Measure

We conclude by summarizing some of the chief reasons that having a conservation is important to a developing theory of cognition. A conservation is a quantity whose number before and after a manipulation, experimental or natural, remains invariant. It is a unity that may be partitioned such that when the partitions are recombined, the unity is restored without loss or gain. Conserved quantities may redistribute themselves so that a loss in one place is always complemented by a gain in another place. In this way, a conservation provides a useful measure, indeed the best measure, for a dynamical systems where almost everything else is changing. It represents a persistence in the midst of change, an invariant. Information specific to a conserved quantity will likewise be conserved (e.g., in phase space, consider the conservation of area of a limit cycle in spite of shape changes; it is information specifying the conservation of momentum over space).

With an invariant information measure, you can quantize the aspects of a task yielding path solution data. With these path action numbers, one can then compare and order paths accordingly to:

(1) compare novice paths to that of experts;

(2) compare the levels of task difficulty under different experimental conditions;

(3) measure improvement in task performance over trials;

(4) classify task control strategies in terms of efficient use of space, time, energy, or momentum changes.

Although students of cognition have measures of each of these, the measures are typically *post hoc* rather than principled, and are not guaranteed to be measuring the same thing. Having a conserved quantity on which to base the above measures would avoid these problems and help the study of cognition to move to a higher scientific plateau.

Acknowledgements

The preparation of this manuscript was supported, in part, by NSF Grants SBR 97–09678, SBR 97–28970.

References

Bain, A. (1873), *Mind and Body: The Theories of their Relation* (London: Henry S. King and Co).

Bernstein, N. (1935), 'The problem of the interrelation between coordination and localization', *Archives of Biological Sciences*, **38**, pp. 1–34. [Reprinted in N. Bernstein (1967), *Coordination and Regulation of Movements* (Oxford: Pergamon Press)].

Braitenberg, V., Heck, D., and Sultan, F. (1997), 'The detection and generation f sequences as a key to cerebellar function: experiments and theory', *Behavioral and Brain Sciences*, **20**, pp. 229–77.

Chalmers, D. (1996), *The Conscious Mind: In Search of a Fundamental Theory* (New York: Oxford University Press).

Dewey, J. and Bentley, A.F. (1949), *Knowing and the Known* (Boston, MA: Beacon).

Feynman, R.P. and Hibbs, A. (1965), *Quantum Mechanics and Path Integrals* (New York: McGraw-Hill).

Finkelstein. D.R. (1996), *Quantum Relativity: A Synthesis of the Ideas of Einstein and Heisenberg* (New York: Springer Verlag).

Gibson, J.J. (1966), *The Senses Considered as Perceptual Systems* (Boston, MA: Houghton Mifflin).

Gibson, J.J. (1979), *The Ecological Approach to Visual Perception* (Boston, MA: Houghton Mifflin).

Herbart, J.F. (1824–1825), *Psychology as a Science, Newly Based Experience, Metaphysics, and Mathematics (Vols. 1 and 2)* (Königsberg, Germany: Unzer).

Jibu, M. and Yasue, K. (1995), *Quantum Brain Dynamics and Consciousness: An Introduction* (Amsterdam/Philadelphia: John Benjamin Publishing Company).

Kadar, E.E., Shaw, R.E. and Turvey, M.T. (1997), 'Path space integrals for modelling experimental measurements of cerebellar functioning', *Behavioral and Brain Sciences*, **20**, p. 253.

Kugler, P.N. and Turvey, M.T. (1987), *Information, Natural Law and the Self-assembly of Rhythmic Movement* (Hillsdale, NJ: Lawrence Erlbaum and Associates).

Kugler, P.N. and Turvey, M.T. (1988), 'Self-organization, flow fields, and information', *Human Movement Science*, **7**, pp. 97–129.

Kugler, P.N., Turvey, M.T., Carello, C. and Shaw, R. (1985), 'The physics of controlled collisions: A reverie about locomotion', in *Persistence and Change,* ed. W.H. Warren, Jr. and R. Shaw (Hillsdale, NJ: Lawrence Erlbaum and Associates).

Landau, R.H. and Mejia, M.J. (1997), *Computational Physics* (New York: Wiley).

Lovejoy, A.O. (1936), *The Great Chain of Being: A Study In the History of an Idea* (Cambridge, MA: Harvard University Press).

Michaels, C. and Carello, C. (1981), *Direct Perception* (New York: Appleton-Century-Crofts).

Peirce, C.S. (1892), *The Monist*, **2**, pp. 321–37.

Penrose, R. (1989), *The Emperor's New Mind* (Oxford: Oxford University Press).

Poincare, H. (1905/1952), *Science and Hypothesis* (New York: Dover Publications Inc.) (A reprint of the Walter Scott Publishing Company, limited translation).

Prigogine, I. (1980), *From Being to Becoming* (San Francisco: W.H. Freeman).

Prigogine, I. (1997), *The End of Certainty* (New York: Free Press).

Rosen, R. (1978), *Fundamentals of Measurement and Representation of Natural Systems* (New York: North-Holland).

Shannon, C. and Weaver, W. (1949), *The Mathematical Theory of Communication* (Urbana, IL: University of Illinois Press).

Shaw, R.E. (1985), 'Measuring information', in *Persistence and Change*, ed. W.H. Warren, Jr. and R.E. Shaw (Hillsdale, NJ: Lawrence Erlbaum and Associates).

Shaw, R., Kadar, E. and Kinsella-Shaw, J. (1995), 'Modelling systems with intentional dynamics: A lesson from quantum mechanics', in *Appalacia II: Origins of self-organization. The Report of the Second Annual Appalachian Conference on Neurodynamics*, ed. K. Pribram (Hillsdale NJ: Lawrence Erlbaum and Associates).

Shaw, R., Kadar, E., Sim, M. and Repperger, D. (1992), 'The intentional spring: A strategy for modelling systems that learn to perform intentional acts', *Journal of Motor behaviour*, **1** (24), pp. 3–28.

Shaw, R.E., Kadar, E.E. and Turvey, M.T. (1997), 'The job description of the cerebellum and a candidate model of its 'tidal wave' function', *behavioural and Brain Sciences*, **20**, p. 265.

Shaw, R.E. and Kinsella-Shaw, J. (1988), 'Ecological mechanics: A physical geometry of intentional constraints', *Human Movement Science*, **7**, pp. 155–200.

Shaw, R.E. and Pittenger, J.B. (1978), 'Perceiving change', in *Modes of Perceiving and Processing Information*, ed. H. Pick and E. Saltzman (Hillsdale, NJ: Lawrence Erlbaum and Associates).

Shaw, R.E., Flascher, O. and Mace, W. (1996), 'Dimensions of event perception', in *Handbook of Perception and Action, Volume I*, ed. W. Prinz and B. Bridgemean (San Diego: Academic Press).

Shaw, R.E. and Turvey, M.T. (1981), 'Coalitions as models for ecosystems: A realist perspective on perceptual organization', in *Perceptual Organization*, ed. M. Kubovy and J. Pomerantz (Hillsdale, NJ: Lawrence Erlbaum and Associates).

Swenson, R. and Turvey, M.T. (1991), 'Thermodynamic reasons for perception–action cycles', *Ecological Psychology*, **3**, pp. 317–48.

Turvey, M.T. (1986), 'Intentionality: A problem of multiple reference frames, specificational information and extraordinary boundary conditions on natural law', *behavioural and Brain Sciences*, **9**, pp. 153–55.

Turvey, M.T. (1990), 'Coordination', *American Psychologist*, **8**, pp. 938–53.

Turvey, M.T. (1992), 'Affordances and prospective control: An outline of the ontology', *Ecological Psychology*, **4**, pp. 173–87.

Turvey, M.T. and Shaw, R.E. (1995), 'Toward an ecological physics and a physical psychology', in *The Science of the Mind: 2001 and Beyond*, ed. R. Solso and D. Massaro (Oxford: OUP).

Turvey, M.T. and Shaw, R.E. (1999), 'Ecological foundations of cognition, I: Symmetry and specificity of animal–environment systems', *Journal of Consciousness Studies*, **6** (11–12), pp. 95–110.

Turvey, M.T., Shaw, R.E., Reed, E. and Mace, W. (1981), 'Ecological laws of perceiving and acting: In reply to Fodor and Pylyshyn (1981)', *Cognition*, **9**, pp. 237–304.

Warren, Jr., W.H. and Shaw, R.E. (1985), 'Events and encounters as units of analysis for ecological psychology', in *Persistence and Change*, ed. W. H. Warren, Jr. and R. E. Shaw (Hillsdale, NJ: Lawrence Erlbaum and Associates).

Wigner, E.P.(1970), 'Two kinds of reality', in *Symmetries and Reflections: Scientific Essays of Eugene P. Wigner,* ed. W.J. Moore and M. Scriven (Cambridge, MA: M.I.T. Press).

Wheeler, J.A. (1982), *Physics and Austerity, Law Without Law* (Anhui, China: Anhui Science and Technology Publications).

Yourgrau, W. and Mandelstam, S. (1979), *Variational Principles in Dynamics and Quantum Theory* (New York: Dover Publications).

Paul Cisek

Beyond the Computer Metaphor

Behaviour as Interaction

Behaviour is often described as the computation of a response to a stimulus. This description is incomplete in an important way because it only examines what occurs between the reception of stimulus information and the generation of an action. Behaviour is more correctly described as a control process where actions are performed in order to affect perceptions. This closed-loop nature of behaviour is de-emphasized in modern discussions of brain function, leading to a number of artificial mysteries. A notable example is the 'symbol grounding problem'. When behaviour is viewed as a control process, it is natural to explain how internal representations, even symbols, can have meaning for an organism, and how actions can be motivated by organic needs.

What's Wrong with 'Computationalism'?

One can say that the general working assumption in brain science today is that *the function of the brain is to convert stimuli into reactions*. This was explicitly stated over a hundred years ago by one of the most influential psychologists of all time, William James:

> The whole neural organism, it will be remembered, is, physiologically considered, but a machine for converting stimuli into reactions. (James, 1890, p. 372)

The modern interpretation of this statement takes the form of what may be called the computer metaphor, or 'computationalism'.[1] This doctrine describes brain function as the computation of behavioural responses from internal representations of stimuli and stored representations of past experience. In its broadest sense, computationalism may be crudely defined in terms of the following analogy: perception is like input, action is like output, and all the things in between are like the information processing performed by computers.

---

[1] The term 'computationalism' may be used to imply any one of a number of theoretical viewpoints which differ on many important issues. A distinction between symbolic computationalism and connectionist computationalism is an example. Here, I will use the term in the broadest sense, referring to any theory which describes brain function as the computation of a response to a stimulus.

Journal of Consciousness Studies, **6**, No. 11–12, 1999, pp. 125–42

Below, I argue that this description is incomplete in an important way, leading brain sciences toward apparent mysteries where none actually exist. This is not to say that computationalism is false, it is merely incomplete and can be easily extended toward a more productive description of brain function without giving up many of its accompanying concepts. Before I suggest how this can be done, I will first briefly look at some aspects of the history of the computational analogy. Where does it come from and how has it come to be the dominant viewpoint in neuroscience, psychology, and philosophy of mind?

I believe that the computer metaphor for the mind earned its popularity by providing convergent answers to several major questions which confronted psychology during the first half of the twentieth century. Below I identify five such questions. Most of these stem directly from the theological foundations of philosophy, and, specifically, from the belief in a distinction between the body and the soul.

The belief in a non-physical soul that is separate from the body predates history, and has been the fundamental assumption underlying the vast majority of human philosophical thought. Today, most psychologists do not believe in a soul, but they work within scientific traditions which grew up in the context of that belief. The influence of this heritage can still be felt in modern psychology, carried within its jargon and even within its academic taxonomy. The seventeenth-century philosophical foundations of psychology held mind–body duality (derived from soul–body duality) as a central theme. This led to three of the questions which I want to discuss.

First, mind–body duality forced an architecture for discussing issues of behaviour. The assumption of a non-physical *Mind* compelled philosophers to conceive of two interfaces between it and the world: *Perception* — which presented the world to the mind; and *Action* — which played out the wishes of the mind upon the world. René Descartes (1596–1650) described these two processes as completely mechanistic, and in animals it was assumed that they bore the sole responsibility for producing even complex behaviour. In Man, however, there existed a non-physical *Mind* which linked these two processes, allowing for free will, rational thought, and consciousness.

When dualism was eventually rejected, the concept of the non-physical *Mind* was replaced with a mechanistic concept of *Cognition*, but the shape of the architecture remained, leading to the *Perception-Cognition-Action* model of behaviour (sometimes called the 'sense-think-act' model). This model established the basic taxonomy within which psychologists classify their work — nearly every question in psychology is immediately labelled either as a perceptual, a cognitive, or a motor control question. For those interested in higher mental functions, the most fundamental question is how Cognition operates, i.e. how it converts perceptions into action plans (Question #1).

Second, mind–body duality led to the mind–body problem and to a series of movements vying to provide its solution. Although the mind was initially seen as separate from the physical world, many insisted that it could still be studied scientifically. Psychology as a science was founded upon this assumption, primarily through the 'introspective' method of Wilhelm Wundt (1832–1920). A series of reactionary movements ensued. First, the introspective method was taken to an extreme by Titchener's *structuralism* — an attempt to discover the elements of consciousness through rigorous introspection by trained human adults. This approach came under a great deal of criticism, the most successful being Watson's *behaviourism* (Watson, 1913) — a rejection of all notions of internal states and an exclusive focus on observable data.[2] This went too far the other way, however, and psychology was left with a desire for a return to concepts of internal states but without resorting to dualism or to a method of pure introspection (Question #2).

An alternative approach to the mind–body problem was the idea that mental states are processes, that they are functional states of the brain. This view is now called *functionalism*, and it has gained great prominence in recent decades. Before functionalism could be truly accepted, however, it had to explain how it can be that mental phenomena are so qualitatively different from the physical brain phenomena which presumably make them happen (Question #3).

Two other issues deserve some discussion. During the nineteenth century, studies of living organisms became separated into those addressing issues on the level of behaviour and those addressing issues of bodily physiology. This was done for very practical reasons of scientific specialization, but it led to the growth of a huge conceptual gap between knowledge obtained within psychology and knowledge obtained through physiological studies of the nervous system. For over a hundred years, there was precious little interaction between the two. In the last few decades, a concerted effort had begun to bring these two disparate fields back together toward a unified science of behaviour, but by then the conceptual gap was very great. It seemed difficult to see how psychological phenomena could be explained with biological elements (Question #4).

Finally, the field of psychology can be said to have suffered for a long time from a kind of 'physics envy'. Many attempts at establishing a science of mind, starting with John Locke in the seventeenth century, strove to discuss mental operation with the same kind of mathematical rigour that was found in physics. There was and is still a widespread attitude that any field aspiring to the status of a 'real science' needs to develop a precise formalism for expressing its concepts. Psychology had always struggled in this regard, and has often been criticized by scientists from other fields. There was a great desire for a formalism, preferably a mathematical one, which could capture psychological phenomena (Question #5).

In the early part of the twentieth century, while psychology faced these five open questions (among others), several new concepts had emerged in seemingly unrelated fields. First, Alan Turing's pioneering work in machine theory resulted in a formal definition of 'computation': According to Turing (1936), all computation is formally equivalent to the manipulation of symbols in a temporary buffer. Second, research aimed at the development of more efficient telephone communication resulted in a formal definition of 'information': According to Shannon and Weaver (1949), the informational content of a signal is inversely related to the probability of that signal arising from randomness. These developments, along with others, launched computer science into what has become one of the most significant technological advancements of modern times.

As computers quickly grew in complexity and in functional sophistication, the potential similarity these machines had to the brain began to be widely recognized.

---

[2] There is a tendency to assume all 'behaviourism' to be so extreme. This is not true — many to whom that label is applied did in fact discuss and study internal phenomena.

Both received information from their external environment, and both acted upon this information in complex ways. Digital computers suddenly joined human brains as the only examples of systems capable of complex reasoning. The analogy between computers and brains was irresistible. Most importantly, once the brain was thought of as analogous to a computer, all of the five questions discussed above suddenly had answers:

First, the computer metaphor provided a candidate mechanism for how Cognition operates — it operates like a digital computer program, by manipulating internal representations according to some set of rules.[3] Second, it allowed discussion of non-dualist internal states — for example, memories and temporary variables. Third, it provided an inspiring metaphor for functionalism — mental entities are like software while physical mechanisms are like hardware. This same metaphor also provided a quick way to bridge the gap between psychology and biology — psychological phenomena are the software running upon the biological hardware. Finally, it provided a mathematical formalism that gave psychology some long-desired rigour — the language of predicate logic and information theory.

All of these answers arrived on the scene within a short time of one another, carried along with the central idea that the brain is like a computer. Because this metaphor provided so many timely solutions, it was very eagerly embraced and tenaciously defended. The computer metaphor for the mind quickly found its way to become the official foundation of modern psychology.

Since digital computers were the only systems capable of complex reasoning whose operation was understood, it was not known which conclusions about their operation should be carried over to theories of the brain and which should not. At first, the analogy was taken quite literally, and notions of symbolic computation and serial processing were seen as inseparable from the concept of functionalism, and necessary ingredients to constructing any human-like intelligence. Meanwhile, the formal definitions of information used in computer science and communications technology found widespread use in psychological theory and practice as well. An example is the work of Paul Fitts, a psychophysicist who quantified the accuracy of human movements in terms of the bandwidth of the channels between perception and action (Fitts, 1954).

In recent decades, computationalism has become the basic axiom of most large-scale brain theories and the language in which students entering brain-related fields are taught to phrase their questions.[4] The task of brain science is often equated to answering the question of how the brain computes. Many debates remain, of course, about such issues as serial vs. parallel processing, analogue vs. digital coding, and symbolic vs. non-symbolic representations, but these are all debated within the accepted metaphor: perception = input, action = output, and cognition = computation.

---

[3] Connectionism is no exception here, albeit its representations are distributed and its rules are encoded in weight matrices and learning laws.

[4] In time, criticisms of the analysis of behaviour in terms of information processing came to be perceived as attacks upon the science of psychology as a whole (Still and Costall, 1991)! I hope that the reader will recognize that the modifications I recommend are not an attempt to disparage psychological science or to bring back behaviourism, but an attempt to move beyond some premature and limiting assumptions of the computer metaphor.

Despite its popularity, certain problems have plagued computationalism from the very beginning. Notable among these are questions of consciousness, emotion, motivation, and meaning. In this article I focus on the question of meaning. This, I will argue, is not a problem for the brain to solve through some dedicated 'meaning assignment' mechanism. Instead, it is only a problem with our description of the brain — a symptom of the shortcomings of computationalism,[5] and an argument for going beyond it.

The question of meaning is a central problem in the philosophy of mind. If the brain is doing computation, defined as a transformation of one representation into another, how does the brain know what these representations mean? To illustrate the problem, an analogy to computers is usually employed. For example, suppose there is a program which takes as input simple queries written in English and responds to them with either 'yes', 'no', or 'don't know', based on a large database of facts. Such programs can be written today, and with sophisticated front-end parsing they may give the illusion of understanding the questions. But the programmer, and anyone who knows how the machine works, will insist that the machine does not *understand* the queries — it only searches for keywords, analyses tenses, applies prepared rules of response, etc. Let's suppose that in the future similar systems may be built, with larger databases and better parsers, that can answer more complex questions. Will they be able to understand? It seems not. It seems we'll just have more and more heuristics and programming tricks, but no actual understanding.

This apparent inability of computers to grasp the semantics of the symbols they manipulate has been discussed under a number of labels. The most famous example is Searle's Chinese Room Argument (Searle, 1980); but it is also known as 'intrinsic meaning' and 'the problem of intentionality' (Dennett, 1978). It underlies the so-called 'frame problem' in artificial intelligence (McCarthy and Hayes, 1969). Stevan Harnad (1990) calls it 'the symbol grounding problem'. That is the label I will use here because I find his presentation of the problem to be the most clear. The question, as posed by Harnad, goes as follows: *'How can the meanings of the meaningless symbol tokens, manipulated solely on the basis of their (arbitrary) shapes, be grounded in anything but other meaningless symbols?'* (Harnad, 1990, p. 335)

Harnad suggests that the issue lies in symbolic vs. non-symbolic representations. He suggests that a symbol system, defined as a system which manipulates arbitrary tokens according to a set of explicit rules, is purely syntactic and thus cannot capture meaning. According to Harnad, the problem lies in the arbitrary nature of the assignment between the symbolic tokens and the objects or states of affairs that these tokens stand for. To ground the symbols, he proposes a hybrid system where the representations of the symbolic system are linked to two kinds of non-symbolic representations: *icons* which are analogs of the sensation patterns, and *categorical representations* which capture the invariant features of these icons. The fundamental symbols are

---

[5] The riddle of meaning is at least in part a symptom of a particularly inappropriate definition of 'information' used by most psychologists and philosophers — the definition given by Shannon and Weaver (1949). It is an important irony that Shannon and Weaver were working on improving the transmission of information on telephone lines, and were openly not concerned with the semantic content of these transmissions. However, because no other definitions were widely recognized at the time, theirs became almost universally accepted as *the* definition of information. See Mingers (1996) for a review of a number of alternate definitions which attempt to bring semantics back into a theory of information.

arbitrary tokens assigned to the non-arbitrary patterns of the icons and categories, and higher-level symbols are composites of these. For example, the word 'horse' is linked to all the images and all the categorical representations involved in the perception of horses, thereby being grounded.

In summary, Harnad is proposing that we solve the symbol grounding problem by backing up out of the premature analogy, made during the beginnings of Artificial Intelligence, that all thought is like symbolic logic. Though I believe that this is moving in the right direction, I suggest that we need to back out further. We need to step back all the way out of the computer metaphor and to consider whether there is a better alternative description of what it is that the brain does.

In the following section, I outline an alternative metaphor for describing the function of the brain. Those who believe that 'information-processing' already captures this function adequately might question the utility of searching for an alternative. I ask these readers to bear with me.

Behaviour as Control

In an attempt to develop a new metaphor, we must first break free from the preconceptions that our current one forces us into. This is not easy, but it may be possible if we step back from modern philosophical debates for a moment and consider issues which at first might appear unrelated. This lets us develop a discussion that is not filtered through the lens of the current metaphor. Later, we can return toward the problems of interest and examine them from a novel perspective. Once this has been done, the reader may decide which viewpoint offers a more parsimonious account of the phenomena that both try to explain.

We begin with a fundamental premise: *The brain evolved.* This is accepted as fact by almost everyone, but its implications for philosophy are rarely acknowledged. The evolution of a biological system such as the brain is not merely a source of riddles for biologists to ponder. It is also a rich source of constraints for anyone theorizing about how the brain functions and about what it does. It is a source of insight that is too often overlooked.[6]

An evolutionarily sound theory of brain function is not merely one which explains the selective advantage offered by some proposed brain mechanism. Lots of mechanisms may be advantageous. What is more useful toward the development and evaluation of brain theories is a plausible story of how a given mechanism may have evolved through a sequence of functional elaborations. The consideration of such a sequence offers powerful guidance toward the formulation of a theory, because the phylogenetic heritage of a species greatly constrains the kinds of mechanisms that

---

[6] In fact, it is often argued that an account of evolution should be secondary: 'any reasonable way to go about finding out how a mechanism evolved would be first to find out how the mechanism works, and then worry about how it evolved' (Crick, 1994). I respectfully disagree. Mechanisms generated by evolution are products of a long sequence of modifications and elaborations, all of these performed within living organisms. Because the modified organisms have to continue living, evolution does not have full freedom to redesign their internal mechanisms. Consequently, the modern form of these mechanisms is strongly constrained by their ancestral forms. Our theories should be similarly constrained. Therefore, one should build an understanding of the fundamental architecture of the brain upon an understanding of the kinds of behaviours and mechanisms present at the time at which that fundamental architecture was being laid down.

may have evolved. Therefore, we should expect to gain insight into the abilities of modern brains by considering the requirements faced by primitive brains, and the sequence of evolutionary changes by which these primitive brains evolved into modern brains.

A contemplation of the most fundamental functional structure of behaviour can start all the way back at the humble beginnings of life. The earliest entities deserving of the term 'living' were self-sustaining chemical systems called 'autocatalytic sets'. There are various theories of how these systems came into existence (Eigen and Schuster, 1979; Kauffman, 1993), developed the genetic code (Bedian, 1982; Crick *et al.*, 1976), and enfolded themselves with membranes (Fox, 1965). Much of the story of how organic molecules organized into cells is not understood. However, although they differ on many important issues, all theories of early life agree that living systems took an active part in ensuring that the conditions required for their proper operation were met. This means that any changes in critical variables such as nutrient concentration, temperature, pH, etc. have to be corrected and brought back within an acceptable range. This is a fundamental task for any living system if it is to remain living. Mechanisms which keep variables within a certain range are usually called 'homeostatic' mechanisms.

Biochemical homeostasis often works through chemical reactions where the compounds whose concentration is to be controlled affect their own rates of production and/or breakdown. For example, compound A might be a catalyst (a chemical activator) for a reaction BA which leads to the breakdown of A. If another reaction PA produces A at a constant rate, then the two reactions operating together will cause the concentration of A to equilibrate at some constant level. Any fluctuations in the concentration of A will cause the relative rates of PA and BA to change so that the concentration of A is brought back to this equilibrium. Because such a 'negative feedback' mechanism exploits reliable properties of chemistry, it is likely to be discovered by the blind processes of variation and selection.

Next, consider a slightly different scenario. Compound B cannot be produced by the organism but must be absorbed from the environment. Suppose that compound B inhibits some cascade of chemical reactions WF which causes the waving of a flagellum (a hair-like appendage used for locomotion). If the concentration of B drops below some threshold level, the locomotion mechanism is released into action and the organism begins to swim randomly. Under the assumption that compound B is non-uniformly distributed in the environment, this motion is likely to improve the situation by bringing the creature to a site of higher concentration of compound B. Once such a site is reached, enough of compound B is absorbed to again inhibit reaction WF and motion ceases. This simple mechanism acts to maintain the concentration of B within some acceptable range, just as the purely chemical mechanism for controlling the concentration of A did above.[7]

---

[7] A functionally analogous mechanism is employed by modern woodlice. The mechanism operates under a simple rule — move more slowly when humidity increases — resulting in a concentration of woodlice in damp regions where they won't dry out. The bacteria *Escherichia coli* use a mechanism only slightly more sophisticated to find food. Their locomotion system increases turning rates when the nutrient concentration decreases, and thus they tend to move up the nutrient gradient. This mechanism, called *klinokinesis*, exploits the reliable fact that in the world of *E. coli*, food sources are usually surrounded by a chemical gradient with a local peak (Koshland, 1980).

The second kind of homeostasis should not be any more surprising than the first. If evolution can exploit reliable properties of biochemistry, then it should also be able to exploit reliable properties of geometry and statistics. That the second kind of homeostasis involves a mechanism which effectively extends its action past the membrane, moving the organism through the environment, does not make it do something other than homeostasis. Both kinds of mechanisms ultimately serve similar functions — they maintain the conditions necessary for life to continue. They may be described as *control mechanisms*.[8]

As evolution produced increasingly more complex organisms, the mechanisms of control developed more sophisticated and more convoluted solutions to their respective tasks. Mechanisms controlling internal variables such as body temperature or osmolarity evolved by exploiting the consistent properties of chemistry, physics, fluid dynamics, etc. Today we call these 'physiology'. Mechanisms whose control extends out through the environment had to exploit consistent properties of that environment. These properties include statistics of nutrient distributions, Euclidean geometry, Newtonian mechanics, etc. Today we call such mechanisms 'behaviour'. In both cases, the functional architecture takes the form of a negative feedback loop, central to which is the measurement of some vital variable. Fluctuations in the measured value of this variable outside of some 'desired range' initiate mechanisms whose purpose is to bring the variable back into the desired range. These mechanisms may be direct, as in the chemical homeostasis example, or involve indirect causes and effects as in the example of klinokinesis (see footnote 7).

The alternative 'control metaphor' being developed here may now be stated explicitly: *the function of the brain is to exert control over the organism's state within its environment.*

This is not a novel proposal. Over a hundred years ago, John Dewey made essentially the same point I am making now. Dewey argued that the concept of stimulus-response is insufficient as a unifying principle in psychology because it only mentions part of the behavioural picture:

> What we have is a circuit, not an arc or broken segment of a circle. This circuit is more truly termed organic than reflex, because the motor response determines the stimulus, just as truly as sensory stimulus determines movement. (Dewey, 1896, p. 363)

The concept of stimulus-response, or reflex arc, focuses attention only on the events leading from the detection of stimulation to the execution of an action, and leads one to ignore the results of that action which necessarily cause new patterns of stimulation. 'The reflex arc theory… gives us one disjointed part of a process as if it were the whole' (Dewey, 1896, p. 370). Surely, nobody has denied Dewey's observation that actions and perceptions mutually affect each other. And yet, the history of brain sciences in the twentieth century suggests that this observation has largely been ignored.

---

[8] The term 'homeostasis' implies some constant goal-state, but we need not be too devoted to that implication. Much of the activity of living creatures is anything but constant. For that reason, I stay away from the term 'homeostatic mechanism' and prefer the more general term 'control mechanism', implying only that control over a variable is maintained to keep it within a desirable range. How that range changes may be determined by various factors, including other, higher-level control mechanisms (Powers, 1973). Furthermore, it should not be assumed that a mechanism which exerts control over some state necessarily involves an explicit representation of the goal state (consider the examples described in footnote 7).

But it wasn't ignored completely. The notion of behaviour as control was fundamental in the early work on cybernetics (Rosenblueth *et al.*, 1943), inspired in part by the theory and practice of engineering feedback systems like Watt's classic centrifugal steam governor. In fact, the word 'cybernetics' was originally intended to imply a control system (Wiener, 1958), though that aspect of its meaning seems to have been neglected in the last few decades. Ashby's (1965) 'Design for a Brain' elaborates these ideas into a theory of control systems capable of maintaining homeostasis and adapting their 'sensorimotor' architecture through rudimentary learning.

The feedback nature of behaviour has often been discussed within psychology as well. The relationship between physiology and behaviour described above had been eloquently discussed by Jean Piaget (1967), whose seminal work on sensorimotor development was highly influenced by cybernetics. William Powers' (1973) book 'Behaviour: The Control of Perception' outlines a model of a behavioural control hierarchy spanning everything from simple reflexes to social interactions — this work has become the foundation of an entire psychological movement called Perceptual Control Theory (Bourbon, 1995). There is also, of course, James Gibson's 'ecological psychology' (Gibson, 1979), about which more will be said below.

Many other theoretical schools of thought share the foundation of the control metaphor, from the theory of 'autopoiesis' (Maturana and Varela, 1980), to the branch of AI research often termed 'situated robotics' (Brooks, 1991; Mataric, 1992; Harvey *et al.*, 1993). It is beginning to be rediscovered in philosophy as well (Adams and Mele, 1989; Van Gelder, 1995). To my knowledge, Hendriks-Jansen (1996) provides the best synthesis of related ideas from numerous disciplines.

While in some fields the control metaphor is often perceived as something novel and revolutionary, in others it has been a founding concept for decades. Feedback control has long been used to describe the physiological operation of the body (Cannon, 1932; Schmidt-Nielson, 1990), including the function of the autonomic nervous system (Dodd and Role, 1991). Only the central nervous system has been considered different, described within the concept of stimulus-response, and only by psychology, neuroscience, AI, and philosophy. In ethology, the study of *animal* behaviour, the feedback control nature of behaviour has been a foundation for years (Hinde, 1996; McFarland, 1971; Manning and Dawkins, 1992).

Why then has the control system metaphor been so neglected within mainstream psychology? Several possible reasons come to mind: (1) The experimental methodology in psychology deliberately prevents the response from affecting the stimulus in an effort to quantify the stimulus-response function. This is appropriate for the development of controlled experiments, but can be detrimental when it spills over into the interpretation of those experiments. (2) There is excess homage paid to the skin, and the structural organization of behaviour (from receptors to effectors) is mistaken to be its functional organization (from input to output). (3) The behaviour of modern humans is so sophisticated that most actions that we tend to contemplate are performed for very long-range goals, where the ultimate control structure is more difficult to appreciate. Thus, because many traditions of brain science began by looking at human behaviour, they were not likely to see the control structure therein. (4) Systems with linear cause and effect are much more familiar and easier to grasp than the dynamical interactions present in systems with a closed loop structure. (5) Interdisciplinary boundaries have split the behavioural loop across several distinct sets of scientific literature, making it difficult to study by any single person. The study of advanced behaviour has become allocated among numerous scientific disciplines, none of which is given the mandate of putting it all together. Even philosophy has only recently begun to look into brain science and biology for insight into mental function. (6) Finally, the various attempts to reintroduce the control metaphor into brain theory must themselves take some of the blame. In an attempt to establish themselves as distinct entities, many of the movements listed above described themselves as revolutionary viewpoints that redefine the very foundations of scientific psychology (for example, Gibson, 1979) or Artificial Intelligence research (for example, Brooks, 1991). Much criticism was levelled at mainstream theories, resulting in impassioned defences. And during such defences, the new movements were portrayed as already familiar and discredited viewpoints (usually as variants of the most extreme form of behaviourism; cf. Ullman, 1980), and thus quickly rejected. But it is not true that these movements redefine psychology — they merely present novel perspectives on existing data, data which continues to be relevant to the study of behaviour.

This essay suggests that the control metaphor is a better way of describing brain function than the computer metaphor. One advantage, of particular interest to philosophy of mind, is that it provides a simple answer to the question of meaning. Briefly, rather than viewing behaviour as 'producing the *right response* given a stimulus', we should view it as 'producing the response that results in the *right stimulus*'. These statements seem pretty similar at first, but there is a crucial difference. While the first viewpoint has a difficult time deciding what is 'right', the second does not.

Animals have physiological demands which inherently distinguish some input (in the sense of 'what the animal perceives as its current situation') as 'desirable', and other input as 'undesirable'. A full stomach is preferred over an empty one; a state of safety is preferred over the presence of an attacking predator. This distinction gives *motivation* to animal behaviour — actions are performed in order to approach desirable input and avoid undesirable input. It also gives *meaning* to their perceptions — some perceptions are cues describing favourable situations, others are warnings describing unfavourable ones which must be avoided. The search for desirable input imposes functional design requirements on nervous systems that are quite different from the functional design requirements for input-output devices such as computers. In this sense, computers make poor metaphors for brains. For computers *there is no notion of desirable input within the computing system*, and hence there is the riddle of meaning, a.k.a. the symbol grounding problem.

Re-examining the Problem of Meaning

Most philosophical inquiries into meaning begin by contemplating the meaning of words and symbols. This is undoubtedly due to the influence that computer analogies and 'language of thought' theories (Fodor, 1975) have historically had over the field. As discussed above, a few decades ago the paradigm of symbolic logic had been perceived as the only mechanistic explanation of complex behaviour available as an alternative to the emptiness of dualism and the ineffectiveness of behaviourism. It thus defined the default premises for modern philosophy of mind. With that foundation, modern philosophical thought revolves around questions of how symbols acquire their meaning.[9] To repeat: '*How can the meanings of the meaningless symbol tokens, manipulated solely on the basis of their (arbitrary) shapes, be grounded in anything but other meaningless symbols?*'.

It is usually assumed that if we can understand the meaning of symbolic tokens, then the meaning of non-symbolic representations (like sensorimotor schemata) will follow trivially. It is also usually assumed that the meaning of perceptual representations can be understood in isolation. For example, Harnad (1990) states that 'motor skills will not be explicitly considered here. It is assumed that the relevant features of the sensory story . . . will generalize to the motor story' (Harnad, 1990, footnote 12). In an attempt to limit the scope of his analysis, Harnad isolates himself from considering the behavioural control loop which, I argue, is where the answer to his problem lies.

Such conceptual isolation is a large part of the reason why meaning appears mysterious. As discussed above, traditional disciplinary boundaries and the need for specialization in science have divided the large problem of behaviour into smaller sub-problems such as Perception, Cognition, and Action. These disciplinary boundaries then spill over into large-scale brain theories, yielding a model of the brain with distinct modules separated by putative internal representations. The Perceptual module is separated from the Cognitive module by an internal unified representation of the external world, and the Cognitive module is separated from the Action module by a representation of the motor plan. These boundaries help to limit the volume of literature that scientists are confronted with, but at the same time they isolate them from potential insights. Those who study the mechanisms of perception and movement control seldom contemplate 'philosophical' issues like meaning. Meaning is left for those who study cognition. And those who study cognition start with a perceptual representation and usually phrase the question of meaning in terms of how meaningful symbolic labels can be attached to that representation, and how symbols can be *about* the things they refer to.

The symbol grounding problem has things backwards. Meaning comes long before symbols in both phylogeny (evolutionary history of a species) and ontogeny (developmental history of an individual). Animals interacted with their environment in meaningful ways millions of years before they started using symbols. Children learn to interact with their world well before they begin to label their perceptions. The invention of symbols, in both phylogeny and ontogeny, is merely an elaboration of existing mechanisms for behavioural control.

Again, let's step back from the philosophical debate surrounding meaning and consider issues that are more fundamental from a biological perspective.

In order to survive, organisms have to take an active part in controlling their situation and keeping it within desirable states. For an organism to exert control over its environment, there must exist predictable relationships between an action and the resulting stimulation ('motor-sensory' relationships). As discussed above, both physiology and behaviour can function adaptively only if there exist reliable properties in the organism's niche which may be exploited toward its survival. Biochemical

---

[9] One might suggest that we can avoid the symbol grounding problem by abandoning symbols in favour of connectionist representations. However, the problem exists for representations in general, be they symbolic or non-symbolic, as long as their role in controlling behaviour is neglected.

control exploits reliable properties of chemistry, diffusion, fluid dynamics, etc., while behavioural control exploits reliable properties of statistics, geometry, rules of optics, etc. With a simple behavioural control mechanism such as klinokinesis, it is easy to see how the system makes use of the statistics of nutrient distribution. With more complex behaviours, the properties of the niche which are exploited by the organism (i.e., brought within its control loop) are more subtle.

Gibson (1979) referred to these reliable properties of the niche as 'affordances'. Affordances are opportunities for action for a particular organism. For example, a mouse-sized hole affords shelter for a mouse, and it does so whether or not a mouse is perceiving it. Thus, an affordance is objective in the sense that it is fully specified by externally observable physical reality, but subjective in the sense of being dependent on the behaviour of a particular kind of organism. For a mouse, the hole affords shelter, while for a cat, it instead affords a potential source of meat. Gibson (1979) suggested that perception of the world is based upon perception of affordances, of recognizing the features of the environment which specify behaviourally relevant interaction. An animal's ecological niche is defined by what its habitat affords.

These relationships between animals and their habitats may be considered precursors to meaning. They are properties of the environment which make adaptive control possible and which guide that control. They make it possible for an organism to establish a behavioural control loop which can be used to approach favourable situations and avoid unfavourable ones. Because these properties tend to come packaged along with semi-permanent physical objects, we can speak of the 'meaning' that these objects have to the organism in question. However, the crucial point is that *the 'meaning of an object' is secondary to the meaning of the interactions which the object makes possible*.

For example, consider a child learning to control her arms. At first, she hardly even distinguishes the meaningless image of the hand on her retina from its background. Not knowing how to control the visual input, the infant at first sends random commands to the muscles. With time, her brain progressively comes to correlate certain random commands with certain motions on her retina — discovering the motor–sensory relationship. This correlation is possible only because the biomechanics of the arm and the laws of optics ensure consistency between the motor commands and the visual motion. After some time, the infant can produce desired visual motions of the hand-shaped retinal image by calling up the appropriate motor commands. She can control her hand. The resulting system might be described as involving representations and transformations between them (control theorists may describe these as pseudoinverse control). But these representations do not require meaning to be somehow *assigned* to them; their meaning is their role in the behavioural control of the arm.[10]

Such behavioural control itself establishes a new domain of consistent motor–sensory relationships, which can be used as the building blocks of higher-order behavioural control. For example, once control over the arm has been learned, it can be used to grasp food and thus to control satiation of hunger. The child's use of her arm for such higher-order behavioural control constitutes her understanding of the meaning of the arm. This is a non-symbolic, pragmatic kind of

---

[10] This is reminiscent of Millikan's (1989) suggestion that the function of a representation is defined not by how it is produced, but by how it is used.

understanding, which only much later leads to the formation of such concepts as 'arm' or 'self' (Piaget, 1954).

In more sophisticated examples of sensorimotor learning, a child might discover that touching part of her environment produces interesting noises. With time the child may distinguish the rattle that lies next to her from the rest of the crib, discovering it as the source of the noises because other parts of the world seem to be irrelevant to these noises. She may learn to grasp the rattle independently of grasping other objects, and learn to shake it this way and that to produce desired sounds. She might thus discover that the rattle affords 'noisemaking'. She might even come up with some kind of crude internal symbol for the rattle. Again, must we worry how meaning gets 'assigned' to this symbol? The symbol is shorthand notation for 'thing I can grasp and shake and make noises with'. The symbol is constructed much later than the sensorimotor strategy which grounds it.

In summary, symbols are merely shorthand notation for elements of behavioural control strategies. The symbol 'rattle' is learned by the child in the context of representations which already have meaning for her by virtue of their role in the behavioural control of her perception. There is no grounding problem because the symbol is only constructed after its meaning has already been established by the affordances that come packaged with the object to which it refers.[11]

The symbol grounding problem has been turned upside-down: The question we're left with is not how meaning is assigned to the symbol 'rattle', but how the symbol is assigned to meaningful interactions with a rattle. How does a shorthand representation emerge within a system of sensorimotor control strategies? What regularities in the environment make the establishment of symbols in the organism possible? Perhaps one example is the fact that the affordances of an external object tend to remain attached to it as a set. For what behavioural abilities is such a shorthand representation useful? One likely use is in predicting events; perhaps another is when constructive imitation among social animals evolves into purposeful teaching (Bullock, 1987). And at what phylogenetic stage of behavioural sophistication are these abilities observed? These are the kinds of questions that research on symbolic thought can productively pursue once it moves beyond the riddle of meaning.

One of the major reasons why meaning may appear as a riddle is its presence in communication. When humans talk over a telephone line, somehow there is meaningful information passed from one to the other, even though nothing more than arbitrary electrical impulses are being transmitted between them. This article is another example of a set of arbitrary symbols which somehow convey meaningful information. How can this be?

Let's consider communication from the evolutionary perspective which has been used throughout. In order to survive, an organism must exert control over its situation, and this control can only be achieved if there exist reliable relationships between actions and their results. While laws of chemistry create opportunity for biochemical homeostasis, laws of geometry, optics, etc. create opportunity for behavioural

---

[11] In a sense, what we have is similar to Harnad's (1990) proposal that the arbitrary symbols such as the word 'rattle' are linked to non-arbitrary and non-symbolic representations. However, these non-symbolic representations are not 'iconic' or 'categorical' representations of perceptual entities, but rather the elements of sensorimotor control mechanisms which operate by making use of consistent properties in the environment.

interaction. A behavioural control loop can be constructed wherever consistencies exist in the environment. And of course, such consistencies also exist in the behaviour of other animals.

Animals respond in complex but predictable ways to various stimuli. When a threatening posture is assumed by one crayfish, another will either back off or respond with its own threat posture. This establishes a domain of interaction between the two creatures where each can attempt to control the other into conceding submission. Often, no actual fighting needs to take place before the dominance hierarchy is achieved. It may be said that the dominant crayfish has successfully exerted control over its opponent. The behavioural control loop can thus extend out through other creatures.

The threatening posture of the crayfish may be called a 'signal'. It demonstrates a threat without having to do any actual physical damage. Other kinds of primitive signals in animal behaviour include the eye spots of certain moths, the white tails of fleeing deer, the exposed teeth of aggressive baboons, etc. All of these are used for exerting control over other individuals, whether they be of the same or of different species.

In social animals, such signalling protocols have been developed to great sophistication. These, it seems, deserve the label of 'communication'. While the earliest symbols (such as the threat posture of crayfish) closely resemble the state of affairs (battle) which they signal, over time the symbol forms may diverge into arbitrary variations. For example, the dance which bees use to communicate the direction and distance of a food source is quite different from the movements needed to arrive at the food source (Frisch, 1967). The dance may be said to involve arbitrary symbolic tokens, in Harnad's (1990) sense. But these tokens are grounded because they play the role of mediating the control that the dancing bee has over the other bees.

Thus, like physiology and behaviour, communication is also an extension of control: one which encompasses other creatures in the environment. To describe communication merely as 'transmission of information' is incomplete. While information is indeed transferred in communication, to miss the control purpose of the transmission is a fatal oversight in an attempt to understand the meaning of the communiqué.

The same case can be made for the meaning of words. As the system for communication grows in complexity and acquires syntax, the meaning of its elements derives from the behavioural control goals of the speaker. In humans, this ability has developed most impressively and meanings have been built upon meanings until even abstract concepts can be expressed. Nevertheless, most communication, including this article, is an attempt at *persuasion*.

Conclusions

A car may be described as a device for converting chemical energy into kinetic energy. This description is not false, and could serve as the foundation for a scientific study of cars. Such study could lead to theories of how parts of the car contribute to its role of energy conversion (for example, the drive-shaft and axle system may be viewed as a coordinate transformation of kinetic energy), but it would fail to provide a complete picture of the car's function. In order to understand the purpose of such things as the steering wheel, one has to understand that the function of a car is to transport people. Energy conversion is merely a useful means toward that end.

Likewise, the 'processing' of 'information' is merely a useful means toward the goals of behaviour. The brain is not merely an input–output device; it is a control system which exerts control over the organism's state in the environment. This task is accomplished through the exploitation of regularities within the environment that define reliable rules for interaction. A viewpoint which isolates the behaving organism from its domain of interaction, as done by the computer metaphor, misses the importance that such regularities have for the establishment of effective control strategies, and the meaning they have within the system. Such a view has forced most modern attempts to understand behaviour to place all explanatory burdens upon mechanisms within the skin, ignoring the contribution that the environment can make toward the guidance of behaviour.

One may argue that any control system is a special case of an input–output system: one where the outputs feed back to affect the inputs. This is true. In fact, it means that the control metaphor is *more precise* than the input–output metaphor. To describe brains as input-output devices is like describing cars as energy-conversion systems without adding something that distinguishes them from chainsaws and chloroplasts. Because the control metaphor is a more precise description, it better constrains the task of explaining behaviour. Control problems, being a subset of input–output problems, present a smaller search-space of possibilities. To step outside that subset is to risk spending time on questions which are not of immediate relevance, even if they apply to some other members of the general set. For example, questions aimed at systems with unbounded time and unbounded memory, such as idealized Turing Machines, may be of limited utility for studies of the biological brain.

The historical interdependence in the development of both functionalism and computationalism has resulted in the two concepts often being viewed as inseparable. However, they are not, and several prominent criticisms aimed at functionalism are really criticisms of computationalism. Notably, Searle's Chinese Room Argument (Searle, 1980) is an argument against the possibility of true understanding within a system that performs input-to-output computations, but as discussed above, that's not a good analogy for what brains do. Brains interact with a world, controlling their situation by performing actions that result in desirable input. They do this by discovering and exploiting the reliable rules of output-to-input transformation that are made possible by the external world — *this is what a pragmatic understanding of the world equates to*. Any system which is capable of performing such control over the environment will perforce contain internal representations that have meaning to it. For Searle's room, and for computer systems which merely accept information, there is no notion of desirable input, no utilization of opportunities for interaction with the world, and thus no understanding of the external world or its projection onto receptor surfaces.

In summary, what I am advocating here comes down to correcting two crucial mistakes that psychology has been led to make: (1) severing the behaving organism from its environment; and (2) decomposing behaviour into Perception, Cognition, and Action modules which are then studied in isolation. Both of these are very old ideas, but they have become particularly entrenched in mainstream psychological thought with the development of computationalism.

To reject these two mistakes by viewing the brain as a control system does not, however, invalidate the progress made under computationalism. For example, the idea of lawful transformations of internal patterns of activity is still useful. We can still refer to this as the 'processing' of 'representations' as long as we focus on the pragmatic value these representations have for the task of control and not dwell solely on how they may describe the world. The idea of specialized brain subsystems is also perfectly reasonable, as long as we delimit these subsystems for functional reasons and not because of conceptual traditions. Finally, many existing models of brain systems, developed in the context of the computer metaphor, are equally compatible with a high-level view of the brain as a control system — this is particularly true of connectionist models.

Making the change in perspective from viewing the brain as an input–output device to viewing it as a control system also leads to a number of important conceptual shifts. A major one is a shift from an emphasis on representations to an emphasis on behaviours, from the analysis of serial stages of processing to an analysis of parallel control streams. This lets one avoid some classic problems in philosophy of mind. First, as discussed above, once neural representations are viewed within the context of the behavioural control to which they contribute, their meaning is not a mystery. Second, motivated action is also not a mystery — when an animal's physiological state no longer meets its internal demands (like a growing hunger), action is generated so as to bring it to a more satisfying state. Third, once one no longer assumes the presence of a complete internal representation of the external world, many forms of the 'binding problem' are no longer difficult. When environmental regularities are allowed to take part in behaviour, they can give it coherence without need of explicit internal mechanisms for binding perceptual entities together (Cisek and Turgeon, 1999).

Finally, the shift away from serial representations leads one to reconsider some classic notions concerning consciousness. Much of psychological theory in the last century has been developed in the context of philosophical viewpoints on the mind–body problem. Because dualism and its variations have thrived at least until the 1950s, they had a great deal of influence on the foundations of psychology. This dualist backdrop led many psychologists to assume that the brain, somewhere within it, presents a model of the world for the mind to observe. Dualism called for a central representation, and computationalism provided that in the form of the internal world model upon which cognition presumably applies its computations. Thus, there has existed for a long time a symbiotic relationship between dualism, a philosophical stance, and computationalism, a psychological viewpoint. And although dualism itself has largely been discredited, some if its influences, such as the assumption of a unified internal world model, remain with us still. Deconstructing that assumption by moving beyond computationalism will have profound effects on what we imagine that the neural correlates of consciousness might be. Perhaps the shift will help us develop a more functional concept of consciousness than the currently prevalent dualistic one, freeing us from the persistence of the so-called 'hard problem'.

The last three decades of brain science have witnessed a progressive backing away from several premature assumptions based on the computer analogy. It was quickly obvious that serial searches among a combinatorial set of possibilities cannot be the way that a human brain reasons, even if today such a process can be made fast enough to beat a chess grand-master. Neuroscience research has made it clear that the brain operates with large numbers of noisy elements working in parallel rather than with a single powerful CPU. Human memory appears to be stored in a distributed manner rather than in the sequential addresses of computer memory. The re-emergence of connectionism has questioned the notion that symbolic logic is the only form of computation to be considered. Artificial life research has demonstrated that robots can be built without accurate sensors and explicit internal representations (Brooks, 1991; Mataric, 1992), and even that such robots can be developed through simulated evolution (Beer and Gallagher, 1992). All these developments demonstrate weaknesses of the ageing analogy that brains are like computers, and motivate us to take a few steps back away from some of the assumptions it originally generated.

This article argues that one more step needs to be taken. We need to step back from the input–output metaphor of computationalism and ask what *kind* of information processing the brain does, and *what is its purpose*? The answer, suggested numerous times throughout the last hundred years, is that the brain is exerting control over its environment. It does so by constructing behavioural control circuits which functionally extend outside of the body, making use of consistent properties of the environment including the behaviour of other organisms. These circuits and the control they allow are the very reason for having a brain. To understand them, we must move beyond the input–output processing emphasized by computationalism and recognize the closed control-loop structure that is the foundation of behaviour.

Acknowledgements
I am grateful to Peter Cariani for his valuable advice on this essay. Supported by a fellowship from the National Institutes of Health (F32 NS10354-02).

References

Adams, F. and Mele, A. (1989), 'The role of intention in intentional action', *Canadian Journal of Philosophy*, **19**, pp. 511–31.

Ashby, W.R. (1965), *Design For a Brain: The Origin of Adaptive Behaviour* (London: Chapman and Hall).

Bedian, V. (1982), 'The possible role of assignment catalysis in the origin of the genetic code', *Origins of Life*, **12**, p. 181.

Beer, R.D. and Gallagher, J.C. (1992), 'Evolving dynamical neural networks for adaptive behavior', *Adaptive Behavior*, **1**, pp. 91–122.

Bourbon, W.T. (1995), 'Perceptual control theory', in *Comparative Approaches to Cognitive Science*, ed. H.L. Roitblat and J.A. Meyer (Cambridge: MIT Press).

Brooks, R. (1991), 'Intelligence without representation', *Artificial Intelligence*, **47**, pp. 139–59.

Bullock, D. (1987), 'Socializing the theory of intellectual development', in *Meaning and the Growth of Understanding: Wittgenstein's Significance for Developmental Psychology*, ed. M. Chapman and R.A. Dixon (New York: Springer-Verlag).

Cannon, W.B. (1932), *The Wisdom of the Body* (New York: Norton).

Cisek, P. And Turgeon, M. (1999) '"Binding through the fovea", a tale of perception in the service of action', *Psyche*, **5**. http://psyche.cs.monsh.edu.au/v5/psyche-5-34-cisek.html

Crick, F.H.C. (1994), 'Interview with Jane Clark', *Journal of Consciousness Studies*, **1** (1), pp. 10–17.

Crick, F.H.C., Brenner, S., Klug, A. and Pieczenik, G. (1976), 'A speculation on the origin of protein synthesis', *Origins of Life*, **7**, p. 389.

Dennett, D.C. (1978), 'Current issues in the philosophy of mind', *American Philosophical Quarterly*, **15**, pp. 249–61.

Dewey, J. (1896), 'The reflex arc concept in psychology', *Psychological Review*, **3**, pp. 357–70.

Dodd, J. And Role, L.W. (1991), 'The autonomic nervous system', in *Principles of Neural Science*, ed. E.R. Kandel, J.H. Schwartz and T.M. Jessell (New York: Elsevier).

Eigen, M. And Schuster, P. (1979), *The Hypercycle — A Principle of Natural Self-Organization* (Heidelberg: Springer-Verlag).

Fitts, P.M. (1954), 'The information capacity of the human motor system in controlling the amplitude of movement', *Journal of Experimental Psychology*, **47**, pp. 381–91.

Fodor, J.A. (1975), *The Language of Thought* (Cambridge, MA: Harvard University Press).

Fox, S.W. (1965), 'Simulated natural experiments in spontaneous organization of morphological units from protenoid', in *The Origins of Prebiological Systems*, ed. S.W. Fox (New York: Academic Press).

Frisch, K.V. (1967), *The Dance Language and Orientation of Bees* (Cambridge, MA: The Belknap Press of Harvard University Press).

Gibson, J.J. (1979), *The Ecological Approach to Visual Perception* (Boston: Houghton Mifflin).

Harnad, S. (1990), 'The symbol grounding problem', *Physica D*, **42**, pp. 335–46.

Harvey, I., Husbands, P. and Cliff, D.T. (1993), 'Issues in evolutionary robotics', in *Proceedings of the Second Conference on Simulation of Adaptive Behavior*, ed. J.-A. Meyer, H.L. Roitblat and S. Wilson (Cambridge, MA: MIT Press).

Hendriks-Jansen, H. (1996), *Catching Ourselves in the Act: Situated Activity, Interactive Emergence, Evolution, and Human Thought* (Cambridge, MA: MIT Press).

Hinde, R.A. (1966), *Animal Behaviour: A Synthesis of Ethology and Comparative Psychology* (New York: McGraw-Hill Book Company).

James, W. (1890), *The Principles of Psychology* (New York: Holt).

Kauffman, S.A. (1993), *The Origins of Order: Self-Organization and Selection in Evolution* (New York: Oxford University Press).

Koshland, D.E. (1980), *Behavioral Chemotaxis as a Model Behavioral System* (New York: Raven Press).

Manning, A. and Dawkins, M.S. (1992), *An Introduction to Animal Behavior* (Cambridge: Cambridge University Press).

Mataric, M.J. (1992) 'Integration of representation into goal-driven behavior-based robots', *IEEE Transactions on Robotics and Automation*, **8**, pp. 304–12.

Maturana, H.R. and Varela, F.J. (1980), *Autopoiesis and Cognition: The Realization of the Living* (Boston: D. Reidel).

McCarthy, J. and Hayes, P. (1969), 'Some philosophical problems from the standpoint of artificial intelligence', in *Machine Intelligence IV*, ed. B. Meltzer and D. Michie (Edinburgh: Edinburgh University Press).

McFarland, D.J. (1971), *Feedback Mechanisms in Animal Behaviour* (New York: Academic Press).

Millikan, R.G. (1989), 'Biosemantics', *The Journal of Philosophy*, **86**, pp. 281–97.

Mingers, J.C. (1996), 'An evaluation of theories of information with regard to the semantic and pragmatic aspects of information systems', *Systems Practice*, **9**, pp. 187–209.

Piaget, J. (1954) *The Construction of Reality in the Child* (New York: Basic Books).

Piaget, J. (1967), *Biologie et Connaissance: Essai sur les Relations Entre les Régulation Organiques et les Processus Cognitifs* (Paris: Editions Gallimard).

Powers, W.T. (1973), *Behavior: The Control of Perception* (New York: Aldine Publishing Company).

Rosenblueth, A., Wiener, N. and Bigelow, J. (1943), 'Behavior, purpose and teleology', *Philosophy of Science*, **10**, pp. 18–24.

Schmidt-Nielsen, K. (1990), *Animal Physiology: Adaptation and Environment* (Cambridge: Cambridge University Press).

Searle, J. (1980), Minds, brains, and programs', *Behavioral and Brain Sciences*, **3**, pp. 417–57.

Shannon, C. and Weaver, W. (1949), *The Mathematical Theory of Information* (Urbana: University of Illinois Press).

Still, A. and Costall, A. (1991), *Against Cognitivism: Alternative Foundations for Cognitive Psychology* (Hemel Hempstead: Harvester Whitesheaf).

Turing, A.M. (1936) 'On computable numbers, with an application to the Entscheidungsproblem', *Proceedings of the London Mathematical Society*, Series 2, **42**, pp. 230–65.

Ullman, S. (1980), 'Against direct perception', *Behavioral and Brain Sciences*, **3**, pp. 373–415.

Van Gelder, T. (1995), 'What might cognition be, if not computation?', *The Journal of Philosophy*, **91**, pp. 345–81.

Watson, J.B. (1913), 'Psychology as the behaviorist views it', *Psychological Review*, **20**, pp. 158–77.

Wiener, N. (1958), *Cybernetics, or Control and Communication in the Animal and the Machine* (Paris: Hermann).

Walter J. Freeman

Consciousness, Intentionality and Causality

According to behavioural theories deriving from pragmatism, gestalt psychology, existential-ism, and ecopsychology, knowledge about the world is gained by intentional action followed by learning. In terms of the neurodynamics described here, if the intending of an act comes to awareness through reafference, it is perceived as a cause. If the consequences of an act come to awareness through proprioception and exteroception, they are perceived as an effect. A sequence of such states of awareness comprises consciousness, which can grow in complexity to include self-awareness. Intentional acts do not require awareness, whereas voluntary acts require self-awareness. Awareness of the action/perception cycle provides the cognitive meta-phor of linear causality as an agency. Humans apply this metaphor to objects and events in the world to predict and control them, and to assign social responsibility. Thus, linear causality is the bedrock of technology and social contracts.

Complex material systems with distributed non-linear feedback, such as brains and their neural and behavioural activities, cannot be explained by linear causality. They can be said to operate by circular causality without agency. The nature of self-control is described by break-ing the circle into a forward limb, the intentional self, and a feedback limb, awareness of the self and its actions. The two limbs are realized through hierarchically stratified kinds of neu-ral activity. Intentional acts are produced by the self-organized microscopic neural activity of cortical and subcortical components in the brain. Awareness supervenes as a macroscopic ordering state, that defers action until the self-organizing microscopic process has reached closure in reflective prediction. Agency, which is removed from the causal hierarchy by the appeal to circularity, re-appears as a metaphor by which events in the world are anthropomorphized, making them appear subject to human control.

I: Introduction

What is consciousness? It is known through experience of the activities of one's own body and inferred from observation of the bodies of others. In this respect, the ques-tion of whether it arises from the soul (Eccles, 1994), or from panpsychic properties of matter (Whitehead, 1938; Penrose, 1994; Chalmers, 1996), or as a function of brain operations (Searle, 1992; Dennett, 1991; Crick, 1994) is not relevant. The perti-nent questions are — however it arises and is experienced — how and in what senses does it cause the functions of brains and bodies, and how do brain and body functions

Journal of Consciousness Studies, **6**, No. 11–12, 1999, pp. 143–72

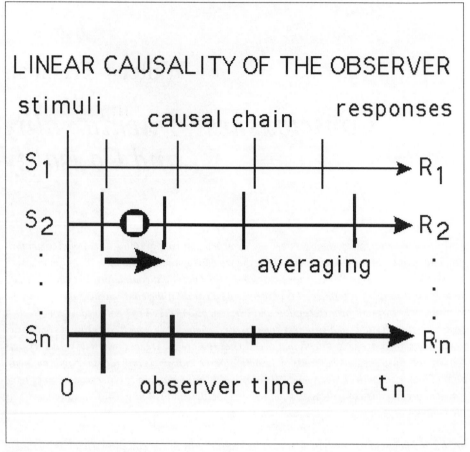

Figure 1
Linear causality is the view of connected events by which causal chains are constructed. The weaknesses lie in requirements to assign points in time to the beginning and the end of each chain, and to intervening events in the chain in strict order, and to repeat pairs of observations in varying circumstances in order to connect pairs of classes of events. As Davidson (1980) remarked, events have causes; classes have relations. In the example, analysis of stimulus-dependent events such as evoked potentials is done by time ensemble averaging, which degrades nonsynchronized events, and which leads to further attempts at segmentation in terms of the successive peaks, thus losing sight of an event extended in time. The notion of 'agency' is implicit in each event in the chain acting to produce an effect, which then becomes the next cause.

cause it? How do actions cause perceptions; how do perceptions cause awareness; and how do states of awareness cause actions? Analysis of causality is a necessary step toward a comprehension of consciousness, because the forms of answers depend on the choice among meanings that are assigned to 'cause': (a) to make, move and modulate (an agency in linear causality); (b) to explain, rationalize and blame (cognition in circular causality without agency but with top-down-bottom-up interaction); or (c) to flow in parallel as a meaningful experience, by-product, or epiphenomenon (noncausal correlation).

The elements of linear causality (a) are shown in Figure 1 in terms of stimulus-response determinism. A stimulus initiates a chain of events including activation of receptors, transmission by serial synapses to cortex, integration with memory,

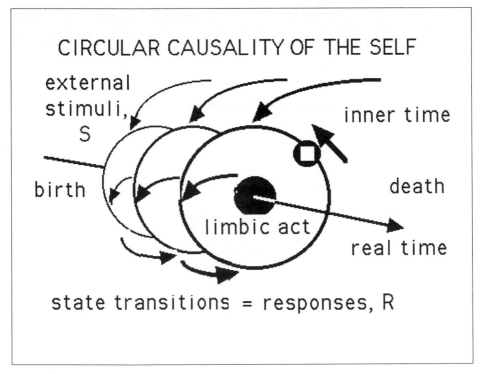

Figure 2
Circular causality expresses the interrelations between levels in a hierarchy: a top-down macro-scopic state simultaneously influences microscopic particles that bottom-up create and sustain the macroscopic state. The state exists over a span of inner time in the system that can be collapsed to a point in external time. Events in real time are marked by changes in the state of the system, which are discrete. This conceptualization is widely used in the social and physical sciences. In an exam-ple used by Haken (1983), the excited atoms in a laser cause coherent light emission, and the light imposes order on the atoms. The laser was also used by Cartwright (1989) to exemplify levels of causality, by which she contrasted simple, direct cause-effect relations not having significant inter-actions or second-order perturbations with higher order 'capacities' (according to her, closely related to Mill's 'tendencies', but differing by 'material abstraction', p. 226), which by virtue of abstraction have an enlarged scope of forward action, but which lack the circular relation between microscopic and macroscopic entities that is essential for explaining lasers — and brains. The notion of an 'agency' does not enter, and multiple scales of time and space are required for the dif-ferent levels

selection of a motor pattern, descending transmission to motor neurons, and activa-tion of muscles. At one or more nodes along the chain, awareness occurs, and mean-ing and emotion are attached to the response. Temporal sequencing is crucial; no effect can precede or occur simultaneously with its cause. At some instant, each effect becomes a cause. The demonstration of causal invariance must be based on repetition of trials. The time line is reinitiated at zero in observer time, and S-R pairs are col-lected. Some form of generalization is used. In the illustration, it is by time ensemble averaging. Events with small variance in time of onset close to stimulus arrival are retained. Later events with varying latencies are lost. The double dot indicates a point in real time; it is artificial in observer time. This conceptualization is inherently lim-ited, because awareness cannot be defined at a point in time.

The elements of circular causality (b) are shown in Figure 2. The double dot shows a point moving counterclockwise on a trajectory idealized as a circle, in order to show that an event exists irresolvably as a state through a period of inner time, which we reduce to a point in real time. Stimuli from the world impinge on this state. So also do stimuli arising from the self-organizing dynamics within the brain. Most stimuli are ineffective, but occasionally one succeeds as a 'hit' on the brain state, and a response occurs. The impact and motor action are followed by a change in brain structure that begins a new orbit. A succession of orbits can be conceived as a cylinder with its axis in real time, extending from birth to death of an individual and its brain. Events are intrinsically not reproducible. Trajectories in inner time may be viewed as fusing past and future into an extended present by state transitions. The circle is a candidate for representing a state of awareness.

Noncausal relations (c) are described by statistical models, differential equations, phase portraits, and so on, in which time may be implicit or reversible. Once the constructions are completed by the calculation of risk factors and degrees of certainty from distributions of observed events and objects, the assignment of causation is optional. In describing brain functions, awareness is treated as irrelevant or epiphenomenal.

These concepts are applied to animal consciousness on the premise that the structures and activities of brains and bodies are comparable over a broad variety of animals including humans. The hypothesis is that the elementary properties of consciousness are manifested in even the simplest of extant vertebrates, and that structural and functional complexity increases with the evolution of brains into higher mammals. The dynamics of simpler brains is described in terms of neural operations that provide goal-oriented behaviour.

In the first half of this essay (Sections 2–6) I describe the neural mechanisms of intention and reafference and learning, as I see them. I compare explanations of neural mechanisms using linear and circular causality at three levels of hierarchical function. In the second half, I describe some applications of this view in the fields of natural sciences. The materials I use to answer the question, what is causality?, come from several disciplines, including heavy reliance on neurobiology and non-linear dynamics. In the words of computer technologists, these two disciplines make up God's own firewall, which keeps neurohackers from burning in to access and crack the brain codes. For reviews on neuroflaming I recommend introductory texts by Bloom and Lazerson (1988) on 'Brain, Mind and Behaviour', and by Abraham *et al.* (1990) on 'Visual Introduction to Dynamical Systems Theory for Psychology'.

II: Level 1, Higher:
The Circular Causality of Intentionality

An elementary process requiring the dynamic interaction between brain, body and world in all animals is an act of observation. This is not a passive receipt of information from the world, as expressed implicitly in Figure 1. It is the culmination of purposive action by which an animal directs its sense organs toward a selected aspect of the world and abstracts, interprets, and learns from the resulting sensory stimuli (Figure 2). The act requires a prior state of readiness that expresses the existence of a goal, a preparation for motor action to position the sense organs, and selective sensitization of the sensory cortices. Their excitability has already been shaped by the past experience that is relevant to the goal and the expectancy of stimuli. A concept that can serve as a principle by which to assemble and interrelate these multiple facets is intentionality. This concept has been used in different contexts, since its synthesis by Aquinas (1272), seven centuries ago. The properties of intentionality as it is developed here are (a) its intent or directedness toward some future state or goal; (b) its unity; and (c) its wholeness (Freeman, 1995).

(a) Intent comprises the endogenous initiation, construction, and direction of behaviour into the world, combined with changing the self by learning in accordance with the perceived consequences of the behaviour. Its origin lies within brains. Humans and other animals select their own goals, plan their own tactics, and choose when to begin, modify, and stop sequences of action. Humans at least can be subjectively aware of themselves acting. This facet is commonly given the meaning of purpose and motivation by psychologists, because, unlike lawyers, they usually do not distinguish between intent and motive. Intent is a forthcoming action, and motive is the reason.

(b) Unity appears in the combining of input from all sensory modalities into gestalts, in the co-ordination of all parts of the body, both musculoskeletal and autonomic, into adaptive, flexible, yet focused movements, and in the full weight of all past experience in the directing of each action. Subjectively, unity may appear in the awareness of self. Unity and intent find expression in modern analytic philosophy as 'aboutness', meaning the way in which beliefs and thoughts symbolized by mental representations refer to objects and events in the world, whether real or imaginary. The distinction between inner image and outer object calls up a dichotomy between subject and object that was not part of the originating Thomist view.

(c) Wholeness is revealed by the orderly changes in the self and its behaviour that constitute the development and maturation of the self through learning, within the constraints of its genes and its material, social and cultural environments. Subjectively, wholeness is revealed in the striving for the fulfilment of the potential of the self through its lifetime of change. Its root meaning is 'tending', the Aristotelian view that biology is destiny. It is also seen in the process of healing of the brain and body from damage and disruption. The concept appears in the description by a fourteenth century surgeon, LaFranchi of Milan, of two forms of healing, by first intention with a clean scar, and by second intention with suppuration. It is implicit in the epitaph of Ambroise Paré, sixteenth century French surgeon: 'Je le pansay; Dieu le guarit' (I bound his wounds; God healed him). Pain is intentional in that it directs behaviour toward facilitation of healing, and that it mediates learning when actions have gone wrong with deleterious, unintended consequences. Pain serves to exemplify the differences between volition, desire and intent; it is willed by sadists, desired by masochists, and necessary for everyone.

Intentionality cannot be explained by linear causality, because actions under that concept must be attributed to environmental (Skinner, 1969) and genetic determinants (Herrnstein and Murray, 1994), leaving no opening for self-determination. Acausal theories (Hull, 1943; Grossberg, 1982) describe statistical and mathematical regularities of behaviour without reference to intentionality. Circular causality explains intentionality in terms of 'action-perception cycles' (Merleau-Ponty, 1942) and affordances (Gibson, 1979), in which each perception concomitantly is the

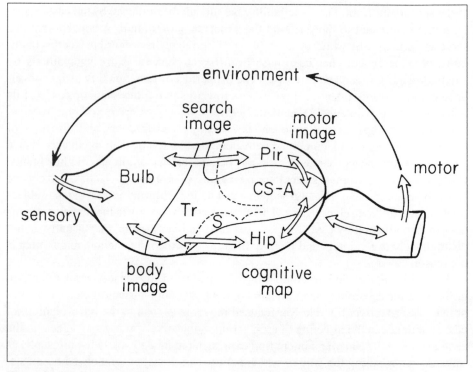

Figure 3
The schematic shows the dorsal view of the right cerebral hemisphere of the salamander (adapted
from Herrick, 1948). The unbroken sheet of superficial neuropil sustains bi-directional interactions
between all of its parts, which are demarcated by their axonal connections with sensory receptors
(olfactory bulb and 'Transitional zone' for all other senses, descending connections to the corpus
striatum, amygdaloid and septum from the pyriform area, and the intrinsic connections between
these areas and the primordial hippocampus posteromedially. This primitive forebrain suffices as
an organ of intentionality, comprising the limbic system with little else besides.

outcome of a preceding action and the condition for a following action. Dewey (1914)
phrased the idea in different words; an organism does not react to a stimulus but acts
into it and incorporates it. That which is perceived already exists in the perceiver,
because it is posited by the action of search and is actualized in the fulfilment of
expectation. The unity of the cycle is reflected in the impossibility of defining a moving instant of 'now' in subjective time, as an object is conceived under linear causality. The Cartesian distinction between subject and object does not appear, because
they are joined by assimilation in a seamless flow.

III: Level 2, Middle:
The Circular Causality of Reafference

Brain scientists have known for over a century that the necessary and sufficient part
of the vertebrate brain to sustain minimal intentional action, a component of intentionality, is the ventral forebrain, including those parts that comprise the external
shell of the phylogenetically oldest part of the forebrain, the paleocortex, and the
underlying nuclei such as the amygdala with which the cortex is interconnected.

These components suffice to support identifiable patterns of intentional behaviour in animals, when all of the newer parts of the forebrain have been surgically removed (Goltz, 1892) or chemically inactivated by spreading depression (Bures et al., 1974). Intentional behaviour is severely altered or lost following major damage to these parts. Phylogenetic evidence comes from observing intentional behaviour in sala- manders, which have the simplest of the existing vertebrate forebrains (Herrick, 1948; Roth, 1987), comprising only the limbic system. Its three cortical areas are sen- sory (which is predominantly the olfactory bulb), motor (the pyriform cortex), and associational (Figure 3). The latter has the primordial hippocampus connected to the septal, amygdaloid and striatal nuclei. It is identified in vertebrates (Freeman 1999) as the locus of the functions of temporal orientation in learning — called 'short term memory', and of spatial orientation — misnamed the 'cognitive map' by O'Keefe and Nadel (1978) following Tolman (1948). These integrative frameworks are essen- tial for intentional action, because even the simplest actions, such as observation in searching for food or evading predators, require an animal to co-ordinate its position in the world with that of its prey or refuge, and to evaluate its progress during plan- ning, attack or escape.

The crucial question for neuroscientists is, how are the patterns of neural activity that sustain intentional behaviour constructed in brains? A route to an answer is pro- vided by studies of the electrical activity of the primary sensory cortices of animals that have been trained to identify and respond to conditioned stimuli. An answer appears in the capacity of the cortices to construct novel patterns of neural activity by virtue of their self-organizing dynamics.

Two experimental approaches to the study of sensory cortical dynamics are in con- trast. One is based in linear causality (Figure 1). An experimenter identifies a neuron in sensory cortex by recording its action potential with a microelectrode, and then determines the sensory stimulus to which that neuron is most sensitive. The pulse train of the neuron is treated as a symbol to 'represent' that stimulus as the 'feature' of an object, for example the colour, contour, or motion of an eye or a nose in a face. The pathway of activation from the sensory receptor through relay nuclei to the primary sensory cortex and then beyond is described as a series of maps, in which successive representations of the stimulus are activated. The firings of the feature detector neu- rons must then be synchronized or 'bound' together to represent an object, such as a moving coloured ball, as it is conceived by the experimenter. This representation is thought to be transmitted to a higher cortex, where it is compared with representa- tions of previous objects that are retrieved from memory storage. An answer to the question of how brains do this, the so-called 'binding problem', is still being sought (Gray, 1994; Hardcastle, 1994; Singer and Gray, 1995).

The other approach is based in circular causality (Figure 2). In this view the experi- menter trains a subject to co-operate through use of positive or negative reinforce- ment, thereby inducing a state of expectancy and search for a stimulus, as it is conceived by the subject. When the expected stimulus arrives, the activated receptors transmit pulses to the sensory cortex, where they elicit the construction by non-linear dynamics of a macroscopic, spatially coherent oscillatory pattern that covers an entire area of sensory cortex (Freeman, 1991, 1992). It is observed by means of the electroencephalogram (EEG) from electrode arrays on all the sensory cortices (Free- man, 1975, 1992, 1995; Barrie et al., 1996; Kay and Freeman, 1998). It is not seen in

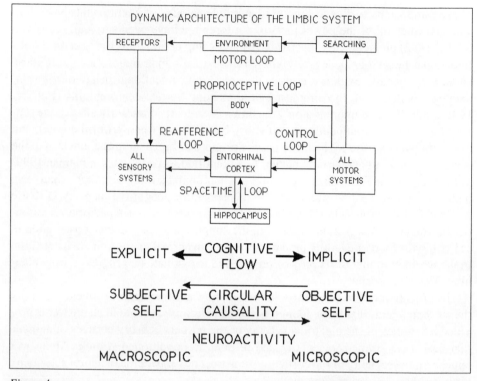

Figure 4

In the view of dynamics, the limbic architecture is formed by multiple loops. The mammalian entorhinal cortex is the target of convergence of all sensory input, the chief source of input for the hippocampus, its principal target for its output, and a source of centrifugal input to all of the primary sensory cortices. The hypothesis is proposed that intentional action is engendered by counterclockwise flow of activity around the loops into the body and the world, comprising implicit cognition, and that awareness and consciousness are engendered by clockwise flow within the brain, comprising explicit cognition. The forward flow (to the right in the brain) commanding action consists of the microscopic activity of brain subsystems, and the back flow (to the left in the brain) regulating the control consists of the macroscopic order parameter that by circular causality regulates and holds or releases the activity of the subsystems.

recordings from single neuronal action potentials, because the fraction of the variance in the single neuronal pulse train that is covariant with the neural mass is far too small, on the order of 0.1 per cent.

The emergent pattern is not a representation of a stimulus, nor a ringing as when a bell is struck, nor a resonance as when one string of a guitar vibrates when another string does so at its natural frequency. It is a state transition that is induced by a stimulus, followed by a construction of a pattern that is shaped by the synaptic modifications among cortical neurons from prior learning. It is also dependent on the brain stem nuclei that bathe the forebrain in neuromodulatory chemicals. It is a dynamic action pattern that creates and carries the meaning of the stimulus for the subject. It reflects the individual history, present context, and expectancy, corresponding to the unity and the wholeness of intentionality. Owing to dependence on history, the patterns created in each cortex are unique to each subject.

The visual, auditory, somesthetic and olfactory cortices serving the distance receptors all converge their constructions through the entorhinal cortex into the limbic system, where they are integrated with each other over time. Clearly, they must have similar dynamics, in order that the messages be combined into *gestalts*. The resultant integrated meaning is transmitted back to the cortices in the processes of selective attending, expectancy, and the prediction of future inputs (Freeman, 1995; Kay and Freeman, 1998).

The same wave forms of EEG activity as those found in the sensory cortices are found in various parts of the limbic system. This similarity indicates that the limbic system also has the capacity to create its own spatiotemporal patterns of neural activity. They are embedded in past experience and convergent multisensory input, because they are self-organized. The limbic system provides interconnected populations of neurons that, according to the hypothesis being proposed, continually generate the patterns of neural activity that form goals and direct behaviour toward them.

EEG evidence shows that the process in the various parts occurs in discontinuous steps (Figure 2), like frames in a motion picture (Freeman, 1995; Barrie, Freeman and Lenhart, 1996). Being intrinsically unstable, the limbic system continually (not continuously) transits across states that emerge, transmit to other parts of the brain, and then dissolve to give place to new ones. Its output controls the brain stem nuclei that serve to regulate its excitability levels, implying that it regulates its own neurohumoral context, enabling it to respond with equal facility to changes, both in the body and the environment, that call for arousal and adaptation or rest and recreation. Again by inference, it is the neurodynamics of the limbic system, with supplementary contributions from other parts of the forebrain such as the frontal lobes and basal ganglia, that initiates the novel and creative behaviour seen in search by trial and error.

The limbic activity patterns of directed arousal and search are sent into the motor systems of the brain stem and spinal cord (Figure 4). Simultaneously, patterns are transmitted to the primary sensory cortices, preparing them for the consequences of motor actions. This process has been called 'reafference' (von Holst and Mittelstädt 1950; Freeman 1995), 'corollary discharge' (Sperry, 1950), 'focused arousal', and 'preafference' (Kay and Freeman, 1998). It predicts and compensates for the self-induced changes in sensory input that accompany the actions organized by the limbic system, and it sensitizes sensory systems to anticipated stimuli prior to their expected times of arrival.

The concept of preafference began with an observation by Helmholtz (1872) on patients with paralysis of lateral gaze, who, on trying and being unable to move an eye, reported that the visual field appeared to move in the opposite direction. He concluded that 'an impulse of the will' that accompanied voluntary behaviour was unmasked by the paralysis. He wrote: 'These phenomena place it beyond doubt that we judge the direction of the visual axis only by the volitional act by means of which we seek to alter the position of the eyes'. J. Hughlings Jackson (1931) repeated the observation, but postulated alternatively that the phenomenon was caused by 'an ingoing current', which was a signal from the non-paralysed eye that moved too far in the attempt to fixate an object, and which was not a recursive signal from a 'motor centre'. He was joined in this interpretation by William James (1893) and Edward Titchener (1907), thus delaying deployment of the concepts of neural feedback in reentrant cognitive processes until late in the twentieth century.

The sensory cortical constructions consist of brief staccato messages to the limbic system, which convey what is sought and the result of the search. After multisensory convergence, the spatiotemporal activity pattern in the limbic system is updated through temporal integration in the hippocampus. Accompanying sensory messages there are return updates from the limbic system to the sensory cortices, whereby each cortex receives input that has been integrated with the input from all others, reflecting the unity of intentionality. Everything that a human or an animal knows comes from the circular causality of action, preafference, perception, and update. It is done by successive frames of self-organized activity patterns in the sensory and limbic cortices.

IV: Level 3, Lower:
Circular Causality of Neurons and Neural Masses

The 'state' of the brain is a description of what it is doing in some specified time period. A state transition occurs when the brain changes and does something else. For example, locomotion is a state, within which walking is a rhythmic pattern of activity that involves large parts of the brain, spinal cord, muscles and bones. The entire neuromuscular system changes almost instantly with the transition to a pattern of jogging or running. Similarly, a sleeping state can be taken as a whole, or divided into a sequence of slow wave and REM stages. Transit to a waking state can occur in a fraction of a second, whereby the entire brain and body shift gears, so to speak. The state of a neuron can be described as active and firing or as silent, with sudden changes in patterns of firing constituting state transitions. Populations of neurons also have a range of states, such as slow wave, fast activity, seizure, or silence. The science of dynamics describes states and their state transitions.

The most critical question to ask about a state is its degree of stability or resistance to change. Stability is evaluated by perturbing an object or a system (Freeman, 1975). For example, an egg on a flat surface is unstable, but a coffee mug is stable. A person standing on a moving bus and holding on to a railing is stable, but someone walking in the aisle is not. If a person regains his chosen posture after each perturbation, no matter in which direction the displacement occurred, that state is regarded as stable, and it is said to be governed by an attractor. This is a metaphor to say that the system goes ('is attracted to') the state along a transient trajectory. The range of displacement from which recovery can occur defines the basin of attraction, in analogy to a ball rolling to the bottom of a bowl. If a perturbation is so strong that it causes concussion or a broken leg, and the person cannot stand up again, then the system has been placed outside the basin of attraction, and a new state supervenes with its own attractor and basin of attraction.

Stability is always relative to the time duration of observation and the criteria for what is chosen to be observed. In the perspective of a lifetime, brains appear to be highly stable, in their numbers of neurons, their architectures and major patterns of connection, and in the patterns of behaviour they produce, including the character and identity of the individual that can be recognized and followed for many years. A brain undergoes repeated transitions from waking to sleeping and back again, coming up refreshed with a good night or irritable from insomnia, but still giving arguably the same person as the night before. But in the perspective of the short term, brains are highly unstable. Thoughts go fleeting through awareness, and the face and body twitch with the passing of emotions. Glimpses of the internal states of neural activity reveal patterns that are more like hurricanes than the orderly march of symbols in a computer, with the difference that hurricanes don't learn. Brain states, and the states of populations of neurons that interact to give brain function, are highly irregular in spatial form and time course. They emerge, persist for a small fraction of a second, then disappear and are replaced by other states.

Neuroscientists aim to describe and measure these states and tell what they mean both to observers of behaviour and to those who experience awareness of them. We approach the dynamics by defining three kinds of stable state, each with its type of attractor. The simplest is the point attractor. The system is at rest unless perturbed, and it returns to rest when allowed to do so. As it relaxes to rest, it has a brief history, but loses that history on converging to rest. Examples of point attractors are neurons or neural populations that have been isolated from the brain, and also the brain that is depressed into inactivity by injury or a strong anaesthetic, to the point where the EEG has gone flat. A special case of a point attractor is noise. This state is observed in populations of neurons in the brain of a subject at rest, with no evidence of overt behaviour or awareness. The neurons fire continually but not in concert with each other. Their pulses occur in long trains at irregular times. Knowledge about the prior pulse trains from each neuron and those of its neighbours up to the present fails to support the prediction of when the next pulse will occur. The state of noise has continual activity with no history of how it started, and it gives only the expectation that its average amplitude and other statistical properties will persist unchanged.

A system that gives periodic behaviour is said to have a limit cycle attractor. The classic example is the clock. When it is viewed in terms of its ceaseless motion, it is regarded as unstable until it winds down, runs out of power, and goes to a point attractor. If it resumes its regular beat after it is re-set or otherwise perturbed, it is stable as long as its power lasts. Its history is limited to one cycle, after which there is no retention of its transient approach in its basin to its attractor. Neurons and populations rarely fire periodically, and when they appear to do so, close inspection shows that the activities are in fact irregular and unpredictable in detail, and when periodic activity does occur, it is either intentional, as in rhythmic drumming, or pathological, as in nystagmus and Parkinsonian tremor.

The third type of attractor gives aperiodic oscillation of the kind that is observed in recordings of EEGs and of physiological tremors. There is no one or small number of frequencies at which the system oscillates. The system behaviour is, therefore, unpredictable, because performance can only be projected into the future for periodic behaviour. This type of attractor was first called 'strange'; it is now widely known as 'chaotic'. The existence of this type of oscillation was suspected by mathematicians a century ago, but confirmation and systematic study was possible only recently after the full development of digital computers. The best known simple systems with chaotic attractors are said to manifest 'deterministic chaos'. They are stationary, autonomous, noise-free, have a small number of components, and have few and fractal degrees of freedom, as, for example, the double-hinged pendulum, and the dripping tap. Large and complex systems such as neurons and neural populations are thought to be capable of stochastic chaotic behaviour, but formal proof is not yet possible at the present level of developments in mathematics, because the systems like brains are interactive with their surrounds and therefore are neither stationary nor autonomous, and they are drenched in noise and infinite in dimension.

The discovery of chaos has profound implications for the study of brain function (Skarda and Freeman, 1987). A dynamic system has a collection of attractors, each with its basin, which forms an 'attractor landscape' with all three types. The state of the system can jump from one to another in an itinerant trajectory (Tsuda, 1991). Capture by a point or limit cycle attractor wipes clean the history upon asymptotic convergence, but capture in a chaotic basin engenders continual aperiodic activity, thereby creating novel, unpredictable patterns that can retain the system's history.

Although the system trajectory is not predictable, the statistical properties such as the mean and standard deviation of the state variables of the system serve as measures of its steady state. Chaotic fluctuations carry the system endlessly around in the basin. However, if energy is fed into the system so that the fluctuations increase in amplitude, or if the landscape of the system is changed so that the basin shrinks or flattens, a microscopic fluctuation can carry the trajectory across the boundary between basins to another attractor. This crossing constitutes a first order state transition.

In each sensory cortex there are multiple chaotic attractors with basins corresponding to previously learned classes of stimuli, including that for the learned background stimulus configuration, and the collection constitutes an attractor landscape. This chaotic prestimulus state of expectancy establishes the sensitivity of the cortex by warping the landscape, so that a very small number of sensory action potentials driven by an expected stimulus can carry the cortical trajectory into the basin of an appropriate attractor. Circular causality enters in the following way. The state of a neural population in an area of cortex is a macroscopic property that arises through the interactions of the microscopic activity of the neurons comprising the neuropil. The global state is upwardly generated by the microscopic neurons, and simultaneously the global state downwardly organizes the activities of the individual neurons.

Each cortical state transition requires this circularity. It is preceded by a conjunction of antecedents. A stimulus is sought by the limbic brain through orientation of the sensory receptors in sniffing, looking, and listening. The landscape of the basins of attraction is shaped by limbic preafference, which facilitates access to an attractor by expanding its basin for the reception of a desired class of stimuli. Preafference provides the ambient context by multisensory divergence. The web of synaptic connections modified by prior learning maintains the basins and attractors. Pre-existing chaotic fluctuations are enhanced by input, forcing the selection of a new macroscopic state that then engulfs the stimulus-driven microscopic activity.

The main reason that all the sensory systems (visual, auditory, somatic and olfactory) operate this way is the finite capacity of the brain faced with the infinite complexity of the environment. In olfaction, for example, a significant odourant may consist of a few molecules mixed in a rich and powerful background of undefined substances, and it may be continually changing in age, temperature, and concentration. Each sniff in a succession with the same chemical activates a different subset of equivalent olfactory receptors, so the microscopic input is unpredictable and unknowable in detail. Detection and tracking require an invariant pattern over trials. This is provided by the attractor, and the generalization over equivalent receptors is provided by the basin. The attractor determines the response, not the particular stimulus. Unlike the view proposed by stimulus-response reflex determinism, the dynamics gives no linear chain of cause and effect from stimulus to response that can lead to the necessity of environmental determinism. An equally compelling reason is the requirement that all sensory patterns have the same basic nonlinear dynamics, so that they can be combined into *gestalts* in conformance with the unity of intentionality, once they are converged and integrated over time.

V: Circular Causality in Awareness

Circular causality, then, comes into play with each state transition in sensory cortices and the olfactory bulb, when fluctuations in microscopic activity exceed a certain threshold, such that a new macroscopic oscillation emerges to force co-operation on the very neurons that have brought the pattern into being. EEG measurements show that multiple patterns self-organize independently in overlapping time frames in the several sensory and limbic cortices, co-existing with stimulus-driven activity in different areas of the neocortex, which structurally is an undivided sheet of neuropil in each hemisphere receiving the projections of sensory pathways in separated areas.

Circular causality can serve as the framework for explaining the operation of awareness in the following way. The multimodal macroscopic patterns converge simultaneously into the limbic system, and the results of integration over time and space are simultaneously returned to all of the sensory systems. Here, I propose that another level of hierarchy exists in brain function as a hemispheric attractor, for which the local macroscopic activity patterns are the components. The forward limb of the circle provides the bursts of oscillations converging into the limbic system that destabilize it to form new patterns. The feedback limb incorporates the limbic and sensory cortical patterns into a global activity pattern or order parameter that enslaves all of the components. The enslavement enhances the co-operation among all of them, which dampens their chaotic fluctuations instead of enhancing them, as the receptor input does in the sensory cortices. Global co-operation in brains is the key to understanding consciousness.

A global operator of this kind must exist, for the following reason. The synthesis of sense data first into cortical wave packets and then into a multimodal packet takes time. After a *gestalt* has been achieved through embedding in past experience, a decision is required as to what the organism is to do next. This also takes time for an evolutionary trajectory through a sequence of attractors constituting the attractor landscape of possible goals and actions (Tsuda, 1991). The triggering of a state transition in the motor system may occur at any time, if the fluctuations in its multiple inputs are large enough, thereby terminating the search trajectory. In some emergent behavioural situations, an early response is most effective: action without reflection. In complex situations with unclear ramifications into the future, precipitate action may lead to disastrous consequences. More generally, the forebrain appears to have developed in phylogenetic evolution as an organ that takes advantage of the time provided by distance receptors for the interpretation of raw sense data. The quenching function of a global operator to delay decision and action can be seen as a necessary complement on the motor side, to prevent premature closure of the process of constructing and evaluating possible courses of action. This view is comparable to that of William James (1879), who wrote: ' . . . the study *à posteriori* of the *distribution* of consciousness shows it to be exactly such as we might expect in an organ added for the sake of steering a nervous system grown too complex to regulate itself', except that consciousness is not provided by another 'organ' (an add-on part of the human brain) but by a new hierarchical level of organization of brain dynamics.

Action without the deferral that is implicit in awareness can be found in so-called 'automatic' sequences of action in the performance of familiar complex routines. Actions 'flow' without awareness. Implicit cognition is continuous, and it is simply unmasked in the conditions that lead to 'blindsight'. In this view, emotion is defined as the impetus for action, more specifically, as impending action (Freeman, 1999). Its degree is proportional to the amplitude of the chaotic fluctuations in the limbic system, which appears as the modulation depth of the carrier waves of limbic neural activity patterns. In partial accord with the James-Lange theory of emotion (James, 1893), it is experienced through awareness of the activation of the autonomic nervous system in preparation for, and support of, overt action, as described by Cannon (1939). It is observed in the patterns of behaviour that social animals have acquired through evolution (Darwin 1872). Emotion is not in opposition to reason. Behaviours that are seen as irrational and 'incontinent' (Davidson, 1980) result from premature escape of the chaotic fluctuations from the leavening and smoothing of the awareness operator. The most intensely emotional behaviour, as it is experienced in artistic creation, scientific discovery, and religious visions, occurs as the intensity of awareness rises in concert with the strength of the fluctuations (Freeman, 1995). As with all other difficult human endeavours, self-control is achieved through long and arduous practice.

Evidence for the existence of the postulated global operator is found in the high level of spatial correspondence in the EEGs simultaneously recorded from the bulb and the visual, auditory, somatic and limbic (entorhinal) cortices of animals and from the scalp of humans (Lehmann and Michel, 1990), though the wave forms of the several sites vary independently and unpredictably (Smart *et al.*, 1997; Gaál and Freeman, 1997), with shared variances under principal components analysis on the order of 50 per cent of the total variance. These levels are lower than those found within each area of 90–98 per cent (Barrie, Freeman and Lenhart, 1996), but they are greater than can be accounted for by any of a variety of statistical artefacts or sources of correlation such as volume conduction, pacemaker driving, or contamination by the reference lead in monopolar recording. The level of coherence above chance holds for most parts of the EEG spectrum and for aperiodic as well as near-periodic waves. The maximal coherence appears to have zero phase lag over distances up to several centimetres between recording sites and even between hemispheres (Singer and Gray, 1995). Attempts are being made to model the observed zero time lag among the structures by cancellation of delays in bi-directional feedback transmission (König and Schillen, 1991; Traub *et al.* 1996; Roelfsma *et al.*, 1997).

VI: Consciousness Viewed as a System Parameter Controlling Chaos

A clear choice can be made now between the three meanings of causality proposed in the Introduction. Awareness and neural activity are not acausal parallel processes, nor does either make or move the other as an agency in temporal sequence. Circular causality is a form of explanation that can be applied at several hierarchical levels without recourse to agency. This formulation provides the sense or feeling of necessity that is essential for human comprehension, by addressing the elemental experience of cause and effect in acts of observation, even though logically it is very different from linear causality in all aspects of temporal order, spatial contiguity, and invariant reproducibility. The phrase is a cognitive metaphor. It lacks the attribute of agency, unless and until the loop is broken into the forward (microscopic) limb and the recurrent (macroscopic) limb, in which case the agency that is so compelling in linear causality can be re-introduced. This move acquiesces to the capabilities of the human observers, who need it and use it in order to comprehend dynamic events and processes in the world. The need has been filled with a variety of terms, including 'dispositions' (Aquinas, 1272), 'tendencies' (Mill, 1843), 'anomalous monism' (Davidson, 1980), 'capacities' (Cartwright, 1989), 'propensities' (Popper, 1990), and 'risk factors' by numerous public health workers.

I propose that the globally coherent activity, which is an order parameter, may be an objective correlate of awareness through preafference, comprising expectation and attention, which are based in prior proprioceptive and exteroceptive feedback of the sensory consequences of previous actions, after they have undergone limbic integration to form *gestalts*, and in the goals that are emergent in the limbic system. In this view, awareness is basically akin to the intervening state variable in a homeostatic mechanism, which is both a physical quantity, a dynamic operator, and the carrier of influence from the past into the future that supports the relation between a desired set point and an existing state. The content of the awareness operator may be found in the spatial pattern of amplitude modulation of the shared wave form component, which is comparable to the amplitude modulation of the carrier waves in the primary sensory receiving areas.

What is most remarkable about this operator is that it appears to be antithetical to initiating action. It can provide a pervasive neuronal bias that does not induce state transitions, but can defer them by quenching local fluctuations (Prigogine, 1980). It can alter the attractor landscapes of the lower order interactive masses of neurons that it enslaves. In the dynamicist view, intervention by states of awareness in the process of consciousness can organize the attractor landscape of the motor systems, prior to the instant of its next state transition, the moment of choosing in the limbo of indecision, when the global dynamic brain activity pattern is increasing its complexity and fine-tuning the guidance of overt action. This state of uncertainty and unreadiness to act may last a fraction of a second, a minute, a week, or a lifetime. Then, when a contemplated act is released, awareness follows the onset of the act and does not precede it.

In that hesitancy, between the last act and the next, comes the window of opportunity, when the breaking of symmetry in the next limbic state transition will make apparent what has been chosen. The observer of the self intervenes by awareness that organizes the attractor landscape, just before the instant of the next transition:

> Between the conception
> And the creation
> Between the emotion
> And the response
> Falls the Shadow
> Life is very long

T.S. Eliot (1936) The Hollow Men

The causal technology of self-control is familiar to everyone: hold off fear and anger; defer closure; avoid temptation; take time to study; read and reflect on the opportunity, meaning, and consequences; take the long view as it has been inculcated in the educational process. According to Mill (1873):

> We cannot, indeed, directly will to be different from what we are; but neither did those who are supposed to have formed our characters directly will that we should be what we are. Their will had no direct power except over their own actions . . . . We are exactly as capable of making our own character, *if we will*, as others are of making it for us (p. 550).

There are numerous unsolved problems with this hypothesis. Although strong advances are being made in analysing the dynamics of the limbic system and its centre pieces, the entorhinal cortex and hippocampus (Boeijinga and Lopes da Silva, 1988; O'Keefe and Nadel, 1978; Rolls *et al.*, 1989; McNaughton, 1993; Wilson and McNaughton, 1993; Buzsáki, 1996; Eichenbaum, 1997; Traub *et al.*, 1996), their self-organized spatial patterns, their precise intentional contents and their mechanisms of formation in relation to intentional action are still unknown. The pyriform cortex to which the bulb transmits is strongly driven by its input, and it lacks the phase cones that indicate self-organizing capabilities comparable to those of the sensory cortices (Freeman *et al.*, 1995). Whether the hippocampus has those capabilities or is likewise a driven structure is unknown. The neural mechanisms by which the entire neocortical neuropil in each hemisphere might maintain spatially coherent activity over a broad spectrum with nearly zero time lag are unknown. The significance of this coherent activity for behaviour is dependent on finding correlates with behaviours, but these are unknown. If those correlates are meanings, then the subjects must be asked to make representations of the meanings in order to communicate them, so that they are secondary to overt behaviours. Knowledge of human brain function is beyond the present reach of neurodynamics because our brains are too complex, especially our mechanisms for social behaviours, language and self-awareness.

VII: Causality Belongs in Technology, Not in Science

The case has now been made on the grounds of neurodynamics that causality is a form of knowing through intentional action. Thus causality is inferred not to exist in material objects, but to be assigned to them by humans with the intent to predict and control them. The determinants of human actions include not only genetic and environmental factors but self-organizing dynamics in brains, primarily operating through the dynamics of intentional action, and secondarily through neural processes that support consciousness, which is commonly but mistakenly attached to free will. While this inference is not new, it is given new cogency by recent developments in neuroscience. What, then, might be the consequences for science, philosophy, and medicine, if this inference is adopted?

The concept of causality is fundamental in all aspects of human behaviour and understanding, which includes our efforts in laboratory experiments and the analysis of data to comprehend the causal relations of world, brain and mind. In my own work, I studied the impact on brain activity of stimuli that animals were trained to ignore or to respond to, seeking to determine how the stimuli might cause new patterns of brain activity to form, and how the new patterns might shape how the animals behaved in response to the stimuli. I attempted to interpret my findings and those of others in terms of chains of cause and effect, which I learned to identify as 'linear causality' (Freeman, 1975).

These attempts repeatedly foundered in the complexities of neural activity and in the incompatibility of self-organized, goal-directed behaviour of my animals with behaviourist models based on input-output determinism. I found that I was adapting to the animals at least as much as they were being shaped by me. My resort to acausal correlation based in multivariate statistical prediction was unsatisfying. Through my readings in physics and philosophy, I learned the concept of 'circular causality', which invokes hierarchical interactions of immense numbers of semiautonomous elements such as neurons, which form non-linear systems, provided they are open in the exchange of matter and energy with their environments. These exchanges lead to the formation of macroscopic population dynamics that shapes the patterns of activity of the contributing individuals. I found this concept to be applicable at several levels, including the interactions between neurons and neural masses, between component masses of the forebrain, and between the behaving animal and its environment, under the rubric of intentionality (Freeman, 1995).

With adoption of this alternative concept, my perspective changed (Freeman, 1995). I now sought not to pin events at instants of time, but to conceive of intervals at differing time scales; not to fill the gaps in the linear chains, but to construct the feedback pathways from the surround; not to average the single responses to monotonously repeated stimuli, but to analyze each event in its uniqueness before generalizing; not to explain events exclusively in terms of external stimuli and context, but to allow for the contribution of self-organizing dynamics.

Circular causality departs so strongly from the classical tenets of necessity, invariance, and precise temporal order that the only reason to call it 'cause' is to satisfy the human habitual need for causes. The most subtle shift is the disappearance of agency, which is equivalent to loss of Aristotle's efficient cause. Agency is a powerful metaphor. For examples, it seems to be common sense to assert that an assassin causes a victim's death; that an undersea quake causes a tsunami; that a fallen tree causes a power failure by breaking a transmission line; that an acid-fast bacillus causes tuberculosis; that an action potential releases transmitter molecules at a synapse; and so forth. But interactions across hierarchical levels do not make sense in these terms. Molecules that co-operate in a hurricane cannot be regarded as the agents that cause the storm. Neurons cannot be viewed as the agents that make consciousness by their firing.

The very strong appeal of agency to explain events appears to come from the subjective experience of cause and effect that develops early in human life, before the acquisition of language, when as infants we go through the somatomotor phase (Piaget, 1930; Thelen and Smith, 1994) and learn to control our limbs and to focus our sensory receptors. 'I act (cause); therefore I feel (effect).' Granted that causality can be experienced through the neurodynamics of acquiring knowledge by the use of the body, the question I raise here is whether brains share this property with other material objects in the world. The answer I propose is that assignment of cause and effect to one's self and to others having self-awareness is entirely appropriate, but that investing insensate objects with causation is comparable to investing them with teleology and soul.

The further question is: Does it matter whether or not causality is assigned to objects? The answer here is: very much. Several examples are given of scientific errors attributable to dependence on the terms of linear causality. The most important, with wide ramifications, is the assumption of universal determinacy, by which the causes of human behaviour are limited to environmental and genetic factors, and the causal power of self-determination is excluded from scientific consideration. We know that linear extrapolation often fails in a non-linear world. Proof of the failure of this inference is by *reductio ad absurdum*. It is absurd in the name of causal doctrine to deny our capacity as humans to make choices and decisions regarding our own futures, when we exercise the causal power that we experience as free will.

VIII: Anthropocentricity in Acts of Human Observation

Our ancestors provided a history of interpreting phenomena in human terms appropriate to the scales and dynamics of our brains and bodies. An example of our limitations and our cognitive means for surmounting them is our spatial conception of the earth as flat. This belief is still quite valid for lengths of the size of the human body, such as pool tables, floors, and playing fields, where we use levels, transits, and gradometers, and even for distances that we can cover by walking and swimming. The subtleties of ships that were hull-down over the horizon were mere curiosities, until feats of intellect and exploration such as circumnavigation of the earth opened a new spatial scale. Inversely, at microscopic dimensions of molecules, flatness has no meaning. Under an electron microscope the edge of a razor looks like a mountain range.

In respect to time scales, we tend to think of our neurons and brains as having static anatomies, despite the evidence of continual change from time-lapse cinematography, as well as the cumulative changes that passing decades reveal to us in our bodies. An intellectual leap is required to understand that form and function are both dynamic, differing essentially in our time scales of measurements and experiences with them. The ontologic and phylogenetic developments of brains are described by sequences of geometric forms and the spatiotemporal operations by which each stage emerges from the one preceding. The time scales are in days and aeons, not in seconds as in behaviour and its neurophysiological correlates.

The growth of structure and the formation of the proper internal axonal and dendritic connections is described by fields of attraction and repulsion, with gradient descents mediated by contact sensitivities and the diffusion of chemicals. Moreover, recent research shows that synapses undergo a process of continual dynamic formation, growth and deletion throughout life (Smythies, 1997). The same and similar terms are used in mathematics and the physical sciences such as astronomy and cosmology, over a variety of temporal and spatial scales, many of which are far from the scales of kinaesthesia to which we are accustomed. On the one hand, morphogenesis is the geometry of motion, which we can grasp intuitively through time-lapse photography. On the other hand, the motions of speeding bullets and hummingbird wings are revealed to us by high-speed cinematography.

The attribution of intention as a property of material objects was common in earlier times in the assignment of spirits to trees, rocks, and the earth. An example is the rising sun. From the human perspective, the sun seems to ascend above the horizon and move across the sky. In mythology, this motion was assigned to an agency such as a chariot carrying the sun, or to motivation by the music of Orpheus, because music caused people to dance. In the Middle Ages, the sun, moon, planets and stars were thought to be carried by spheres that encircled the earth and gave ineffable music as they rotated. The current geometric explanation is that an observer on the earth's surface shifts across the terminator with inertial rotation in an acausal space-time relation. Still, we see the sun rise and set.

Similarly, humans once thought that an object fell because it desired to be close to the earth, tending to its natural state. In Newtonian mechanics it was pulled down by gravity. In acausal, relativistic terms, it follows a geodesic to a minimal energy state. The Newtonian view required action at a distance, which was thought to be mediated by the postulated quintessence ('fifth element') held over from Aristotle, the 'ether'. Physicists were misled by this fiction, which stemmed from the felt need for a medium to transmit a causal agent. The experimental proof by Michaelson and Morley that the ether did not exist opened the path to relativistic physics and an implicit renunciation of gravitational causality. But physicists failed to pursue this revolution to its completion, and instead persisted in the subject-object distinction by appealing to the dependence of the objective observation on the subjective reference frame of the observer.

In complex, multivariate systems with parts that interact by feedback at several levels like brains, causal sequences are impossible to specify unequivocally. Because it introduced indeterminacy, evidence for feedback in the nervous system was deliberately suppressed in the first third of the twentieth century. It was thought that a neuron in a feedback loop could not distinguish its external input from its own output. An example was the reaction of Ramón y Cajal to a 1929 report by his student, Rafael Lorente de Nó, who presented Cajal with his Golgi study of neurons in the entorhinal cortex (Freeman, 1984). Lorente constructed diagrams of axodendritic connections among the neurons with arrows to indicate the direction of transmission, and he deduced that they formed feedback loops. Cajal told him that his inference was unacceptable, because brains were deterministic and could not work if they had feedback. Lorente withdrew his report from publication until Cajal died in 1934. After he published it (Lorente de Nó, 1934), it became an enduring classic, leading to the concept of the nerve cell assembly by its influence on Donald Hebb (1949), and to neural networks and digital computers by inspiring Warren McCulloch and, through him, John von Neumann (1958). The concept of linear causality similarly slowed recognition and acceptance of processes of self-organization in complex systems, by the maxim that 'nothing can cause itself'. The phrase 'self-determination' was commonly regarded as an oxymoron. A similar exclusion delayed acceptance of the concepts of reafference and corollary discharge (Freeman, 1995).

IX: Applications in Philosophy

Description of a linear causal connection is based on appeal to an invariant relationship between two events. If an effect follows, the cause is sufficient; if an effect is always preceded by a cause, then it is necessary. From the temporal order and its invariance, as attested by double-blind experimental controls to parcellate the antecedents, an appearance of necessity is derived. The search concludes with assignment of an agency, that has responsibility for production, direction, control or stimulation, and that has its own prior agency, since every cause must also be an effect.

According to David Hume (1739) and other Nominalists (Freeman, 1999), causation does not arise in the events; it emerges in the minds of the observers. The temporal succession and spatial contiguity of events that are interpreted as causes and effects comprise the invariant connection. It is the felt force of conjoined impressions that constitutes the quale of causality. Since the repetition of these relations adds no new idea, the feeling of the necessity has to be explained psychologically. He came to this conclusion from an abstract premise in the doctrine of Nominalism, according to which there are no universal essences in reality, so the mind can frame a concept or image that constitutes a universal or general term, such as causality, but the concept does not exist in the world, only the unique forms of matter. This was opposed, then as now, to the doctrine of scientific realism. Hume and his Nominalist colleagues were anticipated 500 years earlier by the work of Aquinas (1272), who conceived that the individual forms of matter are abstracted by the imagination ('phantasia') to create universals that exist only in the intellect, not in matter. Early twentieth century physicists should have completed the Humeian revolution in their development of quantum mechanics, but they lost their nerve and formulated instead the Copenhagen interpretations, which reaffirmed the subject-object distinction of Plato and Descartes, despite the force of their own discoveries. Phenomenologists before Heidegger maintained the error, and post-modern structuralists persist in it.

Conversely, John Stuart Mill (1873) accepted 'the universal law of causation' but not Necessity.

> ... the doctrine of what is called Philosophical Necessity' weighed on my existence like an incubus .... I pondered painfully on the subject, till gradually I saw light through it. I perceived, that the word Necessity, as a name for the doctrine of Cause and Effect applied to human action, carried with it a misleading association; and that this association was the operative force in the depressing and paralyzing influence which I had experienced (pp. 101–102).

He developed his position fully in *A System of Logic* (1843).

Kant (1781) insisted that science could not exist without causality. Since causality was for him a category in mind, it follows that science is a body of knowledge about the world but is not in the world. Causality then becomes a basis for agreement among scientists regarding the validation of relationships between events, and the prediction of actions to be taken for control of events in the world. Since it could not be validated by inductive generalization from sense data, but was nevertheless essential to give wholeness and completion to experience [*Apperzeption*], Kant concluded that it must be '*a priori*' and 'transcendental' over the sense data. This led him to designate causality as a category [*Kategorien*] in and of the mind, along with space and time as the forms of perception [*Anschauungsformen*], by which the sense data were irretrievably modified during assembly into perceptions, making the real world [*Ding an sich*] inaccessible to direct observation.

Friedrich Nietzsche (1886) placed causality in the mind as the expression of free will:

> The question is in the end whether we really recognize the will as efficient, whether we believe in the causality of the will: if we do — and at bottom our faith in this is nothing

less than our faith in causality itself — then we have to make the experiment of positing the causality of the will hypothetically as the only one ... the will to power' (p. 48).

Putnam (1990) assigned causality to the operation of brains in the process of observation:

> Hume's account of causation ... is anathema to most present-day philosophers. Nothing could be more contrary to the spirit of recent philosophical writing than the idea that there is nothing more to causality than regularity or the idea that, if there is something more, that something more is largely subjective. (p. 81) If we cannot give a single example of an ordinary observation report which does not, directly or indirectly, presuppose causal judgements, then the empirical distinction between the 'regularities' we 'observe' and the 'causality' we 'project onto' the objects and events involved in the regularities collapses. Perhaps the notion of causality is so primitive that the very notion of observation presupposes it? (p. 75).

A case was made by Davidson (1980) for 'anomalous monism' to resolve the apparent contradiction between the deterministic laws of physics, the necessity for embodiment of mental processes in materials governed by those fixed laws, and the weakness of the 'laws' governing psychophysical events as distinct from statistical classes of events:

> Why on earth should a cause turn an action into a mere happening and a person into a helpless victim? Is it because we tend to assume, at least in the arena of action, that a cause demands a causer, agency and agent? So we press the question; if my action is caused, what caused it? If I did, then there is the absurdity of an infinite regress; if I did not, I am a victim. But of course the alternatives are not exhaustive. Some causes have no agents. Among these agentless causes are the states and changes of state in persons which, because they are reasons as well as causes, constitute certain events free and intentional actions (p. 19).

His premises have been superseded in two respects. First, he postulated that brains are material systems, for which the laws of physics support accurate prediction. He described brains as 'closed systems'. In the past three decades numerous investigators have realized that brains are open systems, as are all organs and living systems, with a continual intake of energy and an infinite sink in the venous return for waste heat and entropy, so that the first and second laws of thermodynamics do not hold for brains, thus negating one of his two main premises. Second, he postulated that, with respect to meaning, minds are 'open' systems, on the basis that they are continually acting into the world and learning about it. The analyses of electrophysiological data taken during the operations of sensory cortices during acts of perception indicate that meaning in each mind is a closed system (Freeman, 1995, 1999), and that meaning is based in chaotic constructions, not in information processing, thus negating the other of his two main premises. In my view, neurons engage in complex biochemical operations that have no meaning or information in themselves, but inspire meaning in researchers who measure them. The degree of unpredictability of mental and behavioural events is in full accord with the extent of variations in the space-time patterns of the activity of chaotic systems, thus removing the requirement for the adjective, 'anomalous', because it applies to both sets of laws for the material and mental aspects of living systems. Moreover, the adoption of the concept 'circular causality' from physics and psychology removes agency. That which remains is 'dynamical monism'.

X: Applications of Causality in Medical Technology

Causality is properly attributed to intentional systems, whose mechanisms of exploring, learning, choosing, deciding, and acting lead to the actualization of the feeling of necessary connection, and of the cognitive metaphor of agency. It is properly used to describe technological intervention into processes of the material world after human analysis of the interrelations of events. Surmounting linear causal thinking may enable neuroscientists to pursue studies in the dynamics of the limbic system that clarify the meanings of statistical regularities in chaotic, self-organizing brain components and change their performance by experimental manipulation. Social scientists may take advantage of the discovery of a biological basis for choice and individual responsibility to strengthen our social and legal institutions by complementing environmental and genetic linear causation. The nature-nurture debate has neglected the third leg of the determinant triad: the self. People can and do make something of themselves. Neurophilosophers studying consciousness in brain function may find new answers to old questions by re-opening the debate on causality. What acausal relations arise among the still inadequately defined entities comprising brains? What is the global operator of consciousness? The mind-brain problem is not solved, but it can be transplanted to more fertile ground.

My proposal is not to deny or abandon causality, but to adapt it as an essential aspect of the human mind/brain by virtue of its attachment to intentionality. This can be done by using the term 'circular causality' divorced from agency in the sciences, and the term 'linear causality' in combination with agency in the technologies, including medical, social, legal, and engineering applications.

For example, medical research is widely conceived as the search for the causes of diseases and the means for intervention to prevent or cure them. A keystone in microbiology is expressed in Koch's Postulates, which were formulated in 1881 by Robert Koch to specify the conditions that must be met, in order to assign a causal relation between a micro-organism and a disease: (1) the germ must always be found in the disease; (2) it must be isolated in pure culture from the diseased individual; (3) inoculation with the isolated culture must be followed by the same disease in a suitable test animal; and (4) the same germ must be isolated in pure culture from the diseased test animal.

These postulates have served well for understanding transmissible diseases and providing a biological foundation for developing chemotherapies, vaccines, and other preventatives. Public health measures addressing housing, nutrition, waste disposal and water supplies had already been well advanced in the nineteenth century for the prevention of pandemics such as cholera, typhoid, tuberculosis, and dysentery, on the basis of associations and, to a considerable extent, the maxim, 'Cleanliness is next to Godliness'. This was intentional behaviour of a high order indeed. The new science brought an unequivocal set of targets for research on methods of prevention and treatment.

The most dramatic development in neuropsychiatry was the finding of spirochetes in the brains of patients with general paresis, for which the assigned causes had been life styles of dissolution and moral turpitude. The discovery of the 'magic bullet' 606 (arsphenamine) established the medical model for management of neuropsychiatric illness, which was rapidly extended to viruses (rabies, polio, measles), environmental toxins (lead, mercury, ergot), vitamin and mineral deficiencies (cretinism, pellagra), hormonal deficits (hypothyroidism, diabetic coma, lack of dopamine in postencephalitic and other types of Parkinson's disease), and genetic abnormalities (phenylketonuria, Tourette's and Huntingdon's chorea). Massive research programs are under way to find the unitary causes and the magic bullets of chemotherapies, replacement genes, and vaccines for Alzheimer's, neuroses, psychoses, and schizophrenias. The current explanations of the affective disorders — too much or too little dopamine, serotonin, etc. — resemble the Hippocratic doctrine of the four humors, imbalances of which were seen as the causes of diseases.

There are compelling examples of necessary connections. Who can doubt that the vibrio causes cholera, or that a now contained virus caused small pox? However, these examples come from medical technology, in which several specific conditions hold. First, the discoveries in bacteriology came through an extension of human perception through the microscope to a new spatial scale. This led to the development by Rudolf Virchow of the cellular basis of human pathology. The bacterial adversaries were then seen as having the same spatial dimensions as the cells with which they were at war. The bacterial invaders and the varieties of their modes of attack did not qualitatively differ from the macroscopic predators with which mankind had always been familiar, such as wolves and crocodiles, which humans eradicate, avoid, or contain in laboratories and zoos. Second, the causal metaphor motivated the application of controlled experiments to the isolation and analysis of target bacterial and viral species, vitamins, toxic chemicals, hormones, and genes. It still does motivate researchers, with the peculiar potency of intermittent reinforcement by occasional success. A recent example is the recognition that pyloric ulcers are caused by a bacillus and not by psychic stress or a deleterious life style, implying that the cause is 'real' and not just mental or 'psychosomatic'. Third, the research and therapies are directly addressed to humans, who take action by ingesting drugs and seeking vaccinations, and who perceive changes in their bodies thereafter. A feeling of causal efficacy is very powerful in these circumstances, and many patients commit themselves without reservation to treatments, well after FDA scientists by controlled studies have shown them to be ineffective. The feeling of urgency on conceptualizing causality to motivate beneficial human actions does not thereby establish the validity of that agency among the objects under study. Feeling is believing, but it is not knowing. The feeling of causal agency in medicine has led to victories, but also to mistakes leading to injury and death on a grand scale.

Koch's postulates describe an approach to a necessary connection of a bacillus with an infectious disease, but not the sufficient conditions. Pathogens are found in healthy individuals as well, and often not in the sick. Inoculation does not always succeed in producing the disease. These anomalies can be, and commonly are, ignored, if the preponderance of evidence justifies doing so, but the classical criteria for causality are violated, or are replaced with statistical judgements. A positive culture of a bacillus is sufficient reason to initiate treatment with an antibiotic, even if it is the wrong disease. Similarly, pathologists cannot tell the cause of death from their findings at autopsy. They are trained to state what the patient died 'with' and not 'of' a configuration of abnormal structures. It is the job of a coroner or a licensed physician to assign the cause of death. The causes of death are not scientific. They are social and

technological, and they concern public health, economic well-being, and the apprehension of criminals.

   Another example of the social value of causality is the statement: 'Smoking causes
cancer.' This is a clear and valid warning that a particular form of behaviour is likely
to end in early and painful death. On the one hand, society has a legitimate interest in
maintaining health and reducing monetary and emotional costs by investing the
strong statistical connection with the motivating status of causality. On the other
hand, the 'causal chain' by which tobacco tars are connected to the unbridled proliferation of pulmonary epithelial tissue is still being explored, and a continuing weakness of evidence for the complete linear causal chain is being used by tobacco
companies to claim that there is no proof that smoking causes cancer but provide only
'risk factors'. Thus the causal argument has been turned against society's justifiable
efforts to prevent tobacco-related illnesses.

XI: The Technology of Mental Illness

The most complex and ambiguous field of medicine concerns the causes and treatments of mental disorders. Diagnosis and treatment for the past century have been
polarized between the medical model of the causes of diseases, currently held in biological psychiatry, and psychoanalysis, the talking cure. Sigmund Freud was
impressed with the phenomena of hysteria in patients who suffered transient disabili-
ties, such as blindness and paralysis, but who presented no evidence of infection or
anatomical degeneration in their brains. He drew on his background in clinical neurology to develop a biological hypothesis (1895) for behaviour, based on the flow of
nerve energy between neurons through 'contact barriers' (3 years later named synapses by Foster and Sherrington). Some axonal pathways developed excessive resistance at these barriers, deflecting nerve energy into unusual channels by 'neuronic
inertia', giving rise to hysterical symptoms. Within a decade he had abandoned the
biological approach as 'premature', working instead with his symbolic model of the
id, ego, and superego, but his ideas were generalized to distinguish 'functional' from
'organic' diseases. Traumatic childhood experiences warped the development of the
contact barriers. Treatment was to explore the recesses of memory, bring the resistances to awareness, and reduce them by client and therapist reasoning together following transference and countertransference.

   The bipolarization between the organic and the functional has been stable for a century. Patients and practitioners have been able to choose their positions in this spectrum of causes according to their beliefs and preferences. Some patients are delighted
to be informed that their disorders are due to chemical imbalances, that are correct-
able by drugs and that are not their fault or responsibility. Others bitterly resent the
perceived betrayal by their bodies, and they seek healing through the exercise of mental discipline and the power of positive thinking. But the balance has become unstable
with two new circumstances. One is the cost of medical care. Health maintenance
organizations are pressuring psychiatrists to see more patients in shorter visits, to dispense with oral histories and the meanings of symptoms for the patients, and to get
them quickly out the door with packets of pills. The power of biological causality is
clearly in operation as a social, not a scientific, impetus, operating to the detriment of
people with complex histories and concerns.

The other circumstance is the growing realization among mental health care specialists that chemical imbalances, poor genes, and horrific experiences of individuals are insufficient explanations to provide the foundations for treatment. Of particular importance for onset, course, and resolution of illnesses are the social relations of individuals, their families, neighbourhoods, religious communities, and milieu of national policies and events. Current conflicts rage over the assignment of the cause of chronic fatigue syndrome to neuroticism or to a virus; of the Gulf War syndrome to malingering or a neurotoxin; of post-traumatic stress disorder to battle fatigue or a character deficit. The dependence of the debates on causality is fuelled by technological questions of human action: what research is to be done, what treatments are to be given, and who is to pay for them? Successful outcomes are known to depend less on pills and counselling than on mobilization of community support for distressed individuals (Frankl, 1973). These exceedingly complex relations, involving faith and meaning among family and friends, may be seriously violated by reduction to unidirectional causes. Patients may be restored to perfect chemical balance and then die anyway in despair. Families may disintegrate into endless recrimination and self-justification, from lack of tolerance of misdirected parental and filial intentions and honest mistakes. So it is with patient-doctor relations. To seek and find a cause is to fix the blame, opening the legal right to sue for compensation for psychological injury and distress. These, too, are legacies of linear causal thinking.

Abnormal behaviour in states of trance or seizure was attributed in past centuries to the loss or willing surrender of self-control or to possession by exotic spirits. In British law the failure of responsibility was codified as legal insanity in 1846 according to the McNaughton Rule: '[To] establish a defence on the grounds of insanity, it must be clearly proven that at the time of the committing of the act, the party accused was labouring under such a defect of reason, from disease of the mind, as not to know the nature and quality of the act he was doing, or, if he did know it, that he did not know he was doing what was wrong.' In the terms of the present analysis, for behaviour to be insane the neural components of the limbic system must have entered into basins of attraction that are sufficiently strange or abnormally stable to escape control by the global state variable. This view encompasses the two facets of causality, microscopic and macroscopic, that compete for control of the self, but it is not an adequate statement of the problem. In fact, the case on which the Rule was based was decided on political grounds (Moran, 1981). Daniel McNaughton was a Scotsman engaged in ideal-driven assassination, and his transfer by the British authorities from Newgate Prison to Bethlam Hospital was designed to prevent him from testifying in public. A similar move for similar reasons was made by the American government in sending the poet Ezra Pound, charged with treason in World War II, to St. Elizabeth's Hospital instead of to Leavenworth Prison. The point is that judgements about which acts are intentional and which are not are made by society, usually by delegation to judges and juries in courts of law, not by doctors, scientists, or individuals in isolation. What biology can offer is a foundation on which to construct a social theory of self-control.

XII: The Science *Versus* the Technology of Self-control

The role of causality in self-awareness is close to the essence of what it is to be human. Nowhere is this more poignant than in the feeling of the need for self-control.

Materialists and psychoanalysts see the limbic self as a machine driven by metabolic deficits and inherited instincts, the id, which carries the ego as a rational critic struggling to maintain causal control, as befits the Cartesian metaphor of the soul serving as the pilot of a boat by adjudicating blind forces. Structure and chemistry are genetically determined. Behaviourist psychologists confuse motivation with intention and view behaviour as the sum of reflexes, caused by environmental inputs and sociobiological processes, while consciousness is epiphenomenal.

Functionalists see mind as equivalent to software that can be adapted to run on any platform, once the algorithms and rules have been discovered. Logical operations on symbols as representations are the causes of rational behaviour, and the unsolved problems for research concern the linkage of the symbols with activities of neurons and with whatever the symbols represent in the world. That research will be unnecessary, if the linkages can made instead to the components of intelligent machines resembling computers (Fodor, 1981). They find it unfortunate that the only existing intelligent beings have evolved from lower species, and that our brains are saddled with the limbic system and all of its irrational baggage. Outputs from the logic circuits in the neocortex, before reaching the motor apparatus, are filtered through the limbic system, where emotions are attached that distort and degrade the rational output. Consciousness is a mystery to be explained by 'new laws of physics' (Penrose, 1994; Chalmers, 1996).

Existentialists hold that humans choose what they become solely by their own actions. The cause of behaviour is the self, which is here described as emerging through the dynamics in the limbic system. The ego constituting awareness of the self discovers its own nature by observing and analyzing its actions and creations, but cannot claim credit for them. In extreme claims advanced by Nietzsche and Sartre, the ego is unconstrained by reality. Freedom is absolute, unforgiving.

In neurodynamics, because of the circularity of the relation of the self and its awareness, the future actions of the self are shaped in the context of its irrevocable past, its body, its given cultural and physical environment, and its present state of awareness, which is its own creation. The finite brain grapples with the infinity of the world and the uncertainty of the interlocked futures of world and brain, by continually seeking the invariances that will support reliable predictions. Those predictions exist as awareness of future possibilities, without which the self cannot prevail. They are experienced in the feeling of hope: the future need not merely happen; to some extent it can be caused.

XIII: Conclusion

The interactions between microscopic and macroscopic levels through mesoscopic dynamics (Freeman, 2000) are the basis for self-organization of brain and behaviour. How do all one hundred billion neurons get together in a virtual instant, and switch from one harmonious pattern to another in an orderly dance, like the shuttle of lights on the 'magic loom' of Sherrington (1940)? The same problem holds for the excitation of atoms in a laser, leading to the emergence of coherent light from the organization of the whole mass; for the co-ordinated motions of molecules of water and air in a hurricane; for the orchestration of the organelles of caterpillars in metamorphosing to butterflies; and for the inflammatory spread of behavioural fads, rebellions, and revolutions that sweep entire nations. All these kinds of events call for new laws such as those developed in physics by Haken (1983), in chemistry by Prigogine (1980), in biology by Eigen and Schuster (1979), in sociology by Foucault (1976), and in neuro-biology by Edelman (1987), which can address new levels of complexity that have heretofore been inaccessible to human comprehension. Perhaps these will serve as the 'new laws' called for by Penrose (1994) and Chalmers (1996), but they need not lead to dualism or panpsychism. They can arise as logical extensions from the bases of understanding we already have in these several realms of science, none of which can be fully reduced to the others.

Consciousness in the neurodynamic view is a global internal state variable composed of a sequence of momentary states of awareness. Its regulatory role is comparable to that of the operator in a thermostat, that instantiates the difference between the sensed temperature and a set point, and that initiates corrective action by turning a heater on or off. The machine state variable has a brief history and no capacities for learning or determining its own set point, but the principle is the same: the internal state is a form of energy, an operator, a predictor of the future, and a carrier of information that is available to the system as a whole. It is a prototype, an evolutionary precursor, not to be confused with awareness, any more than tropism in plants and bacteria is to be confused with intentionality. In humans, the operations and informational contents of the global state variable, which are sensations, images, feelings, thoughts and beliefs, constitute the experience of causation, which is to know intentionality in operation.

To deny this comparability and assert that humans are not machines is to miss the point. Two things distinguish humans from all other beings. One is the form and function of the human body, including the brain, which has been given to us by three billion years of biological evolution. The other is the heritage given to us by two million years of cultural evolution. Our mental attributes have been characterized for millennia as the soul or spirit or consciousness that makes us not-machines. The uniqueness of the human condition is not thereby explained, but the concept of circular causality provides a tool for intervention, when something has gone wrong, because the circle can be broken into forward and feedback limbs. Each of them can be explained by linear causality, which tells us where and how to intervene. The biggest error would be to assign causal agency to the parts of the biological system.

XIV: Summary

Science provides knowledge of relations among objects in the world, whereas technology provides tools for intervention into the relations by humans with intent to control the objects. The acausal science of understanding the self distinctively differs from the causal technology of self-control. 'Circular causality' in self-organizing systems is a concept that is useful to describe interactions between microscopic neurons in assemblies and the macroscopic emergent state variable that organizes them. In this review intentional action emerges from the interactions of the subsystems. Awareness (fleeting frames) and consciousness (continual operator) are ascribed to a hemisphere-wide order parameter constituting a global brain state. Linear causal inference is appropriate and essential for planning and interpreting human actions and personal relations, but it can be misleading when it is applied to microscopic–

macroscopic relations in brains. It is paradoxical to assign linear causality to brains, and thereby cast doubt on the validity of causal agency in the making of choices by humans, merely because the choices are materialized through state transitions in their brain.

Acknowledgements

This research was supported by grants from the National Institutes of Health MH-06686 and the Office of Naval Research N00014-93-1-0938.

References

Abraham, F.D., Abraham, R.H., Shaw, C.D., Garfinkel, A. (1990), *A Visual Introduction to Dynamical Systems Theory for Psychology* (Santa Cruz, CA: Aerial Press).

Aquinas, St. Thomas (1272), *The Summa Theologica*. Translated by Fathers of the English Dominican Province. Revised by Daniel J Sullivan. Published by William Benton as Volume 19 in the Great Books Series. (Chicago: Encyclopedia Britannica, Inc., 1952).

Barrie, J.M., Freeman, W.J., Lenhart, M. (1996), 'Modulation by discriminative training of spatial patterns of gamma EEG amplitude and phase in neocortex of rabbits', *Journal of Neurophysiology*, **76**, pp. 520–39.

Bloom, F.E., Lazerson, A. (1988), *Brain, Mind, and Behaviour* (2nd ed.) (New York: Freeman).

Boeijinga, P.H., Lopes da Silva, F.H. (1988), 'Differential distribution of beta and theta EEG activity in the entorhinal cortex of the cat', *Brain Research*, **448**, pp. 272–86.

Bures, J., Buresová, O., Krivánek, J. (1974), *The Mechanism and Applications of Leão's Spreading Depression of Electroencephalographic Activity* (New York: Academic Press).

Buzsáki, G. (1996), 'The hippocampal-neocortical dialogue', *Cerebral Cortex*, **6**, pp. 81–92.

Cannon, W.B. (1939), *The Wisdom of the Body* (New York: W.W. Norton).

Cartwright, N. (1989), *Nature's Capacities and their Measurement* (Oxford: Clarendon Press).

Chalmers, D.J. (1996), *The Conscious Mind. In Search of a Fundamental Theory* (New York: Oxford University Press).

Crick, F. (1994), *The Astonishing Hypothesis: The Scientific Search for the Soul* (New York: Scribner's).

Darwin, C. (1872), *The Expression of the Emotions in Man and Animals* (London: J. Murray).

Davidson, D. (1980), 'Actions, reasons, and causes' in *Essays on Actions and Events* (Oxford: Clarendon Press).

Dennett, D.H. (1991), *Consciousness Explained* (Boston: Little, Brown).

Dewey, J. (1914), 'Psychological doctrine in philosophical teaching', *Journal of Philosophy*, **11**, pp. 505–12.

Eccles, J.C. (1994), *How the Self Controls Its Brain* (Berlin: Springer-Verlag).

Edelman, G.M. (1987), *Neural Darwinism: The Theory of Neuronal Group Selection* (New York: Basic Books).

Eichenbaum, H. (1997), 'How does the brain organize memories?', *Science*, **277**, pp. 330–2.

Eigen, M., Schuster, P. (1979), *The Hypercycle. A Principle of Natural Self-Organization* (Berlin: Springer-Verlag).

Eliot, T.S. (1936), 'The Hollow Men' in *Collected Poems 1909-1935* (New York: Harcourt Brace).

Fodor, J.A. (1981), *Representations: Philosophical Essays on the Foundations of Cognitive Science* (Cambridge, MA: MIT Press).

Foucault, M. (1976), *The History of Sexuality: Vol. 1. An Introduction* tr. R Hurley (New York: Random Hous)..

Frankl, V. (1973), *The Doctor and the Soul* (New York: Random House).

Freeman, W.J. (1975), *Mass Action in the Nervous System* (New York: Academic).

Freeman, W.J. (1984). 'Premises in neurophysiological studies of learning' in *Neurobiology of Learning and Memory*, ed. G.Lynch, J.L. McGaugh, N.M. Weinberger. (New York: Guilford Press).

Freeman, W.J. (1991), 'The physiology of perception', *Scientific American*, **264**, pp. 78–85.

Freeman, W.J. (1992), 'Tutorial in neurobiology: From single neurons to brain chaos', *International Journal of Bifurcation and Chaos*, **2**, pp. 451–82.

Freeman, W.J. (1995), *Societies of Brains. A Study in the Neuroscience of Love and Hate* (Hillsdale, NJ: Lawrence Erlbaum).

Freeman (1999), *How Brains Make Up Their Minds* (London: Weidenfeld and Nicolson).

Freeman ,W.J., Barrie, J.M., Lenhart, M., Tang, R.X. (1995), 'Spatial phase gradients in neocortical EEGs give modal diameter of 'binding' domains in perception. Abstracts', *Society for Neuroscience*, **21**, 1649 (648.13).

Freud, S. (1895), 'The project of a scientific psychology' in *The Origins of Psychoanalysis*, ed. M. Bonaparte, A. Freud, E. Kris, tr. E. Mosbacher, J. Strachey (New York: Basic Books).

Gaál, G., Freeman, W.J. (1997), 'Relations among EEGs from entorhinal cortex and olfactory bulb, somatomotor, auditory and visual cortices in trained cats', *Society of Neuroscience,* Abstracts, 407.19.

Gibson, J.J. (1979), *The Ecological Approach to Visual Perception* (Boston: Houghton Mifflin).

Gloor, P. (1997), *The Temporal Lobe and the Limbic System* (New York: Oxford University Press).

Goltz, F.L. (1892), 'Der Hund ohne Grosshirn. Siebente Abhandlung über die Verrichtungen des Grosshirns', *Pflügers Archiv,* **51**, pp. 570–614.

Gray, C.M. (1994), 'Synchronous oscillations in neuronal systems: mechanisms and functions', *Journal of Comparative Neuroscience,* **1**, pp. 11–38.

Grossberg, S. (1982), *Studies of Mind and Brain: Neural Principles of Learning, Perception, Development, Cognition, and Motor Control* (Boston: D. Reidel).

Haken, H. (1983), *Synergetics: An Introduction* (Berlin: Springer).

Hardcastle, V.G. (1994), 'Psychology's binding problem and possible neurobiological solutions', *Journal of Consciousness Studies,* **1**, pp. 66–90.

Hebb, D.O. (1949), *The Organization of Behaviour* (New York: Wiley).

Helmholtz, H.L.F. von (1872), *Handbuch der physiologischen Optik. Vol. III* (Leipzig: L. Voss) (1909).

Herrnstein, R.J., Murray, C. (1994), *The Bell Curve* (New York: The Free Press).

Herrick, C.J. (1948), *The Brain of the Tiger Salamander* (Chicago IL: University of Chicago Press).

Hull, C.L. (1943), *Principles of Behaviour, An Introduction to Behaviour Theory* (New York: Appleton-Century).

Hume, D. (1739), *Treatise on Human Nature* (London: J. Noon).

Jackson, J.H. (1931), *Selected writings of John Hughlings Jackson,* ed. J. Taylor, F.M.R. Walshe, G. Holmes (London: Hodder and Stoughton).

James, W. (1879), 'Are we automata?', *Mind,* **4**, pp. 1–21.

James, W. (1893), *The Principles of Psychology* (New York. H. Holt).

Kant ,I. (1781), *Kritik der reinen Vernunft,* ed. W. von Weischedel (Frankfurt am Main: Suhrkamp Verlag) (1974).

Kay, L.M., Freeman, W.J. (1998), 'Bi-directional processing in the olfactory-limbic axis during olfactory behaviour', *Behavioural Neuroscience,* **112**, pp. 541–53.

König, P., Schillen, T.B. (1991), 'Stimulus-dependent assembly formation of oscillatory responses: I. synchronization', *Neural Comp,* **3**, pp. 155–66.

Lehmann, D., Michel, C.M. (1990), 'Intracerebral dipole source localization for FFT power maps', *Electroencephalography and clinical Neurophysiology,* **76**, pp. 271–6.

Lorente de Nó, R. (1934), 'Studies on the structure of the cerebral cortex. I The area entorhinalis', *Journal für Psychologie und Neurologie,* **45**, pp. 381–438.

McNaughton, B.L. (1993), 'The mechanism of expression of long-term enhancement of hippocampal synapses: Current issues and theoretical implications', *Annual Review of Physiology,* **55** pp. 375–96.

Merleau-Ponty, M. (1942), *The Structure of Behaviour,* trans. A.L. Fischer AL, (Boston: Beacon Press) (1963).

Mill, J.S. (1843), *Of Liberty and Necessity,* Ch. II, Book VI. A System of Logic. (Londo: Longmans, Green), 18th ed. (1965).

Mill, J.S. (1873), *Autobiography* (New York: Columbia University Press) (1924).

Moran, R. (1981), *Knowing Right from Wrong. The Insanity Defence of Daniel McNaughtan* (New York: Macmillan).

Nietzsche, F. (1886), *Beyond Good and Evil. A Prelude to a Philosophy of the Future,* tr. W. Kaufmann (New York: Random) (1966).

O'Keefe, J., Nadel, L. (1978), *The Hippocampus as a Cognitive Map* (Oxford: Clarendon).

Penrose, R. (1994), *Shadows of the Mind* (Oxford: Oxford University Press).

Piaget, J. (1930), *The child's conception of physical causality* (New York: Harcourt, Brace).

Popper. K.R. (1990), *A World of Propensities* (Bristol: Thoemmes).

Prigogine, I. (1980), *From Being to Becoming: Time and Complexity in the Physical Sciences* (San Francisco: Freeman).

Putnam, H. (1990), *Realism With a Human Face* (Cambridge MA: Harvard University Press).

Roelfsema ,P.R., Engel, A.K., König, P., Singer, W. (1997), 'Visuomotor integration is associated with zero time-lag synchronization among cortical areas', *Nature,* **385**, pp. 157–61.

Rolls, E.T., Miyashita, Y., Cahusac, P.B.M., Kesner, R.P., Niki, H., Feigenbaum, J.D., Bach, L. (1989), 'Hippocampal neurons in the monkey with activity related to the place in which the stimulus is shown', *Journal of Neuroscience,* **9**, pp. 1835–45.

Roth, G. (1987), *Visual Behaviour in Salamanders* (Berlin: Springer-Verlag).

Searle, J.R. (1992), *The Rediscovery of Mind* (Cambridge MA: MIT Press).

Sherrington, C.S. (1940), *Man on his Nature* (Oxford: Oxford University Press).

Singer, W., Gray, C.M. (1995), 'Visual feature integration and the temporal correlation hypothesis', *Annual Review of Neuroscience,* **18**, pp. 555–86.

Skarda, C.A., Freeman, W.J. (1987), 'How brains make chaos in order to make sense of the world', *Behavioural and Brain Sciences,* **10**, pp. 161–95.

Skinner, B.F. (1969), *Contingencies of Reinforcement; A Theoretical Analysis* (New York: Appleton-Century-Crofts).

Smart, A., German, P., Oshtory, S., Gaál, G., Barrie, J.M., Freeman, W.J. (1997), 'Spatio-temporal analysis of multi-electrode cortical EEG of awake rabbit', *Society of Neuroscience*, Abstracts, 189.13.

Smythies, J. (1997), 'The biochemical basis of synaptic plasticity and neural computation: a new theory', *Proc.Roy.Soc.London B.*, **264**, pp. 575–9.

Sperry, R.W. (1950), 'Neural basis of the spontaneous optokinetic response', *Journal of Comparative Physiology*, **43**, pp. 482–9.

Thelen, E., Smith, L.B. (1994), *A Dynamic Systems Approach to the Development of Cognition and Action* (Cambridge MA: MIT Press).

Titchener, E.B. (1907), *An Outline of Psychology* (New York: Macmillan).

Tolman, E.C. (1948), 'Cognitive maps in rats and men', *Psychological Review*, **55**, pp. 189–208.

Traub, R.D., Whittington, M.A., Colling, S.B., Buzsáki, G., Jefferys, J.G.R. (1996), 'A mechanism for generation of long-range synchronous fast oscillations in the cortex', *Nature*, **383**, pp. 621–4.

Tsuda, I. (1991), 'Chaotic itinerancy as a dynamical basis of hermeneutics in brain and mind', *World Futures*, **32**, pp. 167–84.

von Holst, E., Mittelstädt, H. (1950), 'Das Reafferenzprinzip (Wechselwirkung zwischen Zentralnervensystem und Peripherie)', *Naturwissenschaften*, **37**, pp. 464–76.

von Neumann, J. (1958), *The Computer and the Brain* (New Haven CT: Yale University Press).

Whitehead, A.N. (1938), *Modes of Thought* (New York: Macmillan).

Wilson, M.A., McNaughton, B.L. (1993), 'Dynamics of the hippocampal ensemble code for space', *Science*, **261**, pp. 1055–8.

Ravi V. Gomatam

Quantum Theory and the Observation Problem

Although quantum theory is applicable, in principle, to both the microscopic and macroscopic realms, the strategy of practically applying quantum theory by retaining a classical conception of the macroscopic world (through the correspondence principle) has had tremendous success. This has nevertheless rendered the task of interpretation daunting. We argue the need for recognizing and solving the 'observation problem', namely constructing a 'quantum-compatible' view of the properties and states of macroscopic objects in everyday thinking to realistically interpret quantum theory consistently at both the microscopic and macroscopic levels. Toward a solution to this problem, we point out a category of properties called 'relational properties' that we regularly associate with everyday objects. We see them as being potentially quantum-compatible. Some possible physical implications are discussed. We conclude by touching upon the nexus between the relational property view within quantum physics and some neurobiological issues underlying cognition.

I: Introduction

> Every analysis of the conditions of human knowledge must rest on considerations of the character and scope of our means of communication.
>
> Niels Bohr (1957, p. 88)

How does physics describe reality? Theoretical terms are never directly observable, not even in early mechanics. The speculative character of all scientific concepts, even that of 'position' in early mechanics, was adequately emphasized by Mach. Newton himself was only too acutely aware of the formal nature of his theory's 'description':

> I ... use the word 'attraction', 'impulse', or 'propensity' of any sort toward a centre, promiscuously and indifferently, one for another, considering those forces not physically but mathematically; whereof the reader is not to imagine that by those words I anywhere take upon me to define the kind or the manner of any action, the causes or the physical reason thereof, or that I attribute forces in a true and physical sense to certain centres (which are only mathematical points) when at any time I happen to speak of centres as attracting or as endued with attractive powers (Newton, 1687, Definition VIII, pp. 5–6).

Duhem and Einstein, among others, have additionally pointed out that only the theory as a whole gets verified in experience (Howard, 1990). From these twin considerations it follows that the pragmatic success of a theory alone cannot give its individual

Journal of Consciousness Studies, **6**, No. 11–12, 1999, pp. 173–90

concepts the status of being 'real'. Scientific realism needs further justification. If one grants naïve realism at the level of everyday thinking and practical living, ordinary language remains the sole source of our notions of the 'real'. If so, naming theoretical terms using words of the ordinary language would be a necessary starting step for explicating the nexus we conceive between scientific terms and reality. The need for physical theories to connect with ordinary language was eloquently expressed by Pauli:

> It is true that these [scientific] laws and our ideas of reality which they presuppose are getting more and more abstract. But for a professional, it is useful to be reminded that behind the technical and mathematical form of the thoughts underlying the laws of nature, there remains always the layer of everyday life with its ordinary language. Science is a systematic refinement of the concepts of everyday life revealing a deeper and, as we shall see, not directly visible reality behind the everyday reality of coloured, noisy things. But it should not be forgotten either that this deeper reality *would cease to be an object of physics*, different from the objects of pure mathematics and pure speculation, *if its links with the realities of everyday life were entirely disconnected* (Pauli, 1994, p. 28, italics mine).

Ordinary language considerations, however, enter physical theory at another, more fundamental level. For empirical justification of abstract, mathematical theories not to be circular, the predictive content of the theories needs to be stated in the first instance, not as scientific observations (using the formal terms), but as everyday experiences in the laboratory using ordinary language. Such a description of observation experiences using ordinary language would be evidently *independent of the physical theory*. Ordinary language and the *realist intuitions* underlying it will thus play an important and critical role within scientific realism from the very outset.

Our present view of the everyday world allows for objects to factually have only *one* definite state, out of many logical possibilities. If we toss a coin, logically speaking, both heads and tails are possibilities. In an actual toss, however, only heads *or* tails will show up. We shall refer to this view of the objects of the everyday world as the classically-definite (CDEF) conception of an object and its state. According to this CDEF conception, an object is always (factually) in only one of its many (logically) possible states. In addition, *all* the properties that we can associate with such a factual state of the object will have determinate values at all times, whether measured or not. This CDEF conception is independent of any physical theory and is basic to everyday naïve realism.

It is sufficiently recognized in the literature that such CDEF conceptions fail at the level of quantum mechanical description of atomic objects. However, that these CDEF conceptions also fail in quantum theory when we interpret our laboratory observations as simple experiences, toward empirically verifying a theory, is seldom emphasized, much less dealt with.

In this paper we shall argue the advisability of abandoning altogether the CDEF conceptions while interpreting quantum theory, despite the rather impressive pragmatic successes in applying the theory while retaining the CDEF conception of the everyday world. We shall argue so by raising both theoretical and experimental considerations that call into question how far terms such as wave-particle duality and superposition adequately convey the quantum implications of the corresponding formal terms. Indeed, the central thesis of this paper will be that we cannot even begin to comprehend the essential nature of the quantum mechanical description unless we develop an alternative, quantum-compatible conception of everyday objects in *every-day thinking*. We shall call the task of identifying such a conception of everyday objects and developing appropriate formal ideas based on it as the 'observation problem'.

We then discuss certain interpretive insights of Einstein and Bohr on the grounds that both of them effectively recognized the observation problem. Toward solving the observation problem, a range of properties that we routinely attribute to macroscopic objects in everyday thinking and which seem quantum-compatible is identified. We label them 'relational properties' and discuss in broad terms how they can aid in solving the observation problem. We conclude by touching upon the possible nexus between the relational property view within quantum physics and the neurobiology of cognition, particularly in the context of intentionality and causality.

II: The Two-Slit Experiment, the Classically-Definite (CDEF) Conceptions and the Wave-Particle Duality

> We choose to examine a phenomenon which is impossible, absolutely impossible, to explain in any classical way, and which has in it the heart of quantum mechanics. In reality, it contains the only mystery.
>
> R.P. Feynman (1965, vol. III, p. 1)

Bohr, among all writers on quantum theory, consistently referred to wave-particle duality as an apparent dilemma, a problem to be got rid of, not explained (see, for example, Bohr, 1934/1961, pp. 15, 93, 95, 107). Indeed, our laboratory observations concerning the behaviour of quantum objects are *always* localized, a fact that promotes, if any, a 'particle picture' of the quantum objects. What experimental evidence compels us to ascribe a 'wave nature' to the same objects? A simple step-by-step analysis of the two-slit experiment in several different configurations will show that the so-called wave aspect of the particle at the experimental level is merely an inference resulting from an implicit application of a series of CDEF-based intuitions to the laboratory observations. The basic experimental setup consists of a two-slit screen placed in front of a low-intensity monochromatic light source at an appropriate distance. A battery of detectors (or a photographic plate, effectively an array of detectors) is placed on the other side of the screen.

Configuration 1: Both slits of the screen are kept open, and a battery of detectors is placed at any place on the other side of the screen. No matter how small we make the size of the detector, we *always* see only *one* detector click.

    CDEF conclusion 1: Light is always observed to be a localized 'particle'. Let us call them 'photons', following conventional practice.

Configuration 2: Both slits of the screen are kept open, and two detectors are placed very close to the screen, one at each slit. Predictably, at any instant only *one* of the detectors *always* fires.

    CDEF conclusion 2: Photons always 'go through' one or the other slit.

Configuration 3: The photographic plate is placed far enough from the two-slit-screen (depending on the distance between the slits and the size of the slits), and one slit is closed. The localized light spots continue to appear, but randomly and *everywhere* on the photographic plate with a peak in one region which is dependent on which slit is open.

> *CDEF conclusion 3*: Photons going through *either* slit can arrive at *any spot* on the photographic plate.

Configuration 4: Same as configuration 3, except that now both slits are kept open. Again, localized spots appear randomly on the photographic plate. However, when we let enough of the spots accumulate, a pattern of *alternate dark and bright bands* appears on the screen, familiarly called the 'interference' pattern.

> *CDEF conclusion 4*: when both slits are open, even though single photons (must) continue to be going through one or the other slit, we now see that there are regions where the photons *don't arrive* at all!

Now, using CDEF conceptions, one further *infers* the following: consider a spot P on the photographic plate which falls in the 'dark region' when both slits are open. Of all the single photons that 'go through' the right slit (according to the results of configuration 3), some that would arrive at P if the left slit were closed, do not when the left slit is open!

> *CDEF conclusion 5*: Where a photon going through one of the slits (and all photons must go through one or the other slit, according to the results of configuration 2) would arrive on the screen depends on the status of the *other* slit (open or closed).

Penrose (1989) has called it the 'most mysterious part of the two-slit experiment'. In some unknown manner, the photon reacts to the status of both slits, and this allows us to ascribe an extended or 'wave' nature to the photon. Dirac has famously suggested that the single photon, going through the slits, somehow interferes with itself. There is also the notion that, prior to an actual laboratory observation, the photon has many constituent states that correspond to, in some vague sense, possibilities for going through both slits.

It is important to see that the idea of mysterious 'self-interference' appears necessary only because of the following assumptions:

(a) We accept the 'particle' idea in view of the localized nature of individual observations.
(b) We make further CDEF *inferences* in analysing the outcomes in each experimental configuration.
(c) We combine these inferences across experiments into a single picture of the atomic world.

III: The CDEF Conception at the Everyday Level and Its Incompatibility with the Quantum Formalism

At the core of the quantum mechanical description is a wave function, denoted by Ψ. Its evolution is governed by the Schrödinger equation. If two wave functions, Ψ_1 and Ψ_2, are possible solutions to the Schrödinger equation, then the principle of superposition holds that a linear combination of the two, $\Psi = c_1\Psi_1 + c_2\Psi_2$ is also a solution. However, within a CDEF conception of the world, we find our laboratory observations *always* correspond to either state Ψ_1 *or* Ψ_2, never both. It would be more congenial if we could devise a non-CDEF conception of the observations that would allow us to relate the laboratory observations to Ψ as well.

However, any such inclusive conception would have to revise, not only our notion of the states of atomic objects, but of macroscopic objects also. This point was best illustrated by Schrödinger in his famous 'cat paradox'. A single photon in a state of superposition is connected to a detector, which is wired up to release poison gas inside a box in which a living cat has been placed. It is possible to visualize, in principle, a wave function that describes the joint state of the atom, counter, poison trigger, all the way up to *the cat*. Such a wave function would require that, prior to a laboratory observation, the cat also be in a state of superposition, i.e. a state that does not permit any CDEF-based notion. Nevertheless, independent of quantum theory, CDEF notions would tell us that a cat has to be in *one* of two distinct states, dead or alive, at all times.

We may thus tend to conclude that Schrödinger's paradox cannot point to any problem at the observational level. Instead, we may conclude that the paradox shows the utter necessity to accept that, with quantum theory, the physical descriptions of the world have become purely abstract in the sense that they cannot be brought in contact with our everyday notions of the world, even in an idealized way.

However, the 'cat' in the paradox only serves the role of a macroscopic object. This macroscopic object need not be a biological entity. A flag which would either go up or down a pole, depending on whether or not the detector registers the photon can easily replace the cat. Thus, the issue is solely whether a macroscopic object can be conceived in terms of *physical states* other than mutually exclusive CDEF states, and there is no logical reason to suppose *a priori* that this is impossible. In fact the aim of this paper is to eventually show one such possible conception, although its actual application within quantum theory must await further work.

Another important consideration is that quantum theory is applicable to macroscopic objects also. Thus, it is not possible to solve Schrödinger's cat paradox by retaining the CDEF notions at the macroscopic level, and limiting the implications of superposition to the atomic level. Yet, all extant interpretations begin by adopting precisely such a move. This immediately introduces the famed 'measurement problem'. We have, on the one hand, a description of atomic objects involving superposition that does not permit any CDEF compatible intuitions. Yet, our observations in relation to these superposed states are always CDEF compatible. At present, various interpretations differ only over how to reconcile these two different conceptions of states, while agreeing on the use of the CDEF notions at the level of observation. In solving the problem, which arises only as a consequence of the commitment to the CDEF view at the level of observation, they seem to inevitably admit a role for the conscious observer to account for the appearance of a classically-definite state. This role is antithetical to the very spirit underlying the CDEF conceptions they assume to start with.

For example, in the 'collapse models', the superposed state is taken to be an objectively real state of the quantum system in between measurements, while a physical non-local mechanism is taken to resolve superposed state into a CDEF compatible observed state. Indeed, some physicists have invoked the objective reduction of the collapse-model to link quantum theory with consciousness. Some recent examples are Penrose (1989); Goswami (1989; 1993); Stapp (1993). Another example of a realist interpretation is the 'many-worlds interpretation', in which the state of superposition represents a collection of worlds in each of which the CDEF conception holds. Why does one particular result that we observe take place in our world? The interpretation can only account for it as a fact of our experience. Thus, this interpretation too requires a role for consciousness.

IV: On the Need for an Alternate Non-CDEF Conception of the Everyday World

What grounds do we have for assuming that within quantum theory CDEF notions apply at the level of observations, when quantum theory applies in principle to both microscopic and macroscopic objects? The correspondence principle is taken to provide the justification here. According to this principle, in the limit in which the Planck's constant is negligible compared to the quantities being measured, and this is generally the level of macroscopic bodies, the quantum mechanical description reduces, for all practical purposes, to that of classical description. However, Bohr himself wrote the following with reference to the correspondence principle:

> [I]n the limit of large quantum numbers . . . mechanical pictures of electronic motion may be rationally utilized. It must be emphasized, however, that this connection cannot be regarded as a gradual transition towards classical theory in the sense that the quantum postulate would lose its significance for high quantum numbers. On the contrary, the conclusions obtained from the correspondence principle with the aid of classical pictures depend just upon the assumptions that the conception of stationary states and of individual transition processes are *maintained even in this limit* (1934/1961, p. 85, italics mine).

Einstein too insisted that the fact that quantum theory requires macroscopic objects to be *in principle* in the state of superposition is significant for the issue of interpretation (1969/1949, p. 682).

The practical successes in applying quantum theory within the CDEF conceptions of the observations in the everyday world (via the correspondence principle) has promoted the rather pervasive conclusion that the radical implications of quantum theory are limited to the atomic realm. Our point of departure is to propose that the 'cat paradox' points to the need to go back and devise a non-CDEF, quantum-compatible conception of everyday objects for interpreting the *observations*. Schrödinger himself, in his later years, wrote along lines compatible with such an understanding of the cat paradox:

> It is probably justified in requiring a transformation of the image of the real world as it has been constructed in the last 300 years, since the re-awakening of physics, based on the discovery of Galileo and Newton that bodies determine each other's *accelerations*. That was taken into account in that we interpreted the velocity as well as the position as instantaneous properties of anything real. That worked for a while. And now it seems to work no longer. One must therefore go back 300 years and reflect on how one could have proceeded differently at that time, and how the whole subsequent development would then be modified. No wonder that puts us into boundless confusion! (Schrödinger, Letter to Einstein, 18 November 1950, in Prizibram, 1967, p. 38).

Schrödinger is raising the need to go back and redo *classical physics*, i.e. the physics of macroscopic objects. This is compatible with envisioning the need for an alternative to the CDEF conception of observations and associated macroscopic objects to realistically interpret quantum theory.

Let us allow for the possibility that quantum theory requires an alternative, non-CDEF conception of macroscopic objects. Such a conception would have to be complementary to the present CDEF conception of macroscopic objects. This means that while only quantum conceptions would apply at the atomic level (since classical theory is known to fail at this level), we can choose between two conceptions, one classical and another quantum, at the observational level. This could explain why, while having unmitigated success in using quantum theory practically by choosing to treat the everyday world within the CDEF conceptions, we are nevertheless unable to get an understanding of the state of superposition.

If we are right, in order to consistently interpret quantum theory, the first step would be to discover, *in ordinary language*, words to describe the states of *macroscopic* objects that are compatible with the formal principle of superposition. We shall designate the task of finding such a description of macroscopic objects as the 'observation problem'.

The observation problem aims to realistically interpret laboratory observations in a quantum-compatible manner using the everyday language. The measurement problem aims to realistically interpret the states of atomic objects directly, using quantum theoretical language. Within the CDEF framework, the task of solving the measurement problem remains highly problematic. To develop an alternative to the CDEF conception requires recognizing the observation problem.

The 'measurement problem' presupposes the 'quantum/classical dichotomy' with measurement as the sole link between the two worlds. The 'observation problem' begins simply by asking what it is that we are observing in the laboratory, if we treat them as simple everyday experiences. We thus see the 'observation problem' as being logically prior to the 'measurement problem'. The measurement problem might well dissolve once the observation problem is recognized and appropriately solved.

To take the observation problem seriously is to expect to go beyond describing the observations simply as meter pointer positions, detector clicks or localized spots on photographic plates. Einstein made a similar point:

> We are like a juvenile learner at the piano, just relating one note to that which immediately precedes or follows. To an extent this may be very well when one is dealing with very simple and primitive compositions; but it will not do for the interpretation of a Bach Fugue. Quantum physics has presented us with very complex processes and to meet them we must further enlarge and refine our concept of causality (Einstein, 1931, p. 203).

Einstein is certainly talking about enhancing our notion of causality within physics. However, he developed at length a philosophy of science in which he said the following:

> The whole of science is nothing more than a refinement of everyday thinking. It is for this reason that the critical thinking of the physicist cannot possibly be restricted to the examination of the concepts of his own specific field. He cannot proceed without considering critically a much more difficult problem, the problem of analysing the nature of *everyday thinking* (1936, p. 349).

Einstein evidently expected the refinement of our notion of causality to proceed from the level of everyday thinking. Indeed, Einstein constantly emphasized the 'freely created nature' of all of our concepts.

Even Bohr, who strongly insisted on the permanency of the CDEF conceptions recognized that any new concepts must *first* be shown to be applicable in the world of our experience. Only thus could he write so strongly:

> It would be a misconception to believe that the difficulties of the atomic theory may be evaded by eventually replacing the concepts of classical physics by new conceptual forms . . . the recognition of the limitation of our forms of perception by no means implies that we can dispense with our customary ideas or their direct verbal expressions when reducing our *sense impressions* to order (Bohr, 1934/1961, p. 16, italics mine).

The foregoing discussion brings out the striking difference between the approaches of Bohr and Einstein. It would seem that, contrary to popular opinion, Bohr was far more classical than Einstein was. Jammer has remarked,

> contrary to widespread opinion, [Einstein] rejected the theory not because he, Einstein — owing perhaps to intellectual inertia or senility — was too conservative to adapt himself to new and unconventional modes of thought, but on the contrary, because the theory was in his view too conservative to cope with the newly discovered empirical data (1982, p. 60).

Nevertheless, both Bohr and Einstein, in contra-distinction to the rest of the physics community, devised their interpretations starting from the common point that quantum theory is incompatible with CDEF-notions even at the level of *everyday experience*.

V: Bohr, Einstein and the Observation Problem

If we logically allow for an alternative, quantum-compatible conception of everyday objects, then until we devise such a conception, the observations pertaining to quantum theory should be treated as the *experiences* we have in the laboratory. Both Bohr and Einstein started from this point,[1] and thus could be regarded as having taken the first step in recognizing and solving the observation problem.

> For our theme, however, the decisive point is that the physical content of quantum mechanics is *exhausted* by its power to formulate statistical laws governing observations obtained under conditions specified in plain language (Bohr, 1963, p. 12, emphasis mine; see also p. 61).

> The de Broglie-Schrödinger wave fields were not to be interpreted as a mathematical description of how an event actually takes place in time and space, though, of course, they have reference to such an event. Rather they are a mathematical description of what we can actually know about the system. They serve only to make statistical statements and predictions of results of all measurements, which we can carry out upon the system (Einstein, 1940, p. 491).

Both, however, acknowledged that individual quantum objects are real and *cause* individual events, based on direct experimental evidence (Bohr, 1934/1961, pp. 100, 102, 112; 1957, pp. 16, 24, 73, 87).

How did they reconcile their realist commitments with their apparently instrumental view of the formalism? Both concluded that *more* than just the individual system was underlying the laboratory observations of quantum theory.

For Bohr, the system under observation and the experimental means for observing them formed, as Teller put it, a single epistemic whole:

---

[1]   Stapp (1993) must be credited with emphasizing this aspect in Bohr's interpretation.

From the beginning, the attitude towards the apparent paradoxes in quantum theory was characterized by the emphasis on the features of wholeness in the elementary processes, connected with the quantum of action.

The element of wholeness [has] the consequence that, in the study of quantum processes, any experimental inquiry implies an interaction between the atomic object and the measuring tools which, although essential for the characterization of the phenomena, evades a separate account . . . (Bohr, 1963, pp. 78, 60).

For Einstein, the quantum mechanical Ψ-function captured the state of the ensemble as a 'single whole':

. . . the Ψ-function does not, in any sense, describe the state of one single system. The Schrödinger equation determines the time variations which are experienced by the ensemble of systems which may exist with or without external action on the single system . . . [quantum mechanics] does not operate with the single system, but with a totality of systems . . . (Einstein, 1936, pp. 375–6).

This too must be necessarily an epistemic idea, because the ensemble of quantum systems can be prepared and observed one at a time in an experiment.

Thus, according to both, the 'cause' underlying each laboratory observation involves *more* than just the behaviour of the individual system, whose reality is not doubted. Thus, neither Bohr nor Einstein can be called Machian sensationists or instrumentalists of any kind, because of their realist *commitment*. However, in the absence of an alternative conception, they were obliged to treat their respective holisms epistemologically in order to avoid explicit contradictions with the CDEF-framework. Since their interpretations represent, in our view, the first step toward recognizing the observation problem, we shall discuss some relevant details of their respective interpretations. Both of their interpretations are quite complex, and the present author has treated them in detail elsewhere (Gomatam, 1999).[2] For the present paper, their interpretive ideas will be traced only at a summary level.

VI: Einstein and the Ensemble Holism

Physicists would understand me a hundred years after my time.
Einstein quoted in Pais (1992, p. 467)

Einstein's interpretation is an attempt to justify the assumption of the existence of the atomic object as a localized individual whose description is missing within quantum theory. The interpretation, at the same time, aims to account for the theory's pragmatic successes in describing the observable consequences of the behaviour of an ensemble of these localized, individual systems.

It is the situation in any probabilistic description that each observation can go only toward verifying the behaviour of the ensemble of identically prepared systems.

---

[2] The modern-day Copenhagen interpretation is generally taken to have evolved from Bohr's interpretive ideas. However, in most versions, the Copenhagen interpretation interprets the quantum formalism directly and objectively, as a description of the physics of individual quantum systems in time and space. We have already suggested above that both Bohr and Einstein took an instrumental view of the formalism. I have argued elsewhere (Gomatam, 1998) that although the two differed between themselves, if the version of the Copenhagen interpretation that treats the Ψ-function as objectively real is placed on one side of the so-called 'Bohr-Einstein debate', then both Bohr and Einstein would be on the other side.

Classical statistical descriptions are compatible with presuming a causal link between each individual observation and the real state of each individual system. However, quantum theory is irreducibly statistical. In this situation, Einstein took the laboratory observations as a whole to be related to the state of the *ensemble as a whole*. However, 'the programmatic aim of all physics [is] the complete description of any (individual) real situation (as it supposedly exists irrespective of any act of observation or substantiation)' (Einstein, 1969, p. 667). The view that quantum theory is incomplete follows: One would very much like to say the following:

> Ψ stands in a one-to-one correspondence with the real state of the real system . . . if this works, I talk about a complete description of reality by the theory. However, if such an interpretation doesn't work out, then I call the theoretical description 'incomplete' (Letter to Schrödinger, June 19, 1935; cited in Fine, 1986, p. 71).

Einstein held that 'quantum physics deals with only aggregations, and its laws are for crowds, and not for individuals' (Einstein and Infeld, 1966/1938, p. 286). Yet, he also emphasized that there are quantum individuals, and that they cause the individual laboratory observations (1936, p. 337). It would seem that the holist conception of the ensemble that Einstein is arguing for can be thought of more as a mob than a collection of distinct individuals. In a mob, the individuals act without individual identity, as a part of the mob. Such a view of the individual would relate well with the fact that the CDEF notion of the individuals as localized and separable entities fails within quantum theory. However, the failure of locality and separability may be only symptomatic. More fundamentally, the very idea of conceiving the quantum individuals within the space–time continuum may be at fault: 'the problem seems to me how one can formulate statements about a discontinuum without calling upon a continuum (space–time) as an aid' (Letter to Walter Dällenbach, November 1916, Item 9-072 cited in Stachel, 1986, p. 379). Einstein made the same remark again, in 1954, to David Bohm (*ibid.*, p. 380).[3]

Thus, the need for a non-CDEF conception of the individual object within quantum theory suggests itself in the context of Einstein's ensemble holism.[4] Indeed, given that quantum theory applies to macroscopic objects also, Einstein's ensemble holism is not incompatible with the demand for a new conception of everyday objects. Einstein however acknowledged his failure in general to develop new concepts for quantum theory (*ibid.*, p. 380). He was thus forced to take an essentially negative stand, about the impossibility of treating quantum theory as a complete theory of the individual *qua* individual in the CDEF sense.

---

[3] Stachel's paper is an excellent source and survey of many of Einstein's unpublished ideas in regard to quantum theory.

[4] It might be worthwhile to briefly point out that we see the 'ensemble holism' underlying Einstein's interpretation as being different from and logically prior to the holism due to 'quantum entanglement' that is often discussed in the literature. For a recent discussion of entanglement holism, see Esfeld (1999). The entanglement-holism presumes both the CDEF-interpretation of relevant laboratory observations and the individuality of the entangled particles, neither of which, as we have amply argued, is justified within quantum theory. The ensemble holism of Einstein instead focuses on the perceived missing identity of a single individual within quantum theory. If so, the physical meaning of entanglement of individual particles may yet depend upon ascertaining first the nature of the individual as described by quantum theory.

Bohr, on the other hand, worked within the prevailing CDEF conceptions of macroscopic objects, systematically interpreting all the non-CDEF features of the quantum formalism as indicating the epistemological limits within which these CDEF conceptions could be invoked to describe laboratory observations. In thus being less ambitious, Bohr's interpretation could form the natural next step in the search for a positive interpretation of quantum theory.

VII: Bohr and Quantum Relationality

Bohr envisions that within quantum theory 'an independent reality in the ordinary physical sense can neither be ascribed to the phenomena nor to the agencies of observation' (Bohr, 1934/1961, p. 54). This 'inseparability hypothesis' allows him to treat the laboratory observations as simple experiences and avoid contradiction with CDEF notions within quantum theory. It simultaneously renders the formalism a symbolic procedure, such that the theoretical concepts cannot be directly given physical significance (Bohr, 1963, p. 61). As a result, it is 'the application of the [classical] concepts alone that makes it possible to relate the symbolism of the quantum theory to the data of experience' (Bohr, 1934/1961, p. 16).

Consider a two-slit screen placed in the path of a thermionic source emitting a beam of mono-energetic electrons. A simple analysis of the experiment using classical intuitions shows that we can conceive of two measurable properties of the particle *in relation to the experimental arrangement*: through which slit the particle would pass, and in which direction it would emerge from the screen. The former can be called 'position at the slits' and the latter, 'momentum at the slits'. Classical intuition also suggests that by placing two detectors close to the slits, we can gain information as to the position of the particle at the screen (also called 'which-path information'). By placing a battery of detectors far away, we could ascertain the momentum of the particle at the screen. We cannot, however, do both with optimum precision at the same time. Therefore, the better our position set-up is, the worse will be our momentum set-up; and vice versa. This straightaway suggests the non-visualizable nature of the quantum formalism, and Bohr repeatedly emphasized the failure of pictures of trajectories of the particles in space and time within quantum theory. It is equally important to realize that the properties of the particle are defined here not absolutely, but *in relation* to the experimental arrangement we have made.

Bohr's key interpretive response to this situation is to include the experimental arrangement as a part of the very *definition* of the property under measurement. Thus, the quantum mechanical states express '*relations* between the system and some appropriate measuring device' (Feyerabend, 1961, p. 372, italics added; see also Jammer, 1974, p. 197). Using quantum theory, we interpret each observation as measuring a property that is defined in terms of the *spatio-temporal relation* between the observed system and the experimental arrangement. Two essentially different experimental configurations (the detectors being placed near or far) let the observed system enter into a 'position' or 'momentum' relation with the experimental arrangement. Bohr concludes that we should never expect to simultaneously *define* both properties to arbitrary precision in one experiment. Nor should we expect to be able to combine the results obtained from different individual experiments (Bohr, 1963, p. 4).

Bohr's emphasis, then, is on definition, not measurement. Whereas most physicists would say, due to uncertainty relations, that two quantum mechanical properties such as position and momentum cannot be simultaneously *measured* to arbitrary accuracy, Bohr, as we read him, emphasizes the limits to the definability of the two properties in question. Thus, Bohr treats the non-commutation relations entirely epistemologically, as undefinability relations, rather than as uncertainty relations:

> Quantum mechanics speaks neither of particles, the positions and velocities of which exist but cannot be accurately observed, nor of particles with indefinite positions and velocities. Rather, it speaks of *experimental arrangements in the description of which* the expressions 'position of a particle' and 'velocity of a particle' can never be employed simultaneously (Niels Bohr, *Second International Congress for the Unity of Science*, Copenhagen, June 21–26, 1936).

No doubt, even in pre-quantum theories, the properties are relational. The length of an object, for example, is expressed as a relation between the object and a standard scale. Or, its position is expressed in relation to a chosen co-ordinate reference frame. However, there is a clear difference between the relationality of pre-quantum physics, and the relationality of quantum theory that Bohr's interpretation as we read him implies. In classical theories, a property was expressed by a relation, but inhered in the object. In quantum theory, the property is both expressed by and *inhering in* the relation. The claim of Bohr seems to be that without actually setting the detectors in place physically, we cannot even define what property is being measured for the quantum object, and as soon as we consummate the relation by an actual laboratory observation, the property is asserted to exist. Thus, the experimental arrangement seems to both create the context necessary for speaking of a property of the object and to measure the property.

Bohr's move allows for the introduction of a new quantum notion of property involving relationality to interpret quantum theory. But Bohr himself did not go that route. He instead treated his insight epistemologically, by arguing that the 'inseparability hypothesis' which leads to the relational mode of description, is a result of the limitations of ordinary language we have run into, in the context of quantum theory (Bohr, 1957, p. 25).

Why didn't Bohr attempt an interpretation of quantum theory directly in terms of new 'relational properties'? Bohr steadfastly denied, for some reason, the possibility for devising new non-classical descriptions of everyday objects. He held that 'all new experience makes its appearance within the frame of our customary points of view and forms of perception' (1934/1961, p. 1), and limited such points of view to classical conceptions, i.e. those taken over in pre-quantum theories (see, for example, 1934/1961, p. 16).

Bohr recognized, I believe, that in order to treat it physically, the idea of quantum relationality would have to be first applied at the level of everyday objects:

> Once at afternoon tea in the Institute, Teller tried to explain to Bohr why he thought Bohr was wrong in thinking that the historical set-up of classical concepts would forever dominate our way of expressing our sense experience. Bohr listened with closed eyes and finally only said: 'Oh, I understand. You might as well say that we are not sitting here, drinking tea, but that we are just dreaming all that.' (Bastin, 1971, p. 27).

In the next section we shall present a conception of properties of everyday objects that we routinely invoke in everyday thinking that could provide a point of departure from Bohr's epistemological perspective, to treat his quantum relationality physically.

VIII: The Notion of Relational Properties

I believe that the first step in the setting of a 'real external world' is the formation of the concept of bodily objects and of bodily objects of various kinds.

Einstein (1936, p. 350)

We can identify at least three types of properties that we associate with an everyday object: primary, secondary and subjective properties. 'Length' and 'position' are examples of primary properties. An object being 'blue' or 'my father's gift' are examples of secondary and subjective properties, respectively. Ultimately all properties are subjective in the sense they are our notions of objects. Nevertheless, some properties can be treated as subject-independent more than others. A primary property, for example, can be thought of as a property that will inhere in the object even if all conscious observers were to cease to exist in the universe. By contrast, an object will not have a secondary or subjective property unless at least one conscious observer exists within the universe and perceives the object in that manner.

We can operationalize the above distinction between the three properties in a simple way. 'Length' is an observer-independent property because it can be expressed solely in relation to the length of another physical object, the length scale. The existence of a secondary property in an object would ultimately require appealing to the fact of experience of *any one* conscious observer. For this reason, color, although treated to some extent as an objective property in physics, continues to be considered a 'secondary property'. The existence of a subjective property would require appealing to *a particular* conscious observer.

Following Locke, we have generally presumed that physics should deal only with primary properties to remain objective. Indeed, the formal properties of classical physics are idealizations of primary properties. Progress in physics is often times gauged by how far aspects of non-primary properties can be treated objectively, i.e. as primary properties. However, we discussed in section III the problems in treating the properties associated with the quantum mechanical observables as objective properties in the sense of existing in nature with pre-determined values independent of measurement. Indeed, if primary properties were the only conception of objective properties possible, a subjectivity unavoidably enters quantum theory. It is therefore sometimes wondered whether quantum theory signals the end of the Cartesian divide between mind and matter.

However, the statistical predictions of quantum theory in any given experimental set up cannot be in any way changed by a conscious observer. This indicates strongly the ultimately observer-independent status of quantum mechanical properties. We can then certainly ask: are primary properties the only type of objective properties we can conceive of? We saw that Bohr strongly emphasized the relational mode of quantum mechanical description. Could it be that quantum mechanical properties are relational, yet 'objective' properties? In the reminder of this section, we shall identify a category of properties that we routinely associate with everyday objects, which are indeed objective in that they too are expressed in relation to another object, but differ from primary properties in important ways.

Let us consider a macroscopic object, say a book. From the viewpoint of classical mechanics, such an object would be described by a set of numbers representing mass, position, velocity, etc. Since classical theory is deterministic, these numbers are presumed to exhaust all causally relevant properties of the object. Characterizations of the object as a book, or as a 'gift from my father' are taken to be epiphenomenal descriptions, extraneous to physics.

However, in a consistent interpretation of quantum theory, if the macroscopic world would also have a quantum description, then a macroscopic object must possess more physical properties than those accounted for by classical mechanics. Again, an empathetic consideration of ordinary experience suggests some starting clues. A book can be used as a paperweight or a doorstop. Normally, these would be regarded as different uses of an object. However, they can be considered as involving a new kind of objective property, namely a relational property.

The use of the book as a paperweight requires setting it up in a particular spatio-temporal relation with another physical object, i.e. placing the book on top of a stack of papers. Thus 'paperweight-ness' can be regarded as a potential property of the object, which becomes 'physically real' only when the object is placed in an appropriate spatio-temporal relation with another object. Yet, the property itself objectively belongs to the book. Let us call such a property a 'relational property'. It is similar to and yet different from primary properties that physics has so far studied. Both primary and relational properties are objective in the sense that both are defined in relation to another object, and thus can be said to exist in the object independent of the existence of conscious observers. However, as the ensuing discussion will try to show, the two are different in the sense that while the primary properties are only *expressed by* a relation with another object (such as a scale or a clock), the relational properties are expressed by and *actualized* in a relation with another object.

The classical deterministic description indicates a causal closure in terms of primary properties. For the proposed relational properties to be not epiphenomenal, we can envisage them to be complementary to the primary properties of classical mechanics. Since an 'object' in physics, as Eddington put it, is a conceptual carrier of all of its properties, the two sets of properties would provide two different and complementary physical conceptions of the macroscopic object. The 'causal powers' of the macroscopic object conceived in terms of the relational properties should also be then different from and complementary to the causal powers of the same object conceived in terms of its primary properties.

The term 'complementarity' is used here in the sense of two non-intersecting conceptions, which taken together, provide a more complete conception of the object. This also marks the starting point of Bohr's conception of complementarity. However, within quantum mechanics, Bohr introduced the idea of 'complementarity' epistemologically, as a framework for invoking classical properties. We seek to introduce complementarity ontologically, with respect to relational and primary properties. Bohr's complementarity operated with respect to atomic objects, whereas ours pertains to macroscopic objects. Bohr presented the complementary framework as 'a rational generalization of the causal space-time description of classical physics' (1934, p. 87). We are envisioning that the proposed relational property view of the macroscopic object to engender a non-classical notion of causality.

We now reconsider Schrödinger's cat paradox (section III), from the viewpoint of ontological relational properties within quantum mechanics. A single electron in a state of superposition of one of two possible states (spin up/down; $\Psi = \sum_{i=1,2} c_i \psi_i$) is wired up to a flag on a pole, such that a detection mechanism will send the flag either up or down the pole. The central issue is whether the general superposed state Ψ can be conceived so that the fact that the macroscopic measuring device (flag) can take only one of two mutually exclusive CDEF states (up or down) does not pose a problem for interpretation.

We saw that an object can have many relational properties potentially. The key implication, for our present purposes, is that when we actualize one of these properties (by bringing about the necessary spatio-temporal relationship), others will remain potential. Placing the book in a particular spatio-temporal relation with a stack of papers actualizes its 'paperweight-ness', *while still leaving the other property potential*. In the case of the quantum thought-experiment under consideration, if we let 'spin-up' and 'spin-down' be relational properties, then our interpretation of the observations would have to change. Instead of describing the observation as the flag being either up *or* down the pole (which would make the corresponding properties absolute properties), we would say that the observation reveals which of the two possible *spatio-temporal relations* the system under observation has entered with the measurement device. In such a relational description, the actualization of one of the properties still leaves the other property as potential *in the observed system*. If the observation were to correspond to the spin-up eigenstate, the claim is that the spin-down eigenstate still exists, in some sense, as a potential relational state.

In current theory, the probability amplitude for the non-occurring eigenstate is taken to be zero since $\int_{v}^{v+dv} \psi\psi^* \, dv$ is taken to give directly the probability of finding a particle in the unit volume dv. We instead propose to treat the integral as providing the probability that a detector placed in the localized region dv will register a detection event.[5] Since this disconnects the appearance of the eigenstate from the probability statements, the probabilities for the non-occurring events would reduce to zero at the point of an actual measurement, but the corresponding amplitudes would not become zero. According to the Schrödinger equation, these amplitudes have a non-zero value. From the relational viewpoint, the non-occurring eigenstates remain potential and could be causally efficacious.[6]

---

[5] Bohr made much the same interpretive move, without stating it as such: 'The physical content of quantum mechanics is exhausted by its power to formulate statistical laws governing observations obtained under conditions specified in plain language' (1963, p. 12). Bohr saw this instrumental view of the formalism as a permanent necessity, assuming that 'as a matter of course, all new experience makes its appearance within the frame of our customary points of view and forms of perception' (1961, p. 1). We however introduce this move *provisionally*, arguing the need and the possibility for constructing new, quantum-compatible forms of perception to interpret everyday experience.

[6] I thank Brian Josephson for reminding me that the idea of non-occurring potentialities remaining non-zero at the point of an observation is also to be found in the de Broglie-Bohm-Vigier type of hidden-variable theories. These 'empty waves' are denizens of the 3N-dimensional configuration space. The conception of relational properties (involving a spatio-temporal relation) allows for the objective amplitudes (i.e. potentialities) to have a physical basis in the three dimensional space.

The present paper argues for a redefinition of the problem of interpretation of quantum theory by considering underlying philosophical issues. We have sought to replace the 'measurement problem' by the 'observation problem', and have discussed in broad terms, the implications of a relational property approach as a possible framework for solving the latter. We believe that the relational property view holds the resources for allowing a different use of the formalism complementary to the present one. However, it is beyond the scope of the present paper to expand on all the possibilities of the relational approach. Many more of our classical prejudices concerning objects, their states and properties may likely have to be jettisoned before the relational viewpoint can be fully incorporated within physics. As part of an ongoing effort in this direction, I have discussed elsewhere (Gomatam, 1999) the possibility for introducing 'information' as a basic physical notion within quantum theory.

IX: The Relational Viewpoint within Physics and the Neurobiological Issues Underlying Cognition

According to the dominant trend in current cognitive science known as 'cognitivism', cognition is representational and best explained using the computational framework. Cognitive processes are treated as 'passive', and determined by sensory stimuli originating in the external world. 'Questions of how the brain can a priori create its own goals and then find the appropriate search images in its memory banks are not well handled by cognitivism', since it ignores 'the possible "constructive" role that our brains might play in interpreting the very material of the external stimuli' (Freeman, 1998). Freeman proposes alternatively that perception is an active process, involving the occurrence of events internal to the brain that precede and shape our very perception of the external stimuli. Besides assigning 'an internal origin to the constructs that constitute meaning in the brain and that are the basis of effective action into the world', he additionally proposes, correctly in our opinion, that 'the values for these internally generated constructs are in the success or failure of the *relation* of the organism to its environment, not in its brain' (*ibid.*, emphasis mine). Freeman notes that at present the only idea that comes closest to this view of cognition is the philosophical idea of intentionality in the sense of Brentano (Freeman, 1996, p. 176; Freeman, 1999, section II).

The 'relational view' we have outlined above to interpret quantum theory contains, interestingly, a *physical* idea with the same characteristic. A relational property is defined as involving a spatio-temporal relation with another macroscopic object. The property exists potentially in the object regardless of whether the other object actually exists or not. In this sense, a macroscopic object conceived relationally contains an 'about-ness' to other objects of the world which, when treated as a physical property, can be causally efficacious.

Neuroscience is based on an understanding of the underlying physical and chemical processes in the brain. Presently our understanding of these processes is largely classical. Quantum theory under the relational viewpoint could make it possible to view the brain *as a quantum (i.e. relational) macroscopic object*, having new causal powers in virtue of new physical properties it has, now in *relation* to the environment. Such relational properties, by their very nature, would not be a fixed list of properties of the object. They would come about and vanish with the changing (spatio-temporal) relations of the object with its environment. This feature of relational properties is very much in line with the felt need in neuroscience to identify physical processes originating in the brain which carry their meaning and causal efficacy in relation to the environment.

X: Conclusion

Relativity theory, despite its far reaching idea of the space-time continuum, did not oblige us to change our ways of thinking about space and time at the level of practical living in the everyday world. If quantum theory, as argued in this paper, obliges us to actively reconstruct, *in everyday thinking*, our notions of objects and their causal properties in terms of their mutual spatio-temporal relations, it would represent a much greater and profounder impact on human thinking than all previous physical theories.

Acknowledgements

The author wishes to thank Alan Sommerer and Greg Anderson for many hours of valuable, critical discussions; Alan Sommerer, Greg Anderson, Omduth Coceal, Saul-Paul Sirag and Jean Burns for carefully reading and commenting upon successive versions of this paper; Henry Stapp and Edward MacKinnon for encouraging comments on a preliminary draft; an anonymous reviewer and John Smythies for comments on the concluding sections; Professor Walter Freeman for encouragement and support.

References

Bastin, T. (1971), *Quantum Theory and Beyond* (Cambridge: Cambridge University Press).
Bohr, N. (1957), *Atomic Physics and Human Knowledge* (New York: John Wiley & Sons).
Bohr, N. (1961), *Atomic Theory and Description of Nature* (Cambridge: Cambridge University Press); original work published 1934, pages cited in text refer to 1961 edition.
Bohr, N. (1963), *Essays 1958–1962 on Atomic Physics and Human Knowledge* (New York, London: Interscience Publishers).
Einstein, A. (1931), 'Interview' in Planck (1931).
Einstein, A. (1936), 'Physics and reality', *Journal of the Franklin Institute*, **221** (3), pp. 349–82.
Einstein, A. (1940), 'Considerations concerning the fundaments of theoretical physics', *Science*, **91**, pp. 487–92.
Einstein, A. (1969), 'Remarks to the essays appearing in this collective volume', in *Albert Einstein: Philosopher-Scientist*, ed. P.A. Schilpp (Illinois: Open Court); original work published 1949, pages cited in text refer to 1969 edition.
Einstein, A. and Infeld, L. (1966), *The Evolution of Physics: From Early Concepts to Relativity and Quanta* (MA: Simon and Schuster); original work published 1938; pages cited in text refer to the 1966 edition.
Esfeld, M. (1999), 'Quantum holism and the philosophy of mind', *Journal of Consciousness Studies*, **6** (1), pp. 23–38.
Feyerabend, P. (1961), 'Niels Bohr's interpretation of the quantum theory', *Current issues in the Philosophy of Science; Symposia of Scientists and Philosophers, Proc. of Section L, AAAS*, ed. H. Feigl and G. Maxwell (New York: Holt, Rinehart and Winston).
Feynman, R.P. *et al.* (1965), *The Feynman Lectures on Physics*, vol. III (Reading, MA: Addison-Wesley).
Fine, A. (1986), *The Shaky Game: Einstein, Realism and the Quantum Theory* (Chicago: University of Chicago Press).
Freeman, W. (1996), ' "On Societies of Brains" — Discussion with Jean Burns', *Journal Consciousness Studies*, **3** (2), pp. 172–80.
Freeman, W. (1998), Commentary on a target article by D. Watt, 'Emotion and consciousness: Implications of affective neuroscience for Extended Reticular Thalamic Activating System

theories of consciousness', ASSC Electronic Seminars,
    http://server.phil.vt.edu/Assc/watt/freeman1.html
Freeman, W. (1999), 'Consciousness, intentionality and causality', *Journal of Consciousness Studies*, **6** (11–12), pp. 143–72.
Gomatam, R. (1998), *Toward a Consciousness-Based, Realist Interpretation of Quantum Theory — Integrating Bohr and Einstein*, Ph.D. Dissertation, Department of Philosophy, Bombay University.
Gomatam, R. (1999), 'Quantum information', paper presented at the conference on *Quantum Approaches to Consciousness*, July 1999, Northern Arizona University, Flagstaff, Arizona.
Goswami, A. (1989), 'The idealistic interpretation of quantum mechanics', *Physics Essays*, **2**, pp. 385–400.
Goswami, A. (1993), *The Self-Aware Universe* (New York: Putnam & Sons).
Howard, D. (1990), 'Einstein and Duhem', *Synthese*, **83**, pp. 363–84.
Jammer, M. (1974), *The Philosophy of Quantum Mechanics* (New York: Wiley).
Jammer, M. (1982), 'Einstein and quantum physics', in *Albert Einstein: Historical and Cultural Perspectives*, ed. G. Holten and Y. Elkana (Princeton, NJ: Princeton University Press).
Newton, I. (1687), *Sir Isaac Newton's Mathematical Principles of Natural Philosophy and His System of the World*, trans A. Motte and F. Cajori (Berkeley: University of California Press, 1962).
Pais, A. (1992), *Subtle is the Lord. . . . The Science and Life of Albert Einstein* (Oxford: Oxford University Press).
Pauli, W. (1994), *Writings on Physics and Philosophy* (Berlin, Heidelberg: Springer Verlag).
Penrose, R. (1989), *The Emperor's New Mind* (New York: Oxford University Press).
Planck, M. (1931), *Where is Science Going?* (Woodbridge, CT: Ox Bow Press).
Prizibram, K. (1967), *Letters on Wave Mechanics* (New York: The Philosophical Library).
Stachel, J. (1986), 'Einstein and the quantum: fifty years of struggle', in *From Quarks to Quasars, Philosophical Problems of Modern Physics*, ed. R.G. Colodny (Pittsburgh, PA: University of Pittsburgh Press).
Stapp, H. (1993), *Mind, Matter, and Quantum Mechanics* (Heidelberg: Springer-Verlag).

Giuseppe Longo

Mathematical Intelligence, Infinity and Machines

*Beyond Gödelitis**

We informally discuss some recent results on the incompleteness of formal systems. These theorems, which are of great importance to contemporary mathematical epistemology, are proved using a variety of conceptual tools provably stronger than those of finitary axiomatisations. Those tools require no mathematical ontology, but rather constitute particularly concrete human constructions and acts of comprehending infinity and space rooted in different forms of knowledge. We shall also discuss, albeit very briefly, the mathematical intelligence both of God and of computers. We hope in this manner to help the reader overcome formalist reductionism, while avoiding naïve Platonist ontologies, typical symptoms of Gödelitis which affected many in the last seventy years.

Introduction

When one thinks of the foundations of the mathematical form of 'intelligence', of the ways in which it comprehends and describes the world, it is 'the rule' which comes to mind, or in fact the *regulae ad directionem ingenii*, the fundamental constitutive norms of mathematics and of thought itself. It is in particular during this century that the foundational analysis of mathematics has focussed on the analysis of mathematical deduction. This in its turn has been based on the play of logical and formal rules as described perfectly well by one of the best and most rigorous scientific programmes of our times, namely the formalist programme of the foundations of mathematics (and of knowledge). The influence of this programme in the analysis of human cognition has been enormous: in the '30s, formal computations and machines were rigorously defined, for the purposes of the investigation of formal deduction. And there began the modern adventure of the 'computational mind' or the various functionalist approaches to cognition. Reflections on the notions of mathematical proof and certainty, which started the project, may help in its revision.

---

*A preliminary French version of this paper appeared in *Revue de Synthèse*, 1 (January 1999).

Journal of Consciousness Studies, **6**, No. 11–12, 1999, pp. 191–214

Now, if there is no doubt that the notion of 'proof' does indeed lie at the heart of mathematics and that this proof must follow (or must be able to be reconstructed by) 'rules', yet mathematics is not just about proofs, and moreover it should be noted that these exhibit a wide variety of methods of construction, of reference and meaning. Proofs and constructions are carried on with the greatest of rigour, however there is no reason to think that in mathematics rigour should be understood only as the application of rules *without meaning*. That, as we shall see, is the central hypothesis of the formalist programme: certainty is obtainable only in the absence of meaning in the course of a deduction which, for that very reason, must be 'finitary' and 'potentially mechanisable'; any reference to meaning may lead to semantic ambiguities, may involve 'intuition'. Moreover, there is no doubt that mathematics is 'normative', as far as it follows (or it may be reconstructed by) 'rules', and that it is 'abstract' and 'symbolic'. Yet these complex notions do not need to be identified or reduced to 'formal', in the restricted sense given in this century. A *fortiori*, then, this reduction does not need to apply to general human reasoning and intelligence.

It is not a question of denying the logical component of mathematical deduction, namely the 'if/then' constructions which are always present, any more than the role of deduction conceived of as a pure calculation: many proofs in algebra or in logical systems based on 'rewriting' techniques (or purely syntactic rules of 'deductions-as-calculations') are simply non-self-evident sequences of applications of rules of formal manipulation. As such, these rules may appear to be 'without meaning'— their application in the proof must at no stage make reference to possible meanings (for example, one can develop an equation and reach a surprising result without ever 'interpreting' the equation in possible analytic spaces, but rather using only rules of algebraic computation).[1]

These two components of mathematical thought, namely logic, a system of rules of thought with meaning (Boole, Frege, Russell, etc.), and formalism, a system of rules of calculation without meaning (Hilbert, Bernays, Post, Curry, etc.), therefore make up an essential part of the proof. The problem is that they are not sufficient for the foundational analysis of mathematics. The search for foundations and 'certainty' exclusively within these systems has been a kind of *unilateral diet* of most foundational reflections; for the logicists, who are fundamentally opposed to the formalists, often look for the foundation only in 'signifying' logical rules, while the formalists want only to use calculation free of the ambiguities of signification, even logical signification. Both approaches, then, basically exclude each other as well as other forms of foundation (and rigour), such as reference to 'meaning' broadly construed, geometric meaning, for example. This diet, a kind of philosophical obsession which has tried to reduce the mathematical project to a single conceptual level, has led in particular to the lack of analyses of the role of meaning in the construction and foundation of mathematical concepts (and proofs): the role played, for example, by the concept of mathematical infinity (and its 'constructed meaning', see later), the phenomenal reference to space and time, or to those 'actual experiences' which not only root mathematics in a plurality of forms of knowledge about the world (I will refer for this to Poincaré, Husserl, H. Weyl, Enriques, etc.), but which also enter into the proof itself, as

---

[1]   This is well known in algebra. Relevant examples, related to the theories analysed here may be found in the introductory textbooks: Hindley & Seldin (1986) and Baader & Nipkow (1998).

I shall try to demonstrate. These constitutive elements of mathematical thought and practice must now become the object of genuine scientific analysis and no longer be consigned to the shadowy realm of 'intuition', that black hole of the ontological mystery (a typical symptom of Gödelitis),[2] last recourse of the 'working mathematician' who, fed up with the rules constraining his thinking, all too often resorts to a naive Platonism which allows him/her to dispense with these logico-formal systems which restrict the scope of his/her practices and are incapable of justifying them.

In short, I will try here to point out how (provably) non-formalizable arguments, yet based on robust, abstract, symbolic and human conceptual constructions, step in proofs even in arithmetic, the core theory of the formalist program. By this, one may avoid the dilemma between the Scylla of mechanicist formalism and the Charybdis of the naive platonist answer, largely prevailing in mathematical and philosophical circles of this century.

The presentation is meant for the general reader, in particular for the non-mathematician who is tired or puzzled of hearing that some incompleteness results 'prove' either that all mathematics can be actually computerized, since man, in the act of proving, is not better than a Turing Machine, or, alternatively, that mathematics is 'God-given'. The purpose then is 'reclaiming cognition' to our human being, as we developed one of our most fantastic conceptual constructions, mathematics, in the full rigour of a living process throughout history, yet rooted in deep cognitive process.

I: Incompleteness of Formal Systems: From Gödel to Girard

One of the major successes of mathematical formalism has been its ability to show its own limits — using its own methods of proof! Such a result is both remarkable and rare among the sciences, where 'crises' usually arise 'from the outside', by a change of paradigm and/or method.

Later on, we shall present various theorems of the last thirty years (often called of 'concrete incompleteness' for arithmetic or Formal Number Theory), proved by methods which do exit the given formal system, and for which it is *proven* that one must leave that system, in which the proposition is stated, in order to prove it. Note that this also applies to logical systems (Frege, Russell), since their principles are inscribed into the formal systems in question.

The truth is that a single level of expression, a fixed formal language, does not allow a complete representation of the mathematical structures in question, such as the natural numbers, for example. In other words, one shows that the mere manipulation of arithmetic symbols cannot be 'complete', cannot capture all the properties of a conceptual construction of great mathematical importance and product of innumerable experiences, even in the apparently simple case of (natural or integer) numbers. Or again, that the 'definitive certainty' of reasoning, sought by the original foundationalists, cannot reside in a single conceptual level, that of linguistic formal systems, which should have allowed the codification of the 'multidimensional space' of mathematical construction, independently of meaning as reference to the

---

[2] The virus of the Gödelitis was first isolated, by naming it, by one of the leading authors mentioned below, whose friendship, I hope, will not be affected by the use of the word in this paper.

underlying mathematical structure (e.g. the integer number line or the finite ordinals). Or, also, that some methods or 'conceptual tools', borrowed from other mathematical experiences or grounded on the intended meaning, cannot be 'removed'.

We take 'number theory', or arithmetic, as an example in order to show the way in which even forms of intelligence as apparently 'isolated' as this one, actually make use of also non-formal tools of understanding the world. This, in my opinion, is the true meaning of the various 'incompleteness theorems' in mathematics. In particular, the interaction between the plurality of linguistic levels and the concept of infinity in mathematics will be at the centre of our analysis.

But what do we mean by logical/formal calculus, and the axiomatic method of Hilbert? First we must create a language of *well-formed formulae* using a precise syntactical structure, which is defined *independently of meaning*, or, more exactly, by juxtaposing letters and symbols in a specified manner ('*A* **and** *B*' is well-formed; '*A* **and** ' is not). Then we must identify certain well-formed formulae as axioms, and fix some rules of deduction, where the passage effected by the rule from hypotheses to consequences is based *exclusively* on their syntactic structure. If, for example, we posit the formulae A and $A \Rightarrow B$, then B follows mechanically — without reference to the possible meanings of A, $\Rightarrow$, and B. (Of course, if one interprets $\Rightarrow$ as *implies*, then we recover the classical *Modus Ponens*; though this 'meaning', which is human and historical, is unnecessary for the formal deduction.) This, then, is the level of formal language, or object of study, at which one may analyse proofs.

In particular, Hilbert's programme relies on a distinction between the *mathematical* (or theoretical) level, called 'object level', expressed in a language of formulae, and a *meta*mathematical (or metatheoretical) level, i.e. the 'mathematics' thanks to which one is able to talk about the object level. On top of this, a 'third level' (or new conceptual dimension) was added by Tarski in the 1930s: the semantic structures which allow the interpretation of formulae and formal operations. (Although the mathematicians of the previous century had already laid the foundations for such a distinction, for example, Argand/Gauss re the complex numbers, and Beltrami/Klein re non-Euclidean geometries.) Thus, to take an example which deals with the geometric significance of algebraic formulae:

- $(x = \sqrt{-1})$ is a *formula* in the formal language of algebra, i.e. at the object or mathematical level of study;

- '$x = 2$ and $x = 5$ are contradictory' is a *metalinguistic phrase* — it affirms a *property of formulae*, the fact they are contradictory; and lastly,

- the interpretation of the x in $(x = \sqrt{-1})$ by a point in the cartesian plane furnishes a *geometrical semantics* for that algebraic formula, and, more generally that of all complex numbers (the Argand-Gauss interpretation).

In the case of arithmetic, according to Hilbert, it was necessary to prove, working at the metamathematical level, that the object-level language, entirely formalized by axioms and rules of deduction, allows all the formalizable properties of the integers to be proven (that is the hypothesis of the 'completeness' of the formal system). In other words, axioms and finitary rules can wholly account for mathematical deduction; that, in a certain sense, they wholly account for mathematical intelligence and certainty, or more precisely, that they are able to reconstruct these *a posteriori*, and give them a rigorous logical/mechanical foundation by providing them with a formal framework.

But why is arithmetic granted such importance? For several reasons: firstly, the natural numbers constitute an 'elementary' starting point; at the same time, their theory is at the very heart of mathematics. (Remember that Cantor and Dedekind gave definitions of the real numbers in terms of integers.) Moreover, Frege had remarkably set the logical foundations of mathematics in terms of the natural numbers in the 1880s, in the texts *Ideography*, and the *Foundations of Arithmetic*.

It is worth recalling the enormous clarification this programme brought about, as much by its methodological rigour as by the fact that it subsequently made possible the reification of the logical/arithmetic intelligence thus defined, in the form of prodigious electronic machines. For, once the logic has been translated into a formal system and the meaning forgotten, the machine simply needs to be taught to compare sequences of letters; when A and $A \Rightarrow B$ are given, just check *mechanically* if the two occurrences of A are 'identical' and then write B. 'Pattern matching' (letter-by-letter comparison), or the search for a common syntactic structure of formulae which are not trivially correlated (a process called 'unification' or identification modulo certain syntactic transformations) is indeed at the very heart of mechanical reasoning and, in our own times, proof by computers. With the appearance of machines, the idea that the formal level (calculation using signs without meaning) could express human intelligence in its entirety took on its modern form, going far beyond the pretensions of most of the original foundationalists, who aimed more modestly at the rigorous but *a posteriori* reconstruction of mathematics, constituting its only formal foundation. Certain thinkers even went so far as to affirm that 'intelligence is to be defined as that which can be manifested by means of the communication of discrete symbols' (Hodges, 1995) by means of meaningless manipulations of these discrete symbols.

Nevertheless, this program, which aimed at the codification of mathematics and its formal foundation, did not take long to fail, precisely because of arithmetic itself, and Gödel's (justifiably) notorious Incompleteness Theorem. It should be said, however, that the relevance of this theorem is limited by the absence of 'explanation' it furnishes of the incompleteness phoenomena. The subsequent discoveries, to which I would like to introduce the reader, can help us to understand better this game of intelligence which interests us, at the level of mathematical proof. So before proceeding to other, newer, and more informative results, I will try to add a few words *à propos* this classic, which has become the object of innumerable presentations and reflections, some of which have become quite popular, like those of Hofstadter and Penrose. Who knows if by trying — insofar as possible — to be concise, and by working in a level of informal rigour, we will be able to avoid those transcendent and ontological elements which (mis)lead many to believe that 'in mathematics, there are propositions which are true, but not provable', while at the same time leaving this notion of truth vague and mysterious. All we ask is that the reader pay careful attention to each word in the next few dozen lines for if we affirm that 'one proves that the proposition $\mathcal{G}$ is unprovable', for example, this phrase must be understood with particular care: what is said here is simply that, once a certain system of axioms and rules of deduction are fixed, one may prove that, inside this system, $\mathcal{G}$ cannot be proven, i.e. deduced using its axioms and rules. Some close attention is required, since in these contexts, one often uses words and phrases which refer to themselves ('proving unprovability', or 'this phrase is unprovable'); yet, what I am saying is absolutely informal and literal: the mathematical proof itself being long and extremely technical.

Recall in the first place that the First Incompleteness Theorem of Gödel (1931) states solely (I repeat, *solely*) that there exists a proposition or formal arithmetic phrase which is *undecidable* in the framework of arithmetic, *under the assumption* that the latter is consistent (i.e. not self-contradictory). It is an 'undecidability' theorem, in the sense that it gives us a proposition, call it $\mathcal{G}$, which can not be proven by the formal theory, and moreover, whose negation can also not be proven by the formal theory. In the statement of the theorem and in its proof, absolutely nothing is said about the truth of $\mathcal{G}$ or its negation (which of them is true?). The Second Incompleteness Theorem then proves that, *within arithmetic*, one can demonstrate the logical equivalence of the proposition $\mathcal{G}$ and the formalised statement of consistency. As a consequence, since $\mathcal{G}$ is unprovable if arithmetic is consistent, neither can the consistency of arithmetic itself be proven by 'arithmetic tools'. More precisely, no finitary metamathematics, which is as such numerically 'codifiable' (see below), can prove the consistency of arithmetic.[3] This unprovability property (of its own consistency) moreover, can be extended to *any* mathematical theory, which is sufficiently expressive to allow the codification of its own meta-theory (note the interaction between the theoretic and metatheoretic levels here.)

The above is one of the keys of Gödel's proof: the remarkable idea of *numerically codifying arithmetic formulae*, by means of a laborious but conceptually simple technique (later called 'gödel-numbering'). Once this has been achieved, formulae which describe properties of numbers, for example $(x + 4 = 1 + x + 3)$, can 'speak' about formulae, in the sense that they can be applied to the numeric codes of formulae. If, for example, $(x + 4 = 1 + x + 3)$ has numeric code 76, then it implies in particular that $(76 + 4 = 1 + 76 + 3)$, which is a fact about an instance of 'itself', where 'it' is understood as 'formula #76'. In fact the codification of a (well-formed) formula does not depend in any way on its 'meaning', but only on its syntactic structure: the finite sequence of symbols of which it is composed. But, recall, syntactic structure is sufficient for the formal analysis of deduction.

By this, and in a very specific sense, arithmetic formulae can 'speak' about themselves, or even about their own properties. Consistency, for example, is a property of formulae, *and therefore* a property of numbers, once given that numbers, as codes, can be put in one-to-one correspondance with formulae. From then on, *meta*mathematics, which studies the properties of formulae, becomes a sub-field of arithmetic. Briefly, as a mathematical theory, arithmetic describes its own metatheory. This is a mathematically difficult observation, which has opened the door to so much beautiful mathematics — and to so much extravagant speculation: arithmetic speaks or refers to itself or, even, it is 'conscious' of itself, or about the infinite regressions of self-references, as if looking in facing mirrors, etc. Some of the most severe

---

[3] A formal theory is consistent if, in its language, one may write an unprovable sentence. Thus, by the Second Theorem for Arithmetic, consistency is unprovable exactly when it is assumed to be true. Or, consistency is unprovable (in arithmetic) if and only if arithmetic *is* consistent. In Gödel's theorems, one has to make a subtle distinction between consistency and 'omega-consistency': a technical nuance which is outside the scope of a discussion like the present one. A detailed presentation of Gödel's theorems may be found in Smorinsky (1978).

pathologies related to Gödelitis, for which, of course, the immense Gödel has no responsibility whatsoever.

Now, arithmetic *is* consistent, in the sense that one cannot derive a contradiction from its axioms and rules (and thus the proposition $\mathcal{G}$ holds, as it is provably equivalent to consistency). The proof of consistency, however, must be made *outside* of arithmetic, as a consequence of the two theorems of Gödel; in other words, one can not do it with 'purely formal reasoning', which is mechanisable and hence codifiable within arithmetic. Again, one can not do it in a *formal theory*, i.e. a theory within which one avoids the *meaning* of axioms and rules, or, as we will try to explain later on, within which an ordinary axiom can not 'speak of infinity' nor of the standard order structure of numbers, or where the rules do not mix the theoretic, metatheoretic, and semantic levels. As already mentioned, if one assumes consistency or proves it, then it becomes banal to observe that $\mathcal{G}$ is true: the Second Incompleteness Theorem actually proves the equivalence between the two, inside arithmetic. Indeed, there is no method to *affirm* the truth of $\mathcal{G}$, other than specifying a notion of truth for formulae and *proving* it to be true. Thus, in order to know that $\mathcal{G}$ is true one must assume or prove consistency, to which it is provably equivalent.[4]

Now whither the famous ontological mystery, as claimed by Platonists who reject the formalist program? '$\mathcal{G}$ is true, but not provable' means nothing but '$\mathcal{G}$ is not provable in the framework of arithmetic, if one assumes consistency' and, under this *necessary* assumption, one can prove its truth; or, more precisely, '$\mathcal{G}$ cannot be proven except by techniques stronger than arithmetic — those which allow one to prove consistency' (and we shall see what these may be). In mathematics, when one affirms that something is true, *in one way or another*, one must (explicitly define 'truth' and) *prove that it is true*, and that is all. This is what one should analyse, i.e. *how* one may prove things outside a specific formalism, before stooping to theological arguments. One should research what non-mechanisable, non-arithmetical forms of reasoning (dare we say 'forms of intelligence'?) allow the necessary kind of proof.

As a matter of fact, the statements *and the proofs* of Gödel's theorem do not ever mention any notion whatsoever of 'truth', either magical or mathematical. They are an absolutely remarkable game of codes for formulae, formal fixpoints equations, explicit computations (Gödel invented 'programming in arithmetic', an amazingly difficult and original challenge). At most, one may understand the statements by observing that they prove a 'gap' between formal proofs and various possible notions of truth (Tarski's, Kripke's — there are many) over the (standard) model of arithmetic (the natural numbers). That is, that formal provability differs from any reasonable *definition* of truth for arithmetic formulae, in particular if this definition assumes that any proposition is either true or false (the so called *tertium non datur*). The further shift, from this *gap between* a *notion of truth* and *formal provability*, to the God given

___

[4] There is a simple 'classical' argument which *proves* the truth of $\mathcal{G}$, *under the assumption* that arithmetic is consistent and once the First Incompleteness Theorem has been shown. It is just a naive paraphrasis of one implication in Gödel's Second theorem, which derives formally $\mathcal{G}$ from consistency, i.e. *within* arithmetic. In summary, by the previous note, $\mathcal{G}$ is unprovable if and only if (consistency is unprovable if and only if arithmetic *is* consistent if and only if) $\mathcal{G}$ is true. Note then that a biconditional, '$\mathcal{G}$ is unprovable if and only if $\mathcal{G}$ is true' (remarkable, isn't it?), is *not* equivalent to an 'and'; it implies an 'and' if one 'assumes' or proves consistency (or $\mathcal{G}$). In truth, though, one should also say what 'truth' means for a formula of arithmetic, exactly.

'set of true, but unprovable sentences', is just some sort of medieval confusion between mathematical provability and ontological arguments. As a matter of fact, in mathematics, it is not the existence of entities or objects that matters, but the objectivity of mathematical/conceptual constructions. The results discussed below will better explain that this gap is actually between formal (arithmetical) provability and other forms of conceptual constructions, proper to mathematics.

Consistency of arithmetic was soon proven by Gentzen, in 1934. However, his proof presupposes an extremely strong form of induction (that of 'transfinite' order), which made it unconvincing to many: the very essence of arithmetic being ordinary induction (that of finite order).[5] This is the reason it went largely unnoticed, even though its technique of 'cut-elimination' went on to become a pillar of Proof Theory. This proof can be given in a set theory which includes an axiom affirming that 'there exists an infinite set'. This axiom does not have any sense unless one 'understands' the meaning of 'infinity': its codification by an arithmetical predicate is impossible since arithmetic cannot 'say anything' about infinite sets (nor even ' single out' the finite sets or the standard integer numbers). Gödel also gave a proof of the consistency of arithmetic in 1958, in an interesting logical calculus (an extension of Church's calculus of the '30s, a 'Theory of Types and Proofs' called system T), which also uses transfinite induction.

We owe a more enlightening proof of the consistency of arithmetic to (Tait and) Girard, in 1970. It is given in a framework, called system F, similar to, but much more expressive than, Gödel's system T; system F is still an extension of Church's calculus, but with second-order impredicative types.[6] Girard proves, '*à la* Tait', a 'normalisation' theorem, which implies consistency, using, among others, a principle called 'second-order comprehension' which combines, in an unavoidable way, theory, metatheory and semantics. I will try to explain this very informally, by abusing of easy and short ways to render difficult concepts and techniques. At a certain point in the proof, one takes an infinite set of terms (this is a *metatheoretical* operation: one collects terms of the object level, which one perceives 'from on high', i.e. metatheoretically, an 'easy' operation for us human beings, along the proof, as we do it in our ordinary language). Then, one puts them in the place of a set-valued variable inside a *term* — which amounts to saying, one works at the *theoretical level* here. Thus in the course of the proof itself, one mixes metalanguage, and even semantics, with the language of terms, since the operation can be done only if one agrees to interpret formal, set-valued variables as actual sets (the so called 'semantic convention' of the axiom of second-order comprehension).[7]

---

[5] Arithmetical Induction is nothing other than the assertion that if one can show (or if one assumes) $A(0)$ and, writing '$\forall$' for 'for all', one shows (or assumes) that $\forall y \, (A(y) \Rightarrow A(y+1))$, then one can conclude $\forall y \, A(y)$. In an equivalent fashion, if $\forall z \, ((\forall x < z \, A(x)) \Rightarrow A(z))$, then one can again deduce $\forall z \, A(z)$. Transfinite induction allows an infinite number of hypotheses, i.e. it may be informally understood as interpreting z (and x) above as 'infinite numbers' (called transfinite ordinals). Gentzen used transfinite induction over a restricted set of formulae, not all formulae of arithmetic — a key point.

[6] A mathematical notion is impredicatively given (is *impredicative*), when one uses a totality in order to define, by this notion, an element of that very totality.

[7] For a unifying approach to both Gödel's and Girard's normalisation theorems, see Girard *et al.* (1989). Type Theory, the frame of Gödel's 1958 work as well as Girard's, is elegantly and deeply related to Category Theory, *the* theory of mathematical structures (see Lambek & Scott, 1986, and Asperti & Longo, 1991). A very interesting categorical understanding of normalisation may also be found in

This passage of proof (it is not the only one) is not codifiable within arithmetic, and Girard proves it, by showing that his theorem implies the consistency of arithmetic and hence cannot be proven inside arithmetic. The reasoning is impeccable, comprehensible, and human: one needs to understand the blend of meta/theory/semantics to carry on the proof and no purely formal/mechanical/finitary account of it can be given. In a certain sense it corroborates what Wittgenstein already claimed in the 1930s: that the distinction between mathematics and metamathematics is fictitious. More accurately, it is a fine and technically convenient conceptual distinction, for the purposes of the proof-theoretic analysis of mathematics, but as humans, we may move freely, by means of the interaction of language and meaning, between one level and another, just as we do every day in real mathematics (or in ordinary language: e.g., 'I never say true sentences with more then 34 words' — a typical mixture of the levels above). Treatment by a single linguistic level does not allow this type of interplay, no more than the reasonings which concern it: the Tait-Girard and Gödel Theorems (and an observation of Tarski) prove it. Now, digital machines do not function above the formal linguistic level, the level which can be codified by sequences of zeroes and ones. Our human language is, on the other hand, a dynamic construction, built in a permanent resonance with meaning, and so is mathematics and its proofs. By this, they are in a position to capture at once language, metalanguage, and meaning, which is *demonstratably* undoable by formal/mechanical means, i.e. by giving a finite coding technique by a discrete set of meaningless signs. If one wants to keep them purely formal, i.e. mechanically manipulable, this immediately gives the distinct levels of metalanguage and semantics. If you try to encode these levels — by further formal signs — the game starts over again, by further metalanguage and semantics.

In truth, this is the central point: by definition, a digital calculator must codify (everything) in symbols, roughly 0 and 1, and the encoding can depend neither on meaning nor implementation. Hodges' definition, mentioned above, applies very accurately to *mechanic* deduction: '(mechanical) intelligence . . . is effectively defined as that which can be expressed by the communication of discrete symbols', and this codification following the requirements of the functionalist hypothesis *must not* depend on the specific *hardware* which realises it. The goal of high-level programming languages is precisely to be transferable from one computer to another, from one programming environment to another, without problems. That is only possible because their level is exclusively formal/theoretic, codifiable with finite sequences of symbols, and neither depends nor should depend on any (ordinary) meaning any more than on contexts. Human intelligence, on the other hand, depends on the structure of our brain, the fact that it is housed in *our* cranial cavity, and the complexities of its biological and cultural history: it is a rich blend of invariant, general laws and contextual meanings. It bears no rigid distinction between 'language' and 'metalanguage'; moreover, the meaning of the processing which occurs at any point in the brain depends also on *where* the elaboration takes place, on its geometry, on the *type* of the preceding neurons. And so on, up to the actual position of one's

---

Cubric *et al.* (1998). Among the innumerable applications and developments of the normalisation theorems, a technically intriguing one may be found in Castagna *et al.* (1995), which has had some fall-out in the mathematics of programming, as the work of many in Type Theories (the author's papers are downloadable from http://www.dmi.ens.fr/users/longo).

hands — as much because of the rôle the hands played in the evolving complexity of our cerebral cortices, as for the more historical fact that one understands body language (in particular, hand-waving), no matter who is the Italian speaking! For some, this contextual dependence, rooted in evolutionary, social and cultural history, can represent a limit, when in fact it is about a richness: the 'gesture' of the mathematician, who tries to explain the 'construction of a limit', refers to a deep and shared conceptual construction (e.g. the notion of actual infinity used) and it is inscribed irreducibly into the proof. This gesture does not make reference to an 'ontology', but to a constitutive route through the history of mathematical knowledge, it is an essential part of the metaphors that yield the conceptual invariant. The actual challenge is to understand how we get to a (relatively) stable invariant, such as the concept of infinity, yet grounded in our material, contextual lifes.[8] In some cases, the conceptual invariant results from a stability gained through intersubjective exchange, rich in meaning: its formal representation is a remarkable 'attempt' at capturing its expressiveness, but it is essentially incomplete.

II: Infinity and Proofs

In proving the consistency theorems mentioned above, the use of the notion of infinity turns out to be inevitable, and this fact is made even more explicit in other, more recent 'incompleteness theorems'. If I continue to speak of the use of *actual infinity* in the theory of integers (arithmetic), it is deliberately to 'play into the enemy's hand': it would seem too easy to maintain that the proof of theorems about infinite-dimensional differentiable manifolds (which are very abstract spaces), require mathematics to be able to speak of infinity; but what proof theory has taught us is that even our good old positive integers sometimes require the concept of (actual) infinity, if one takes as a frame the usual set-theoretic approach.[9]

There are other properties stemming from arithmetic, which are not codifications of metatheoretic properties like consistency, but actual properties of numbers (like, for all x there exists a y such that $(6 + x = y + 2)$ — just a little more complicated), which can be shown to be *unprovable* by finitary techniques, i.e. by deductions that can be codified inside arithmetic. But one *can* prove that these formulae are true for the natural numbers, by proofs which use 'infinity' in an essential way, in a set-theoretic perspective. Our goal here is to reflect on the manner in which these theorems place the mathematical proof in a plurality of forms of intelligence, not just 'formal', in particular the concept of infinity is applied.

The Paris-Harrington theorem (PH) and the 'Friedman Finite Form (of Kruskal's theorem)'· (FFF) are two arithmetic statements of the type 'for all x there exists a y such that blah, blah, blah . . .'. Here 'blah, blah, blah . . .' is a property of numbers which may be complicated, but not overly so (see Paris & Harrington, 1978; for (FFF), see Harrington *et al.*, 1985). Both (PH) and (FFF) *imply* the consistency of arithmetic, by a proof *within* arithmetic; but, given that consistency, when formalised as a proposition of arithmetic, can not be proven in arithmetic, neither can either of

---

[8] Some more references and discussion may be found in Longo (1997; 1999b).

[9] Yet, other approaches may be followed, see the forthcoming footnotes: we hint, in the text, to the mainstream set-theoretic approach, but different proofs of the same unprovable statements, may be non-arithmetizable for different reasons.

the two statements. Moreover, both statements describe more or less 'concrete' properties of numbers (partitions or 'colouring', as for (PH), inclusions of finite trees, as for (FFF)).[10]

Even though unprovable in arithmetic, (PH) and (FFF) are *true*. For a mathematician who is speaking of interesting propositions (FFF, in particular, is very interesting — it is a variant of a well-known theorem by Kruskal which is rich in applications), that means that (s)he can prove them *and can mean nothing else*; alternatively, that (s)he possesses convincing techniques for deducing the truth of these propositions, relative to the structure in question: the natural numbers. These techniques, and I am now thinking particularly of (FFF), base themselves essentially on the order-structure of the natural numbers, a geometric or an infinitary property, and on sequences of finite and infinite 'trees': they involve making 'instruments of proof' out of our mathematical experience of reasoning about the well-ordered sequence of natural numbers, and confronting simple planar (infinite) structures, trees, by inclusion, and describing the difference between finite and infinite.[11]

As with the (provable) truth of $\mathcal{G}$, there is no miracle in the truth of (FFF): it is quite simply the 'laborious conquest' of a proof handling (countable) infinities, organised in particular as the totally (well-)ordered set of the natural numbers, or as a partially ordered set of 'nodes' and 'branches' of trees.[12] As indicated in the footnote, that is a relatively easy proof, but formally underivable: but ... why must foundation be solely formal/mechanical? Do we not have other things to say about mathematical proof?

---

[10] A partly informal and simplified statement of (FFF) may be given as follows. In mathematics, trees have a 'root', 'branches' and 'nodes' and may be included one into the other, in a roughly ordinary sense. Then (FFF) says: 'For any n, there exists an m such that for any sequence of finite trees $T_1, T_2, ...,$ T_m, such that each T_i has at most n(i + 1) nodes, there exist j and k such that j < k < m + 1 and T_j is included in T_k.' Finite trees may be coded by numbers, thus the statement is a formal statement of (first-order) arithmetic, a $\prod_2^0$ statement of the arithmetical hierarchy, to be precise.

[11] In a comparison, observe that the proposition $\mathcal{G}$ of Gödel, in section I, affirms that there 'does not exist a proof of $\mathcal{G}$' (nor of its negation) or that $\mathcal{G}$ (more accurately: its numeric code) is a solution in x of the following equation ($x = $ *the code of 'there does not exist a proof of the proposition coded by x'*). This is a very fine game between metatheory (the notion of proof), theory (the formal proposition $\mathcal{G}$) and semantics (the integer numbers, where x must be found). Yet, it is 'artificial' or *ad hoc* (it is the arithmetic coding of a metatheoretic statement), it is not an 'interesting' property of numbers *per se* or of finite codable structures, such as trees, say. Finally, the statement and its informal meaning suggest why $\mathcal{G}$ must be unprovable, if arithmetic is consistent: this is exactly what $\mathcal{G}$ says! The difficulty of the theorem entirely lies in the construction of $\mathcal{G}$ and not in the proof of its unprovability nor in its truth, if consistency is assumed. Indeed, the 'apparent evidence' of $\mathcal{G}$ has had a major misleading role in many philosophical reflections. This is not so for statements such as (FFF), whose truth is far from 'evident' and yet it is an 'ordinary' theoretical expression about numbers as codes of finite trees (no self-reference, no metatheory involved). Its formal unprovability is extremely hard to be proved; it uses an ordering of trees by inclusion and transfers its properties, by isomorphic immersion, into the order of (very large) transfinite ordinals. This, following Gentzen, implies the consistency of arithmetic.

[12] The truth of (FFF) is not so obvious, as it requires a 'simple', yet smart proof, under classical, but strong assumptions. Briefly, the set-theoretic proof of (the truth of) (FFF), via Kruskal's theorem, uses an 'oracle' on a $\sum_1^1$ set, an impredicative and extremely non-effective infinitary construction. This proof is relatively easy: one or two pages which any mathematician today could reconstruct without much pain (indeed, some fun). The analysis of this 'easyness' has been one of our motivations. On the other hand, as we already said, the proof of its unprovability (and its essential impredicativity) is very difficult; it is a major technical breakthrough obtained by Friedman. (The complete statement, the proofs and other technical remarks about (FFF) may be found in Harrington *et al.* (1985) — in particular, two papers by Smorinsky in that volume brilliantly explain the role of impredicatively given sets, the place where syntax and semantics are entangled.)

What is the *constitutive process* of the 'certitude' of mathematical proof, even outside formalizations? The formalist program is only a component, necessary and important, but provably incomplete, of the analysis of proof. Moreover, on the analysis of the formal consistency, it achieves nothing without appealing to larger and larger infinities (see section III): a conceptual abyss which must in turn be 'founded'.

So I speak here of the 'laborious conquest' of the notion of infinity, since clarity about the infinite is a conceptual conquest which has been needed in mathematics for centuries; the analysis of this constitutive process is, for me, an essential subsequent component of foundational analysis.

We know in effect how the Greeks hesitated when faced with infinite sequences of converging points ($\sqrt{2}$; Zeno's paradox) and how they laboriously, even anguishedly, distinguished between potential infinity, and a more indefinite infinity, negatively defined by Euclid and Aristotle (*apeiron*), though the latter provided some degree of clarity on the difference between the two. Not until Thomas Aquinas, Duns Scotus, and the late Middle Ages did a real change of scientific paradigm take shape. Infinity became a positive attribute, characteristic of God and possibly created by God (infinity *in actu*, as opposed to infinity *in fieri*). As for the mathematics of the sixteenth to eighteenth centuries, we recall that it developed in a milieu of uncertainty, but with a forever increasing audacity with regard to the use of the notions of actual infinity, and of limit, which no longer represented a negative concept, but on the contrary, a manner of calculation (cf. Pascal, Galileo, Cavalieri, Newton, or the metaphysics of Leibniz). See Zellini (1980); Gardies (1984). This culminated in the great casualness yet conceptual clarity of Cantor's treatment of infinity. His 'paradise of infinities' constitutes another turning-point, one which has truly marked our mathematical era. Cantor introduced infinity in the context of operations: addition, multiplication, and iteration of limits on infinities of infinities. Such juggling with infinities of excessive scale even led to paradoxes. A century of mathematical work accompanied by an increasing refinement of techniques and also an increasing solidity of definitions was needed. Today a mathematician really knows what it means to 'give a proper definition', above all in the difficult cases which imply infinitude, thanks mostly to the stubborn effort of logicists and formalists! This allows us in the present to promote infinite trees, the well-order of the natural numbers, and ordinals beyond them, to the title of daily instruments of proof; and this, without falling into the same errors and paradoxes into which the audacious 'founding fathers' stepped. This mathematical praxis, finally, allows a rigorous *definition*, in different contexts, of infinity, i.e. exactly that which was undefined, according to Euclid and Aristotle. (Cantor really was a mathematician who dared to conjecture and prove the most surprising observations about infinity; truly, he dared to 'think beyond infinity'.)[13]

---

[13] For us today, Cantor's transfinite Arithmetic is nothing particularly difficult. After counting $0,1,2,3,\ldots$, we use ω to denote the 'limit' of this process, its 'closure' on the horizon. Then we continue counting $\omega + 1, \omega + 2, \omega + 3, \ldots, \omega + \omega = \omega \times 2$. Similarly, $\omega \times 2, \omega \times 3, \ldots \omega \times \omega = \omega^2$. By now, the rule of the game should be clear, and one continues to apply exponents: $\omega^2, \omega^3, \ldots, \omega^\omega$. The limit of $\omega^\omega, \omega^{\omega^\omega}, \ldots$, should simply be '$\omega$ to the ω', ω times. This 'ordinal' number is called ε_0. It is the least solution of the equation $x = \omega^x$. If one succeeds in proving (or if one assumes) that these ordinals are 'well-ordered', then one can prove that arithmetic is consistent (this is Gentzen's 1934 proof). The statements (PH) and (FFF) which we have mentioned imply the well-ordering of ε_0, and much more, for evidently one can also create $\varepsilon_1, \varepsilon_2, \ldots, \varepsilon_\omega$ and so on and so forth, up to 'huge' ordinals (whose *individual* definition requires a reference to the entire *collection* of ordinals: the 'impredicatively defined' ordinals). In fact,

In other words, the order of the numbers, in space or time, their succession and extension into a discrete and well-ordered structure, i.e. their ordered extention beyond the infinity of the numbers themselves, the planar structure of possibly infinite trees, are all infinitary or geometric properties which allow the proof of these finitary arithmetic propositions, which *require* treatment beyond formal systems codifiable within arithmetic. Evidently, one must *speak* of these structures, and the proofs which rest on them, with human words of finite length, but one can not do so in a 'complete' manner with *formal calculi* which can be manipulated without reference to meaning: this, no more no less, is what all these results tell us; the *historical praxis* of infinity, the order according to which we organise, mentally or in the plane, numbers and trees (possibly infinite), are part of the foundations, are the extremely solid roots of unmechanisable methods of proof. This amounts to saying that in these proofs, in the feat of making hypotheses and passing from one line to the next, these lines certainly consisting of finite words of our language, 'meaning' steps in, in a *provably* essential way. That is, along the proof, one must *understand* the concept of actual infinity or the geometric structure of well-ordering; or that one should be speaking of infinite sets of numbers or infinite trees in the plane, by situating oneself in a mental geometry, where one may 'pick up' the least element of an arbitrary non-empty subset of the natural numbers (an absolutely infinitary operation, in set-theoretic terms). Although at the end of the day, these proofs should themselves be described by finite words, *for passing from one phrase to another*, in at least one step along the proof, it is therefore necessary to grasp that *behind* these words lie the significance of infinity or of orders, *provably uncodifiable* in arithmetic. Or, in other words, neither codifiable nor manipulable by sequences of finite symbols without meaning, by a Turing Machine, say, as its computation could be equivalently expressed within arithmetic. Thus, these recent incompleteness results show that even 'simple' properties of the integers are provably true using tehniques which can not be represented at the solely theoretic level, codifiable as such with zeroes and ones, or with other mechanisable techniques.[14] The difference with respect to Gödel's

---

this is a game children and mathematicians often play: give me a number, and I give you a bigger one. But the game is not arbitrary as it 'takes meaning' in an interesting geometric structure, the well-ordering of the natural numbers. The game only *extends* this order, by extending the operations of sum, product, exponentiation, and 'limiting' beyond ω. The deed of giving a name to infinity, in a coherent manner, has been inscribed in the conceptual area we call mathematics. The ordinal ω is not of this world, but neither is it a convention nor a mere symbol; rather, it consists of a *construction principle*, a 'disciplined gesture' grounded in a historically rich mathematical practice (cf. Longo, 1999a).

[14] The firm, but naive, formalist may still say: yes, but then, this proof, still given by finite sets of words, may be 'formalized' in a 'suitable' Set-Theory or in second-order arithmetic! Of course, one may add the independent statement as a new axiom, but this is cheating and just moves forward the problem (by Gödel's technique one may give further independent statements). The point is that sufficiently expressive, but not *ad hoc* theories, such as proper second order ones, are not 'formal', in the rigorous sense of the hilbertian tradition: they are infinitary and use tools from infinitary logics or impredicative definitions. The set-theoretic treatment mentioned above, by the use of a $\sum_1^1$ formula or the derivation in second-order arithmetic (provably) require impredicative notions. By this, syntax and semantics get mixed (first, by the so-called second order convention in the comprehension axiom) and, as for arithmetic, validity, as truth in all models, is no longer 'effective', i.e. the set of valid propositions is not recursively enumerable (thus, they cannot be derived by any machine whatsoever, by Church Thesis). For this, and for the related use of totalities in order to define elements, the firm, but coherent and competent formalists reject proofs in these impredicative frames (see, for example, the life-long work of Feferman (for a collection of papers see Feferman, 1998, or Simpson, 1998).

G. LONGO

theorems should be clear. As already mentioned, that classic is only an undecidability theorem: it gives no hint on how to prove $\mathcal{G}$ or the consistency of arithmetic, to which it is provably equivalent. This helped to fall into the mysticism of an unproved truth, that 'the mathematician could see by looking over the shoulder of God' (Barrow) — a very convenient position, indeed. In the latter, concrete, cases, the authors had to *prove* the truth of the arithmetic statement in question, actually on standard numbers; of course, this, jointly to unprovability, implies undecidability, but it is stronger.

Thus, such proofs have no need of reference to ontological miracles evoking 'inaccessible mathematical truths'. We humans are absolutely not constrained to reason with no reference to meaning and only with finite formal sequences codifiable in zeroes and ones, which is exactly what computers do, or, equivalently, to deduce only from 'pattern matching' (if A is syntactically identical to the A in $A \Rightarrow B$, or it may be made so, then write B). Our rigour is not simply formal/linguistic: for example, we construct a praxis of 'infinity' in different conceptual frameworks, and we make it into a rigorous mathematical (or geometrical) concept, admittedly a difficult acquisition which required centuries of work.[15]

It is difficult to speak of the above, for one of the goals of this century's mathematical logic, has been exactly to 'avoid infinity' in foundational analysis, even if one must consider it pertinent to the practice of mathematics: it is too dangerous to be foundational. Yet, infinity is a central element which today we can rest on, thanks to, in part, results obtained in mathematical logic. And infinity constitutes precisely one of these mathematical concepts, which, in order to have meaning, needs a plurality of references to other forms of 'intelligence', considered even in their historical evolution, by the very fact that it has been given in different forms of knowledge.

In conclusion, with reference in particular to the incompleteness phenomena, we would argue that *in mathematics a concept is proposed, a method is chosen, structures are built, theorems are proven, specifying where and by what means, in a manner which is certainly not arbitrary, as 'context of proofs and meanings' are provided. In mathematics there exist no propositions which are 'true and unprovable' and, at the same time, mathematics is not simply mechanisable calculations, since each time it is a question of proving, if necessary using infinitary methods or meaningful derivations, if and within what framework such and such a proposition is unprovable, and if and in what framework it is provably true.* Those who claim, in a

---

[15] There is another way to understand how meaning steps into the proof of these recent incompleteness results. The three statements mentioned above (normalisation, (PH), (FFF)) may be all formalized in arithmetic in the form '$\forall x \exists y.P(x,y)$', where '$\exists y$' means 'there exists y' and where P is a decidable predicate. In each case, '$\forall x \exists y.P(x,y)$' is unprovable in arithmetic. Yet, from (the proof of) its truth over the natural numbers, one may easily derive that 'for all n, arithmetic proves "$\exists y.P(n,y)$"'. There is a subtle but crucial distinction here, which is a 'semantical' one: in order to prove, in arithmetic, the statement '$\forall x \exists y.P(x,y)$', the variable x has to range *only* on the natural numbers. That is, the proof of the universal statement 'for all x . . .' may be given just if interpreting x as a *generic* natural number, with no use of formal induction (similarly as one would prove a statement such as 'for all r, real number, . . . such and such a function is continuous . . .', where induction on r is not possible and the proof should be given for a 'generic' real). In other words, the fact that n in '$\exists y.P(n,y)$' must be a generic natural number and *not a formal variable* (which could be interpreted also in non-standard models) is crucial to the proof and forbids first order induction over arithmetic formulae. As a matter of fact, this 'meaningful' property (n is a natural number), easy for us, provably cannot be formalized in arithmetic (a consequence of the so called 'overspill lemma' in the model theory of arithmetic); in Set Theory, its proof requires infinitary assumptions.

mystic tone, that there are propositions which are 'true but unprovable' must give an example of one, by singling it out: but, in doing so, they will also have to prove it, that is to say, prove its truth within a well-defined construction. That is just what had to be done, by Gentzen or normalisation, for Gödel's $\mathcal{G}$ (or consistency of arithmetic, to which it is formally equivalent), for (PH) and (FFF) with regard to the natural numbers, as well as for the 'Continuum Hypothesis' and the 'Axiom of Choice', each in a different set theoretic construction, as hinted next.[16]

III: Infinity and Metaphors

The foundation of infinity has been largely advocated by (formal) Set Theory. However, I believe that today there can be no *a priori* foundation or *a posteriori* formal justification for mathematical infinity other than 'metaphors'. Actual infinity does not belong to our sensory experiences, be it direct or indirect; nor even to the practice of counting or of classical geometry based on shapes: for that, potential infinity will do, both with regard to adding '+1' and endless motion. The historical praxis of actual infinity, which I have mentioned several times, constitutes a further progress: the limit of this unending '+1' must be conceived, the horizon closed, while positing a point at the 'limit'. This practice is, as I have stated, established on the basis of innumerable reflections situated between the mystical and the emotional: starting with the Greeks, continuing with the controversies surrounding the infinite grace of Mary (Duns Scotus's actual infinity), the discussions of perspective in painting (the vanishing point at infinity of Renaissance painting, which is an entirely artificial construct, one among many other possible kinds), the anchorage in monads and the metaphysical infinity of Leibniz which aimed to give sense to the infinitesimal calculus. And it is precisely this 'constitution' of the concept of actual infinity, so rich in emotivity sedimented over the centuries and *rendered 'objective' by mathematical practice*, which has become an essential part of its mathematical specification, by the very fact of its entry into proofs. Nevertheless, mathematical infinity is not the same as the

---

[16] An alternative proof of (FFF) can be derived form the work in Rathjen & Weierman (1993). The authors give an infinitary, non arithmetizable proof, yet much more 'constructive' than the set-theoretic one, mentioned here (their proof is 'intuitionistically' acceptable, at least by 'open minded' intuitionists.) The proof uses the constructive theory of Inductive Definitions and avoids a (classically) crucial passage 'per absurdum' of the set-theoretic approach. Infinity also is used in a slightly different or finer way: the infinitary ordinal structure mentioned above shows up, in the inductive proof, in a minimal way, with no use of $\sum_1^1$ sets, but 'just' of least impredicative ordinals.

  This paper focuses on the notion of infinity, as a crucial tool of the existing analysis of proofs. However, a further approach should be more closely explored, if possible. Instead of forcing (formal) induction, by stretching it along the ordinals, we should just rely on the order-structure of natural numbers and use generic elements (see the proof by generic n in footnote 15). A different philosophical attitude, giving up the absolutely central role of formal induction (since Peano and Frege), can lead to a more 'finitary' approach, yet non formalizable, as referring to an essentially geometric argument (the order of natural numbers, which is, formally, fully expressed only by second order impredicative principles: every non-empty subset as a least element). But spelling this out may be a difficult, if ever possible, project, which would allow us to propose an alternative to the deep analyses given in the set-theoretic frame and, perhaps, even to the more constructive approach by Inductive Definitions. These approaches 'explain' natural numbers by a (very interesting) detour via infinity. Infinity is fine and good, a beautiful and very human conceptual construction, but perhaps, in arithmetic, it may be replaced by a geometric insight into numbers and an analysis of proofs also by generic elements (which involve meaning along the deduction).

different mental experiences of infinity mentioned above, since it is *the invariant concept* which we posit after all of these various experiences. At the same time, however, its foundation is to be found precisely in the (vectorial) sum of these experiences, each one being a metaphor, an opening onto other forms of knowledge and other forms of intelligence.[17]

In reality, it has been a vain effort to try to give a foundation to infinity using methods within mathematics, let alone within some formal system or another. As we have already recalled, it was Cantor who first defined and treated infinity with a wholly mathematical rigour, objectifying it in notation and in computation, and, furthermore, unifying it in a theory of (possibly infinite) sets. Frege was to give this theory a rigorous logical form, which was corrected later on and formulated into a formal Set Theory, under the influence of the Hilbert school. But, ever since the start of this adventure, which was to change the face of mathematics, the infinite had posed serious problems.

Cantor had developed his theory with the aim of analysing the continuum of the real numbers: he proved that these are strictly more numerous than the natural numbers; and his proof suggests how one can carry on building ever larger infinities, indefinitely. He then spent many years trying to prove the *Continuum Hypothesis*: that the infinite number of real numbers, their *cardinality* as he termed it, is the 'immediate successor' to that of the natural numbers. He failed in his attempt, as did his successors, Zermelo, Bernays, von Neuman and many others. All of these formalizers of Cantorian Set Theory were also unable to prove the derivability of another key property of infinity: the *Axiom of Choice*, which affirms that one can choose an element for each set belonging to a (possibly infinite) collection of sets.

These properties of infinity, now well-defined as a mathematical concept, seem to escape formal treatment by set theory: thanks to a result of Gödel in 1938 and to a theorem of Cohen's (1963), it would be proven that formal Set Theory had not managed to say anything about the Continuum Hypothesis or the Axiom of Choice. Or, in other words, these propositions are undecidable in the Formal Set Theory, as is Gödel's proposition $\mathcal{G}$ in arithmetic. One might then ask whether they are true. One must then *prove* if and within what framework they are true, just as for Gödel's proposition: if one then proceeds to a certain construction of a universe of sets, this construction being due to Gödel, one proves that the Continuum Hypothesis and the Axiom of Choice are true in this universe; if one goes on to another construction, namely Cohen's, both of them are shown to be false in this other set-theoretic universe. Those who believe in mathematical propositions which are true in God's spirit, who believe in absolute and undemonstratable mathematical truths, would then have to say whether they believe that, for God, the Continuum Hypothesis and the Axiom of Choice are true or false. All that we humans can say is that, for example, Gödel's

---

[17] The notion of metaphor used here is close to that of Lakoff and Núñez (2000): it is in fact the meaning of a certain concept, in our case the concept of mathematical infinity, which, as I will try to explain, is constructed with reference to a plurality of metaphors. Their notion of 'conceptual blend', at the core of the unity of our forms of knowledge, as a the permanent 'transfer of meaning and conceptual practices' nicely underlies or allows the 'vectorial sum' of different constructions I refer to here. Before getting acquainted with the approach proposed by Lakoff and Núñez, I was influenced by the use of metaphors in the conceptual constructions of mathematics, by a remarkable book on the philosophy of mathematical physics: Chatelet (1993).

construction is 'simpler', in the sense which is suggested to us by certain regularities in the world, such as the principles of 'minimal' structures: but to do this, it is necessary to specify what one understands by 'minimal' in this precise context, to propose an infinitary mathematical construction, and to prove, within it, the truth of the Continuum Hypothesis and the Axiom of Choice. Now, although it is not 'minimal', even Cohen's construction is not arbitrary: it uses a notion of 'generic element' which is very relevant in mathematics (and in arithmetic, as we noticed in a footnote).

To summarize briefly the question of infinity in formal set theory, the independence of the Continuum Hypothesis and the Axiom of Choice proves that, if one remains at the level of the formal system, without it taking on or before it even takes on a meaning in a mathematical construction, then this formal theory, though created in order to be able to speak of these crucial properties of mathematical infinity, remains once again completely silent. But the formal theory would have to be at least formally consistent (in particular, if one adds to it the Continuum Hypothesis and the Axiom of Choice as axioms). A consequence of (certain extensions of) Gödel's incompleteness theorems is that this consistency cannot be proven — without assuming the ability to construct further infinities, extremely large cardinal numbers, which go beyond the formal theory whose consistency one aims to prove. That is, if one wants to prove the consistency of a formal Theory of Sets which may express a given infinite (cardinal) alpha, say, then one *needs* 'to assume the existence' of a cardinal number beta, strictly larger than alpha.

There is nothing metaphysical about this, nor is any ontology of mathematical infinity involved: what one is stating is simply that, if we are capable of making, or we assume that we are capable of making certain conceptual constructions of 'very large' infinities (by iterating power-set operations, limits and much more), then, using these same constructions, we can prove the consistency (*construct models*) of such and such a set theory; and this is what certain difficult results of Set Theory of recent decades have led to.

Recall now that, conscious of the difficulties involved with infinity, a number of mathematicians/logicians from the beginning of the century, Hilbert and Brouwer to name only two, had sought to exclude actual infinity from foundational theories. Hilbert had recognised the centrality of infinity in mathematics; more importantly, he had affirmed the indispensable character of this notion for mathematical thought, and claimed that the mathematician should work in Cantor's 'paradise of infinities'. However, Hilbert also claimed that, in order to guarantee the certainty of reasoning, a finitistic foundational analysis must be its basis, since 'operations on the infinite can only be guaranteed using the finite as a basis'. That is, the mathematics of infinite and ideal objects had to be saved by a finitistic metamathematics, the frame for consistency proofs. (Two writings of 1925, published as appendices to the French edition of *The Foundations of Geometry*, as well as in van Heijnoort (1967) are fine elaborations of such a programme.)

By contrast, as we have seen, infinity reappears, not only in consistency proofs, but, in view of the recent incompleteness theorems, even in proofs of 'proper' statements of theories of the finite par excellence such as arithmetic. And this is so in an essential manner, at least within the well established frames mentioned here, Set Theory and intuitionistically acceptable theory of Inductive Definitions, see footnote 16. In other words, not only are we unable to guarantee the consistency of theories which speak of infinity using the finite, as in the case of different set theories (a key aim of the formalist program), but we can even need infinity to prove, by induction and/or within Set Theory, consistency of arithmetic as well as certain finitary statements of arithmetic, such as (PH) or (FFF) mentioned above.

Nothing too bad so far, precisely because the certainty of our use of infinity is, in my opinion, extremely 'robust', or as much as mathematics itself: it stems from the interaction and mutual support of numerous mental and historical experiences which can even come from outside mathematics. Its conceptual solidity stems from its rootedness in a plurality of conceptual constructions which have allowed us to conceive of, propose and define the mathematical invariant gradually across history. I return to this plurality one last time, since it is the nub of the whole issue: mathematical infinity is not a metaphor, but our very proposal of an invariant (stable) concept constructed on the basis of (and as an invariant with regard to) a plurality of mental experiences including religious metaphors, the vanishing point in perspectival painting, distant points of convergence — to name but a few conceptual practices which work with (and ground by a praxis) infinite mathematical sets and structures.

Gödel's splendid 1931 theorem left us with a dramatic metaphysics of mathematics: because it concerns 'only' a theorem of undecidability (it says nothing about how one proves the undecidable proposition in arithmetic, or the consistency which is formally equivalent to it), for decades people have not stopped talking about absolute 'truths' of mathematics, of 'looking over God's shoulder', instead of going and studying the many beautiful proofs of consistency elaborated since 1934. Since undecidability does not help us to understand what methods of proof might exist outside arithmetic (by simply stating that certain propositions cannot be proven in arithmetic), the debate has become trapped in a Manichean conflict: on the one side those who say that the limits of man (the 'human computor', in the act of proof) are the same as those of the machine; on the other, those who have hymned the praises of the manifold mysteries of mathematics, precisely because, in sticking uniquely to Gödel's theorems, which says nothing on how to prove the unprovable statement (consistency), one passively accepts the notion of demonstrability (and even mathematical rigour!) as being exclusive to formal systems, indeed the mechanisable ones: the rest belongs to the obscure realm of intuition. A closer analysis of more recent undemonstrable statements of arithmetic, but demonstrably true, to which I have just alluded, may allow the 'foundation' of a crucial practice, namely the rigorous use of the mathematical concept of infinity. A use to which, for now, I can give no other foundation than a practical, historical, and yet extremely solid one, which makes reference to metaphorical meanings, in a blend of conceptual practices.

IV: Metaphors and Analogies: Between Intelligence, Emotions and Affection

Before introducing a daring discourse on forms of mathematical intelligence and their relation to 'affectivity', I would like to open a parenthesis. In the act of disseminating the foundations of mathematics into a variety of forms of knowledge and intelligence, in this attempt to analyse their evolutionary and historical genesis, I, as logician/mathematician, am explicitly violating one of the most established dogmas of the philosophy of science of this century, the dogma which forbids any confusion between 'genesis' and 'foundation', 'creativity' and 'deduction', 'logic of discovery' and 'internal rationality' of a discipline. But it is precisely 'this dogma of the principial fracture between epistemological elucidation and historical explanation . . . between epistemological origin and genetic origin . . . [which must] be overturned completely', as Husserl stated in his too rarely read *The Origin of Geometry* (1933–36). This is a crucial question concerning the whole of scientific knowledge, because in each of these forms of knowledge there exists the difficult interplay between epistemological autonomy — on the level of logic and internal justification, and the genesis of the knowledge — on the level, amongst other things, of its relationship to other forms of knowledge and the course of its evolution and history.

There is no such a thing as the set of 'universal Laws of Thought', with no genesis and at top of which lie the perfectly formal rules of mathematics. The hierarchy which has developed in our cultures represents a distorting mirror of human intelligence, whether this be in the idea of the logical formalism as the only (or ultimate) form of rationality, of the only form of scientific method (which are, in the last analysis, those of formalized mathematics) or in the idea of unbreachable compartmental boundaries. Instead, we have to discern the elements of continuity and the links between the different forms of intelligence, the different human forms of relationship to the world, and some of the different forms of scientific knowledge. For example, Damasio (1994) gives the point of view of the neurophysiologists on the question. Indeed he sets out to explain the neurophysiological discoveries on the basis of which affectivity and intentionality are shown to be integral parts of rationality. For him, that is where Descartes' error lies, namely in his separation of the rational soul from the emotional soul which, for the neurophysiologist, is completely impossible.

I think, thus, that we have to enrich this analysis of meaning and of human intelligence as grounded in our active existence of living beings, from our biological reality up to the dialogue between humans in history. Furthermore, I would like to add that intentionality and affectivity are not only essential 'stimuli' for intelligence, a point on which everybody would agree, but that they affect the very content of intelligence. Or rather, that intentionality, affectivity and the emotions are not simply the possible bait for or the possible break of the 'rational' machine, but that they help to determine its direction, and therefore its content. Personally, I see things as follows.

The understanding of a fact, up to and including the conjecture of the mathematician, is based on analogies, metaphors, and consequently on choices of direction in the representation and the contents of the conceptual construction, whose meaning is rich with affectivity: one proposes, chooses and understands an analogy, a metaphor in view of or because they have an emotional or affective content; one is therefore led by intentionality. In other words, one *chooses* to 'build a certain bridge' between different kinds of knowledge, between different kinds of 'intelligence' (and that is what constitutes analogy and metaphor), on grounds which are both affective and emotional or intentional (they have aims). That is why the very content of a 'rational' practice, which is based on metaphors and analogies, is rich in intentionality and emotions, because through the direction given to the metaphor or the analogy 'bridge', they contribute to its determination. Now, this infinity, a key concept and unavoidable notion of a mathematics which remains the stronghold of rationality, can only be understood as the invariant concept of a plurality of conceptual experiences, both practical and emotional, running from religious metaphors to the metaphor of depth in painting, via the convergence of parallel lines at the horizon, and the limit of iterated movements. It is a kind of vector resulting from a set of vectors, a construction which is therefore different from all the given vectors, but nevertheless always dependent on them, be this in terms of direction or contents.

To conclude, Descartes provided us with an important intellectual clarification by helping us to found the modern scientific method and by purging 'reasoning', amongst other things, of the residues of magic, of the empty logic and syllogisms of medieval times, and of pervasive religious mysticism. Now that we have understood and, on the whole, know fairly well how to put all of that into practice in scientific frameworks, we are in a position to balance the schism with which he endowed us with his bequest of a 'method' and 'rules *ad directionem ingenii*'. This will allow us to go further, to understand better with the aid of scientific analysis, and to put our finger on those particularly difficult points where rationality and affectivity become confused and yet form the basis of one another reciprocally, in other words where a plurality of forms of human intelligence mix together, even in mathematics. This is particularly needed as its formal foundation, beautifully developed in Mathematical Logic, is essentially incomplete; in particular, it lacks the analysis of meaning as embedded in human intentionality.

V: Induction, Machines, and the Lord Almighty

In this research programme, mathematics can lend valuable assistance, because if we manage to break its permanent siege-status, in other words the absolute and separate role ascribed to it, if we manage to destroy the ivory tower in which Platonists and formalists want to shut it up, it will be able to provide us with a good example of, among other things, a relevant cognitive practice. A relatively simple example, since even when profound and difficult, mathematics nevertheless remains conceptually simple: elegance and conceptual necessity are its watchwords, and among its *raisons d'être*. Elegance and necessity are combined with constructions of a profoundly human nature. To such a degree that (but don't say this too loudly) the Lord Almighty and computers are largely incapable of doing mathematics. The former is unable to keep the planets, which are nevertheless His greatest creation, in orbits which are sections of cones, as Kepler recommended. More precisely, the orbits of our own planets do not integrate a system of differential equations. Occasional omnipotent flicks are needed in order to sustain our orbits about the sun, as intuited by Newton.[18] This, if I dare say it, is more a pragmatic than a mathematical solution (surely not a limitation of His omnipotence: He just decided to organise matters in a different way from Kepler's expectation). And let us hope that He continues to make such large gestures, because there is no stability theorem for the solar system.[19] Yet, we soundly try in every possible manner to describe its movement approximatively using the difficult instruments of the best language we possess for speaking about the movement of the

---

[18] Newton discovered that the orbits of the planets affect one another reciprocally by mutual gravitational attraction, and in particular that effects of 'gravitational resonance' could even endanger the stability of the solar system — his profound religiosity provided him with that solution, the only one which can garanty stability for good (see the next footnote).

[19] See the results in Laskar (1990): the orbit of the Earth is provably impredictable beyond 100 million years. Similarly for the solar system, as a whole, beyond 1 million years; just nothing when considering the expected life of the Sun (5 or more billion years?) — the point is that Pluto's orbit is 'very chaotic'.

bodies, curves and geometries of space, namely the mathematics of dynamical systems; but we only manage to obtain qualitative descriptions of the behaviour of systems of chaotic determinism.

As for digital machines, as hinted above, they cannot demonstrate the consistency of arithmetic, via normalization say, or (FFF), because meaning (or an 'impredicative' blend of syntax and semantics) is *provably* essential to the proof and, thus, formal-artithmetizable reasoning does not suffice; but much more should be said, as they are incapable of even giving proofs by arithmetical induction which scarcely reach beyond the banal. I shall explain what this means. By arithmetic induction, as we have already recalled, we mean the following finitary rule: if one proves $A(0)$ and if one proves that, 'for all x, $A(x) \Rightarrow A(x + 1)$', then one can deduce from this that 'for all x, $A(x)$'. What could be more mechanisable? Well, there are proofs which, though hardly complex, use this rule in an '*a priori*' non-mechanisable fashion, as it is relatively rare for one to manage to make the inductive step or to proof that 'for all x, $A(x) \Rightarrow A(x + 1)$', where A is exactly the proposition you want to prove. In a number of significant cases, a proposition B has to be found, stronger than A or such that $B \Rightarrow A$, and for which, by contrast, one manages to prove that 'for all x, $B(x) \Rightarrow B(x + 1)$'. Proposition B is called the 'inductive load'. B can be much more complex than A. There exists no *a priori* criterion for choosing B, excepting a few vague heuristic indications according to which the inductive load must 'contain everything which is needed', on the basis of the hypotheses, the structure of the proof one is constructing, and the thesis aimed for. In actual fact, the choice of B among an infinite number of possibilities is based on analogies or on setting bridges or embedding in broader mathematical frames. An analogy, for example, with a proof already found, which can be algebraic even when one is working in geometry, or, indeed, an analogy with an induction on the number of dimensions inspired by another, very different proof based on the length of the formulae etc., etc. Analogies and bridges between different forms of knowledge, normally within mathematics, I would say, but not always, for the analogy, just like the metaphor, can easily take us outside this.[20]

Of course, none of this affects the foundational programme of the formalist, who in such cases can always reconstruct *a posteriori* the logical/formal framework of the proof in which he will quite simply replace A with B where this proves necessary. But, this nevertheless constitutes an insurmountable obstacle for the logico-computational hypothesis, since there exists no machine which is capable of choosing B in the context of a new proof, out of infinitely many possible induction loads, while an analogy or an intentional, meaningful choice may suggest it.

The question of the 'inductive load' now represents a crucial point of interactivity for interesting programmes of proof by machines. There exist in fact many fine

---

[20] Induction is particularly relevant also in view of 'gödel-numbering' techniques. By these, one may encode in arithmetic many mathematical structures, which apparently do not need to lie within numbers. For example, in Longo & Moggi (1984), some sort of computable functionals (functions of functions of functions . . .) are hereditarily encoded into arithmetic and can be easily defined by arithmetizable formulae (with some further work). Yet, even the proof that they are well defined in higher types requires a huge inductive load, calling for topological spaces and continuous functions. But this is very common in mathematics. Another amazingly heavy inductive load (by the 'candidates of reducibility') may be found in Tait-Girard's normalisation proofs, even in the arithmetizable fragment, i.e. the proof of '∃ y.P(n,y)' of footnote 15.

systems of automated calculation and deduction which are eminently interactive: the mathematician isolates enormous calculations and huge database searches which he then has the computer execute quickly and perfectly; he distils from these a number of terrifying lemmas, whose proof needs numerous mechanisable passages, and then he transfers these into the system, intervening finally in those crucial choices in the proof of a theorem, such as the choice of the inductive load, of the hypothesis rich in meaning, etc. Other automated proof assistants check, *a posteriori*, proofs or properties of programs, a major help for some work in algebra and in programming. Finally, freed from the myths surrounding it, the computer, with its special powers of deductive/formal calculation, in certain cases quite literally gives wings to human calculations and proofs, thanks to an interaction between man and well-constructed machine. It is an interaction which leaves to man the use of analogy, metaphor and meaning, in other words to that ability to make connections within the network of integrated knowledge and forms of intelligence which makes up the specific unity and force of human thought.

Conclusions

To conclude, the formalist hypothesis argues that only calculations using signs without meaning can allow the *a posteriori* reconstruction of any mathematic reasoning and the elaboration of its logical/formal foundation; its proponents nevertheless recognise the plurality of the forms of deduction (the famous 'mathematical creativity') which must be able to be reinscribed *a posteriori* exclusively onto the formal level. The logico-computational hypothesis, which is much stronger, assumes that logical/formal intelligence, based on the manipulation of formulae conceived of as sequences of discrete symbols lacking meaning, themselves codified with, for example, zeroes and ones, allows the representation of all forms of intelligence, thus not only the *a posteriori* reconstruction of the formal skeleton of mathematics, but also the perfect simulation of the advances of those reasoning in each of these fields. Now, if the various Incompleteness Theorems mentioned here lead to the failure of the former of these two programmes, they imply *a fortiori* the same result for the latter; what is more, this second programme fails even when confronted with a question as 'banal' as that of the inductive load in arithmetic.

How then do the many defenders of these two programmes face up to this fact? The formalists, who are quite conscious of the metamathematical relevance of the Incompleteness Theorems, argue that it is more or less a question of metamathematical 'tricks' (implying *ad hoc* metatheory), and that the proofs of all 'interesting' propositions can be reconstructed formally. Now, deciding what is interesting is a matter of opinion, and I believe that Girard's Normalisation theorem is both interesting and rich in applications, particularly for the mathematics of computer science, although the Tait-Girard style proof requires metatheory. The same can be said of FFF (Kruskal's theorem has lots of applications), in which, and even more explicitly, the variety of our forms of knowledge enters into its set-theoretic proof, in particular through the concept of infinity.

As for the defenders of the logico-computational hypothesis, if I have understood properly, they either just ignore these findings, or give to them interpretations borrowed from more learned formalist arguments — which are nevertheless restricted exclusively to mathematics; in other words, they continue to believe that the machine's limits are man's limits, or that the 'interesting' things that man knows how to do, the machine also knows how to do, and they consign everything else to the category of uninteresting or nonexistent things. Some, such as Searle (1992), call these theses 'eliminationist' — an expressive, if somewhat sombre term. Finally, others bravely argue that the digital computer will, one day, go beyond working only on a formal/theoretical level. This is possible: I do not bet on future, when living clones of jupitarians will be our next generation computers . . . Yet, so far, this possibility contradicts the two key hypotheses of functionalism (which deal with both the design of digital computers and their languages), namely the codifiability (on the theoretical level only) of any form of intelligence into discrete symbols, themselves in turn codifiable into formal arithmetic or into similar theories, and the independence of this codifiability (but not of the code itself, obviously) with regard to any specific implementation. As I have already noted, the formal codifiability (the uniqueness of the conceptual level on which the zeroes and ones are found) and the independence 'of what one knows how to do' with regard to specific contexts and implementations (the fundamental idea of the programmability of computers, namely 'software portability') exclude precisely the network of connections characteristic of human thought which is based on the unity of its specific hardware and its software: our 'modularised' brain with its history. For any monist, this network and this unity cannot be fragmented into metalanguage, language and semantics, and then, in a machine, into 'software' and 'hardware' (the soul and the body?), in order to carry on 'meaning independent' computations, represented at the linguistic/theoretical level only. Some theorems I mentioned proved this for us, by showing the incompleteness of this artificial split of human reasoning.

By contrast, it is just this unity, this indivisible ego — or divisible purely for reasons of temporary mathematical commodity, or for the construction of machines, this contextual dependence, this specific hardware, living in the world and in history — which allows us to make these 'bridges', metaphors or analogies between different forms of intelligence. These analogies and metaphors are essential elements of human reasoning, including mathematical reasoning; what is more, they are governed by intentionalities and emotions. It is on precisely these constitutive elements of 'meaning' that cognitive and foundational analysis should also be concentrating today, by focusing on the remarkable conceptual invariance and symbolic abstraction, so typical of mathematics, but which (provably) cannot be defined as purely formal.

References

Asperti A. and Longo G. (1991), *Categories, Types and Structures* (Boston, MA: MIT Press).

Baader F. and Nipkow T. (1998), *Term Rewriting and All That* (Cambridge: Cambridge University Press).

Barwise J. (ed. 1978), *Handbook of Mathematical Logic* (North-Holland).

Castagna G., Ghelli G. and Longo G. (1995), 'A calculus for overloaded functions with subtyping', *Information and Computation*, **117** (1), pp. 115–35.

Chatelet, G. (1993), *Les enjeux du mobile* (Seuil).

Cubric D., Dybjer P. and Scott P. (1998), 'Normalization and the Yoneda embedding', *Mathematical Structures in Computer Science*, **8** (2), pp. 153–92.

Damasio, A. (1994), *Descartes' Error* (New York: Putnam/Grosset).

Feferman S. (1998), *In the light of logic* (Oxford: Oxford University Press).

Gardies G. (1984), *Pascal entre Eudoxe et Cantor* (Vrin).

Girard J-Y., Lafont Y. and Taylor P. (1989), *Proofs and Types* (Cambridge: Cambridge University Press).

Harrington, L. *et al.* (ed. 1985), *H. Friedman's Research on the Foundation of Mathematics* (North-Holland).

van Heijnoort (1967), *From Frege to Gödel* (Cambridge, MA: Harvard University Press).

Hindley J.R. and Seldin J.P. (1986), *Introduction to Combinators and Lambda-Calculus* (Cambridge: Cambridge University Press).

Hodges, J. (1995), in *The Universal Turing Machine*, ed. R. Herken (Springer-Verlag).

Lakoff, G. and Núñez, R. (2000), *Where Mathematics Comes From: How the Embodied Mathematics Creates Mathematics* (New York: Basic Books).

Lambek J. and Scott P.J. (1986), *Introduction to higher order Categorical Logic* (Cambridge: Cambridge University Press).

Laskar, J. (1990), 'The chaotic behaviour of the solar system', *Icarus*, **88**, pp. 266–91.

Longo G. (1997), 'Géométrie, Mouvement, Espace: Cognition et Mathématiques', *Intellectica*, **2** (25), pp. 195–218.

Longo, G., (1999a), 'The mathematical continuum, from intuition to logic' in *Naturalizing Phenomenology: issues in contemporary Phenomenology and Cognitive Sciences*, ed. J. Petitot *et al.* (Standford University Press).

Longo G. (1999b), 'Mémoire et Objectivité en Mathématiques', Colloque *Le réel en Mathematiques*, Cérisy, Septembre 1999 (actes à paraître).

Longo G. and Moggi E. (1984), 'The hereditary partial recursive functionals and recursion theory in higher types', *Journal of Symbolic Logic*, **49** (4), pp. 1319–32.

Paris J. and Harrington L. (1978), 'A mathematical incompleteness in Peano Arithmetic', in Barwise (1978).

Rathjen M., Weierman A. (1993), 'Proof theoretic investigations on Kruskal's theorem', *Annals of Pure and Applied Logic*, **60**, pp. 49–88.

Searle, J.R. (1992), *The Rediscovery of Mind* (Cambridge, MA: MIT Press).

Simpson S. (1998), *Subsystems of Second Order Arithmetic* (Springer-Verlag).

Smorinsky, C. (1978), 'The incompleteness theorems', in Barwise (1978).

Zellini P. (1980), *Breve storia dell'Infinito* (Adelphi, 1980 trad. francaise pour Seul).

J.S. Nicolis and I. Tsuda

Mathematical Description of Brain Dynamics in Perception and Action

'. . . perception depends dominantly on expectation and marginally on sensory input.'

Walter Freeman (1995)

*A given but otherwise random environmental time series impinging on the input of a certain biological processor passes through with overwhelming probability practically undetected. A very small percentage of environmental stimuli, though, is 'captured' by the processor's nonlinear dissipative operator as **initial conditions**, and is 'processed' as **solutions** of its dynamics. The processor, then, is in such cases instrumental in compressing or abstracting those stimuli, thereby making the external world to collapse from a previous regime of a 'pure state' of suspended animation into a set of stable complementary and mutually exclusive eigenfunctions or 'categories'. The characteristics of this cognitive set depend on the operator involved and the hierarchical level where the abstraction takes place. Depending on the context, the transition from one state to another occurs in such a cognitive operator. The chaotic itinerancy may play a crucial role for this process. In this paper we model the dynamics which may underlie such a cognitive process and the role of the thalamo-cortical pacemaker of the (human) brain. In order to model them, conceptualization by the notion of 'attractor ruin' in high-dimensional dynamical systems is necessary.*

I: Why Dissipative Chaos?

Suppose you release a small animal (e.g., a kitten) on the floor. Even if it is satiated, even if it does not look after a partner, even if it does not run away from an enemy, the healthy animal is still going to happily occupy itself in a ceaseless game-like exploratory activity. What is the survival value, if any, underlying such seemingly purposeless behaviour? Very simply, the animal is creating a variety of behaviours by means of cortical chaotic signals which should eventually drive the motor activity. Chaotic dynamics is, in general, characterized by the sensitive dependence on initial conditions (Ruelle, 1989). This characteristic is measured by the Lyapunov exponents (see for example, Ott, 1993), which indicate an expanding or a contracting rate averaged over all the time course or with respect to the stationary probability distribution. In

Journal of Consciousness Studies, **6**, No. 11–12, 1999, pp. 215–28

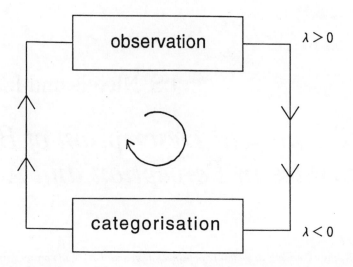

Figure 1

To 'observe' you need *a priori* categories, but to form categories you need observations. The processor creates a variety of frames to observe the environment via his positive Lyapunov exponents — thereby forming *basins* from the responses it receives, and categorizes via his negative Lyapunov exponents — thereby relegating specific subsets of stimuli to attractors. The process involves a *nonlinear recurrence* — a feedback loop — so the 'chicken/egg' syndrome does *not* arise: observation and categorization are performed *in unison*.

chaotic dynamical systems, an expanding phase and a contracting phase of nearby orbits successively appear in its time evolution. In the time average, one obtains at least one positive Lyapunov exponent for chaotic orbit. On the other hand, any act of information processing involves two separate phases: an expanding phase and a contracting phase. It may be executed by the organism involved either in succession or (usually) in unison by means of a recursive feedback loop (Fig. 1).

(a) During the *expanding phase*, the organism-processor via its motor activity makes a subset of alternatives manifest themselves both in number and *a priori* probabilities by observing the environment.

(b) During the *contracting phase*, the processor, via its sensors, contracts (compresses) the basin of attraction created earlier onto one (out of many) coexisting attractor.

It is imperative then for a biological processor to possess coexisting chaotic strange attractors in order to comply with both of these requirements (steady states and limit cycles provide only for the contracting phase and are useful only as classifiers provided that the processor has already been given, or has created, the set of alternative responses, i.e., reactions to 'raw stimuli' from the environment).

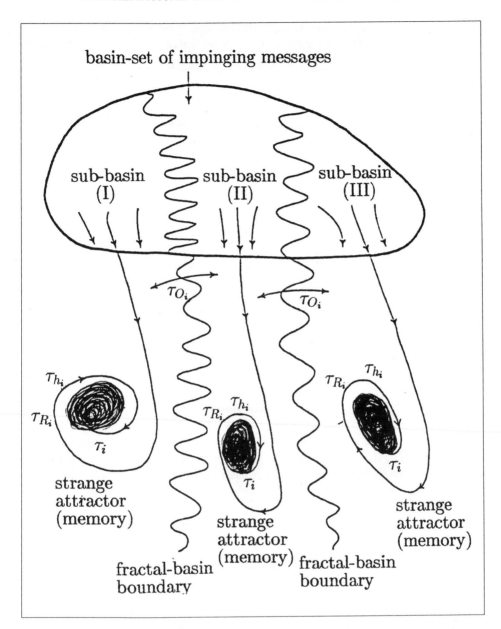

Figure 2 Sketch of a cognitive channel working after the dynamics of chaotic itinerancy (see text); within each attractor a 'micro'-intermittency may go on as well (as, for example, in the Lorentz system).

τ_i = relaxation time on an attractor.
τ_{R_i} = residence time before the interruption of the thalamo-cortical pacemaker.
τ_{h_i} = holding time, after the interruption of the thalamo-cortical pacemaker.
τ_{0_i} = transient time between attractors after leaving attractor *i*.

The thalamo-cortical pacemaker is responsible for the jumpings among the co-existing memories-attractors (a multifractal-*inhomogeneous* attractor). The jumpings can be viewed as *chaotic itinerancy*. The processor is partitioning a set of raw undifferentiated external stimuli onto coexisting attractors and establishes a 'grammar' via chaotic itinerancy. Each 'categorisation' entails an irretrievable loss of information: $I_i = \ell n \dfrac{N}{D_i}$ bits where N is the embedding dimension of phase space and D_i the correlation dimension of the attractor involved. The *fractal* character of the basin boundaries makes the act of partitioning of external stimuli onto the individual attractors-memories-categories, *ambiguous*: As the basin boundaries are 'entangled', a given stimulus — known with limited accurary — as it 'moves' towards a given attractor, inevitably enters the domain of jurisdiction of other attractors. So only after a long transient (of unknown value) is going to 'land' on an attractor — not known in advance.

Furthermore if the coexisting attractors are weakly destabilized to become an attractor ruin, they will intermittently attract and repel amongst themselves a given stimulus — giving the dynamics of ambiguity and multistability in perception and categorisation (e.g. a 'vase' which after a while appears as 'two faces looking at each other' and then back again a 'vase' ... and so on.) So, under conditions of fractal basin boundaries and semistable attractors, we can devise and implement dynamical models for *multi-stability* and *ambiguity* in perception — with a very *simple* hardware giving rise to extremely *complex* software.

Most people tend to take *a priori* probabilities for granted, literally as given *a priori*. Perhaps few realize that behind an *a priori* probability lies much action, much trial and error, and most important, a *converging* process. The very definition of *a priori* probability presumes that $P_i = \lim_{N \to \infty} N_i / N$ where N is the total number of trials (stimuli) and N_i is the number of outcomes under scrutiny. It is not enough just to allow $N \to \infty$. The limit must exist. This requirement of convergence — which lends dynamical overtones to the definition of probability — is more clearly manifested in the way a Markov chain of Λ states and a matrix of transition probabilities P_{ij} from the state i to j converges to its *a priori* stationary probability u_i for all states i. This convergence takes place via a cascade of iterations given by the linear map

$$u_{i,t+1} = \sum_{j=1}^{\Lambda} P_{ji} u_{j,t}$$

Note that in last analysis an *a priori* probability is itself an attractor. Actually, in ergodic theory (see, for example, Arnold and Avez, 1967), a possible dynamical origin of stochastic phenomena is investigated. In our scheme (Fig. 2), the *a priori* probabilities involved are the normalized measures of the individual sub-basins of attraction. In a chaotic attractor in general, *a priori* probabilities are the integrals on specific segments of its invariant measure (Fig. 3a,b).

We have considered this simple example just to show that a chaotic strange attractor provides, among other things, the *deus ex machina* that via its positive Lyapunov exponents creates *variety* by *amplifying* initial uncertainties in some directions in state space and provides *order* by *constraining* initial uncertainties along directions characterized by the negative Lyapunov exponents. More generally, for 'linguistic' processes we can claim that the subtle interplay between unpredictability (variety) and reliability (order) that underlies any cognitive scheme is typical of the behaviour

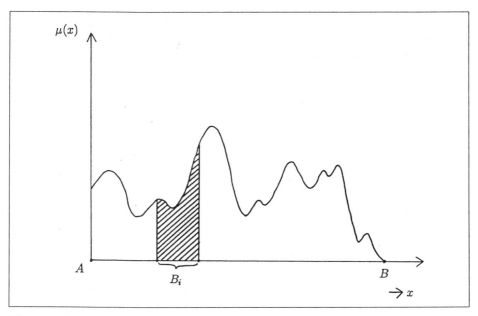

Figure 3 (a) Relating the concept of a priori probability with chaos

$\mu(x)$: The invariant measure of a given chaotic attractor.

$$\mu[B_i] \equiv P(B_i) = \int_{[B_i]} \mu(x)dx = \textit{a priori} \text{ probability of the appearance of the trajectory within the segment } B_i$$

AB: Domain of attraction, $\int_{[AB]} \mu(x)dx = 1$

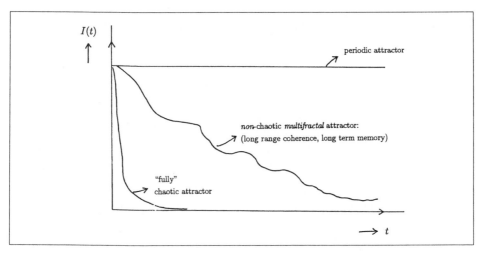

Figure 3 (b) The transinformation: The information exchanged between two 'symbols' t time units apart — within the attractor, as a function of time t.

$$I(t) = \sum_{i=1}^{\Lambda} \sum_{j=1}^{\Lambda} \mu[F^t(B_i) \cap B_j] \log_2 \left\{ \frac{\mu[F^t(B_i) \cap B_j]}{\mu[F^t(B_i)]\mu[B_j]} \right\} \text{ bits}$$

F: The flow — the dynamical model $F(x;\mu)$

of a class of low-dimensional nonlinear dissipative dynamical systems (see Hirsch and Smale, 1974, for nonlinear dynamical systems of low dimension). This reduction to low-dimensionality can provide cognitive models. Let us think of a dissipative dynamical system which is generally described by

$$\frac{dx}{dt} = F(x;\mu) \quad \overline{divF} < 0,$$

where the inequality indicates a dissipation of the phase volume, hence the name 'dissipative dynamical system'. In these systems, the ensuing unpredictability, due essentially to the nonlinear character of the underlying process, is manifested either

(1) Through a sensitive dependence on control parameter(s) μ, giving rise to instabilities, such as bifurcations, broken symmetries, and multiplicity of solutions beyond some instability point μ_c.

or,

(2) Through a sensitive dependence on initial conditions $x(0)$, giving the local (uniform or nonuniform: multifractal) exponential separation of nearby orbits, or the chaotic behaviour of the variables involved.

In compensation, there exist *three* mechanisms for moderating this unpredictability and establishing some order. Namely,

(1) The drastic reduction of the number of degrees of freedom in the vicinity of a bifurcation and the emergence of essentially only a few dominant 'order parameters' in a reduced state-space description (the 'centre manifold'). These order parameters may subsequently interact in a nonlinear fashion, giving rise to low-dimensional dissipative chaos.

(2) The existence of (multiple) attractors possessing invariant measures in the dynamical system governed by the interplay among the order parameters.

(3) A (possible) renormalization or fractal scale-invariance in the evolution of the system, as a result of which a study of the statistics of the time series involved in a restricted window of the variables and the parameters may give results which hold invariant for many other windows or scales of variables and parameters. In short, information is produced not only via cascading bifurcations giving rise to broken symmetry, but also via successive iterations giving rise to ever-increasing resolution.

Beyond a certain resolution interval, ever-present microscopic fluctuations are no longer smeared out but get amplified, passing from lower to higher levels. It is *these* fluctuations that essentially account for new information generated — or rather 'revealed' — by the evolving nonlinear dissipative system.

For an attractor simulating the dual aspect of a cognitive system (or a 'processor'), there exist two basic requirements: large dynamical storage capacity *and* good compressibility. Stable steady states and stable limit cycles of (information) dimension zero and one, respectively, are very poor as *dynamical* information storage units. In other words, they cannot improvise from within, as it were; their algorithm is too inflexible, with too few degrees of freedom. Nevertheless, they are ideal as information compressing gadgets or 'photographic' categorizers.

Strange chaotic attractors, on the other hand, via a harmonious combination of expanding and contracting trajectories in state space, e.g., by possessing positive (λ_+) and negative (λ_-) Lyapunov exponents can, in principle, satisfy both requirements. They may possess a considerable (information) dimension, as well as higher-order dimensions, making them suitable for dynamical and selective information storage, while being 'attractors' $\left(\left|\sum \lambda_-\right| > \sum \lambda_+\right)$, they serve also as information compressors (Oono, 1978; Shaw, 1981; Nicolis, 1982; Nicolis and Tsuda, 1985; Matsumoto and Tsuda, 1985; 1987; 1988).

The chaotic transition among 'attractors' (more precisely, below we will call them *attractor ruins*) in multi-basin system, however, leads us another type of chaotic behaviour, namely chaotic itinerancy (see also Fig. 2). *Chaotic itinerancy* (Ikeda *et al.*, 1989; Kaneko, 1990; Tsuda *et al.*, 1987; Tsuda, 1991a, b) is a universal behaviour in high-dimensional dynamical systems. The appearance of chaotic itinerancy allows processings of a multiplex time-series, which generates a variety of orbits and categorizes the orbits, associated with an alteration of positive and negative Lyapunov exponent in high dimension (Fig. 1).

Let us imagine many coexisting attractors in a high-dimensional phase space. As far as they retain an attracting character, each attractor is separated from the rest, by a basin of attraction. Accompanied with this structure, an asymptotic behaviour is represented by one of such attractors, depending on the initial conditions. What happens when the system destabilises? If instability is strong enough, many chaotic modes appear and consequently the system goes toward a turbulent state, that is, a quite noisy macroscopic state. Then, even a 'trace' of original attractors disappears. If instability is, however, not so strong, an intermediate state between order and disorder can appear. The dynamics may be regarded as an itinerant process which assures a transition among states which were at the beginning described as an attractor but now are no longer an attractor. In this case, a crucial characteristic is that a 'trace' of attractor remains in spite of the generation of unstable directions in the neighbourhood of attractors.

This itinerant process often becomes chaotic. A destabilized attractor could be called an *attractor ruin*. Here, a unique global attractor is a collection of attractor ruins and itinerant orbits connecting them, which could be called an *itinerant attractor*. Although the study on the mechanism of such an instability just recently started, it is expected that the structure of phase space in the neighborhood of attractor ruins is complex, such as possessing fractal basin boundaries (Grebogi *et al.*, 1987). Indeed, in the multiple basins regime, a riddled basin is often observed before the appearance of chaotic itinerancy (Kaneko, 1997). Thus the orbits connecting attractor ruins can be chaotic, which, by possessing a topology of a riddled basin boundary, implies a long-range correlation (see Fig. 2).

II: High-dimensional Mental State Expressed as a Global Attractor of EEG

Let us think of the dynamical situation of multiple-strange attractor system (Fig. 2). One can expect the instability by which a chaotic transition among strange attractors is allowed. When the intermittent jumpings take place via thalamo-cortical interactions in the phase space partitioned by multiple basins of attraction, a successive alteration of dimensionality occurs in such a way that the trajectories dynamically remain in the vicinity of a given strange attractor, that is, in low-dimensional subspace, before the thalamic stimulations effectively enter, and after the effective stimulations the trajectories jump to another strange attractor along high-dimensional manifold. This itinerant behaviour generates the feature of the Lyapunov spectrum that a lot of exponents tend to accumulate around zero, which brings about high dimensionality of the overall dynamical attractor.[1] This high dimensionality assures the production of variety of orbits, namely the generation of a rich 'language', while assuring a time division multiplexing categorizer.

A cognitive implication of chaotic *itinerancy* can now be addressed. Suppose a situation arises that one meets an ill-posed problem, where the information given is not sufficient to solve the problem, or inconsistent data are given. For instance, A promises his friend B to meet at the airport, but he fails. Then, A will imagine several situations to account for the failure, e.g., B has forgotten the promise; a shuttle bus conveying B has been accidentally delayed; B might have changed his mind about coming and tried to contact A but failed, and so on. A's ideas change from one to another, associated with a temporal change of associative memory. If the conditions necessary to reach a conclusion by rational inference are not completely given — this is a common situation — the temporal series of guesses and the dynamic association of memory will be 'chaotic'. It should be, however, noted that it is *not* random, because one has a tendency to draw a context-dependent inference while conceiving a possible causation. This process can be represented by a chaotic transition among attractor ruins.

Another example is the Shannon test, which was originally addressed as a tool for estimating the information content of a natural language. Again consider two persons, A and B. There is a text known to A only, and B tries to guess successive letters. A marks each trial telling B 'YES' or 'NO'. For further guessing, B can record the letter when told it was correct. Let us think about the process of guessing by B. For the first letter, he must guess by tossing a coin in his mind. This process is completely random. But, as the game goes on, he becomes able to guess the next letters, according to his further guessing for the word, based on the collection of knowledge of correct letters. As the game proceeds further, he becomes able to guess the next words, according to his further guessing for the sentence, based on the collection of knowledge of correct words. In this way, the context is formed in his mind, and context-dependent inference proceeds, based on a dynamic association of memory. During the game, the consequence of inference might be equivocal, or be inconsistent with the knowledge already acquired. B's mind is not, however, fixed to this state, but alternates from one idea to another. Perhaps by chaotic activities of collective modes of electric postsynaptic membrane potentials, the collapse of old ideas and the generation of new ideas are achieved. Then, it could be that old ideas are recovered in his

[1] A most interesting case refers to *non*-chaotic multifractal attractors — exactly at the phase transition between periodicity and full chaos. In such a case *all* Lyapunov exponents are *non*-positive *but*, the deviations around the average values are considerable (Nicolis *et al.*, 1983; Nicolis and Katsikas, 1994). In such cases, the attractor possesses a *long range coherence* which makes it an ideal model for linguistic systems displaying long term memory (The 'transinformation' has *a long tail*: Nicolis and Katsikas, 1994). In relation with this topic, it is interesting to note that the presence of negative Lyapunov exponents with a small absolute value can change the differentiability of the invariant manifold. This appears, in general, in chaos-driven contracting systems, where the invariant set is expressed as a continuous (or singular-continuous) but nowhere-differentiable function (Moser, 1969; Rössler *et al.*, 1992; Rössler *et al.*, 1995; Tsuda, 1996; Tsuda and Yamaguchi, 1998).

mind, hence becomes the itinerant transition. According to divergence and convergence of ideas, this inference process becomes *chaotically* itinerant due to a successive change of the Lyapunov exponents between positive and negative values.

III: Neurophysiological Evidence About the Role of the EEG and Modelling of the Role of the Thalamo-cortical Pacemaker

Brain-like structures have evolved by performing signal processing — initially by minimizing 'tracking errors' on a competitive basis. Such systems are highly complex and at the same time notoriously 'disordered'. But, the neural activities should be more or less synchronized, though they look highly random, otherwise we cannot measure them by electroencephalogram (EEG). The functional trace of the cerebral cortex of the (human) brain is a good example. The EEG appears particularly fragmented during the execution of mental tasks as well as during the recurrent episodes of REM sleep.

A stochastically regular or a highly synchronized EEG on the other hand characterizes a drowsy (relaxing) or epileptic subject respectively and indicates — in both cases — a very incompetent information processor (Iasemidis and Sackellares, 1996). We suggest that such behavioural changeovers from the docile state to the regime of pressing mental activity are produced via bifurcations which trigger the thalamocortical non-linear pacemaking oscillator to switch from an unstable limit cycle or a simple *homogeneous* chaotic attractor of *low* dimensionality to chaotic itinerancy specifically characterized by an *inhomogeneous multifractal* (strange) attractor of *higher* dimensionality and high *selectivity*.

Our analysis aims at showing that the EEG's charateristics are not accidental but inevitable and even necessary and therefore functionally significant. An information processor (analogue or digital) is a cognitive gadget which tracks and identifies the parameters of an unknown signal or 'pattern'. Here the signal is usually contaminated by various kinds of perturbations which can be viewed as noise of various levels such as quantum noise, thermal noise, and high frequency components of macroscopic system compared with ordered motions. In order to accomplish this task, the processor has to perform three distinct operations in the following sequence: (a) produce from 'within' a wide variety of spatio-temporal patterns, (b) cross correlate (i.e. 'compress') each of those patterns with the incoming one, and (c) on the basis of some pre-established 'hypothesis-testing' criteria select or filter-out the pattern which forms the largest cross correlation with the unknown signal. To track a signal timing is of the essence. (The simplest tracker in use in communication engineering practice is the phase-locked loop [PLL]). This means that the existence of self-sustained non-linear dissipative oscillators at the hardware level of the processor is a prerequisite for the cognitive operation.

Functionally stable oscillators, in contradistinction to static (switching 'on–off') devices, offer indeed a number of evolutionary advantages as follows:

(a) Time keeping.
(b) Dynamic information storage (dynamic memory).
(c) Display of extremely broad spectrum of complex behaviours for very simple stimuli(chaotic itinerancy among chaotic attractor ruins).

Parsimony — which undoubtedly possesses survival value — requires that the locally generated dynamical patterns in the processor should not always be 'on'. They rather should emerge upon request (i.e. upon triggering from externally impinging stimuli) from a set of available dynamical elements and some basic and rather simple recursive rules for combining those elements.

Below we present a sketch for a dynamic model of a brain processor. Individual neuronal oscillators at the cerebral cortex constitute the above mentioned set of dynamical elements. The thalamocortical oscillator on the other hand, whose projection to some dimension is the recorded EEG, is the adaptive agent which performs two distinct operations:

(a) It provides pacemaking activity resulting in the formation of internal, synchronized (phase-locked) or coherent (spatio-temporal) neuronal patterns, which stand for the hardware of attractors. By making such neuronal groups coherent the pacemaker helps them elevate themselves above the ambient noise level and also distinguish themselves from coexisting neighbouring neuronal formations — within brief time intervals.[2]

(b) It generates the recursive rules governing the sequential appearance of these coherent patterns on a time-division-multiplexing scheme via chaotic itinerancy with multifractality of the pacemaking activity (see also Fig. 2).

IV: A 'Biological' Processor Possessing Multiple Coexisting Attractors

The processor amounts to a nonlinear dissipative dynamical system. Suppose there are coexisting attractors (steady states, limit cycles, tori or strange attractors) with basins divided by fractal separatrices — either fully connected or disconnected.

From a biological processor we expect the formation of multiple dynamical memories (attractors) about which we have a number of demands, if they have to emulate at all biological structures. Such demands are: *input sensitivity, fault tolerance, content addressability, large storing capacity*, and *compressibility. A dynamical* memory has to be contrasted from a *static* one. In the latter, *all* possible environmental stimuli are prestored in the processor. For large memories such a technique becomes impractical. Hence, instead of memorizing all possible 'answers' to a set of environmental inputs one is rather storing an *algorithm* which once triggered generates on the spot all possible strings which may match the incoming message provided they share the same attractor, i.e., they belong to the same basin.

Synoptic description of the state of the art in EEG research is given elsewhere (see e.g. Nicolis, 1982; 1986; 1991; Nicolis & Katsikas, 1994 and references therein). Here we will comment only on the essentials. Specific thalamic nuclei are capable of self-sustained oscillatory activity. Via fibres emanating from these nuclei and projecting on various cortical areas as well as fibres leading back from cortical areas to other nonspecific thalamic nuclei, a thalamocortical–corticothalamic feedback oscillatory activity — the so called thalamo-cortical pacemaker — is established. A

---

[2]   The neurons — members for each coherent set — belong to *different* areas in the cortex: since different kind of neurons decode for different attributes of a given object (i.e. texture, colour, shape, dimension, motion, etc.) the *phase locking* of those elements by the thalamocortical pacemaker allows for the memorization of a coherent object.

macroscopic manifestation of this activity is the EEG — summing up instantaneous subthreshold dendritic potentials from any lead on the scalp.

It appears that the thalamo-cortical pacemaker acts as a scanner of the population of cortical neurones on a time division multiplexing basis: in the presence of external stimulation, this nonlinear oscillator takes up, on a forced intermittent basis, individual subsets of cortical postsynaptic membrane dendritic potential and entrains them thereby forming semicoherent neuronal groups. This is accomplished by plastic modifications of the synapses. In the absence of external stimulation (where all sub-basins are 'empty') the pacemaker is a *homogeneous* chaotic attractor — spending on each modality 'equal' time — a mere $\sim 10^{-1}$ sec. (see basic references in Nicolis, 1986, and also in Babloyantz and Destexhe, 1986). A similar scenario has already been addressed by Freeman (1987; 1994; 1995a,b,c) for the biological significance of the chaotic activities observed in local EEG of the olfactory bulb. The difference between the two is the time scales of basic oscillations involved.

The sequence of jumps then from attractor ruin to attractor ruin may be simulated as a semi-Markov process of itinerant motion. The pacemaker here plays two roles: on the one hand, it is responsible of stimulating the attractors one by one and, on the other hand, is instrumental in making them commute. One might ask what happens to these attractor-memories when the pacemaker leaves them and 'grabs' another subgroup of cortical neurones. Do they dissolve into oblivion? The answer is that the pacemaker simply 'wakes up' these memories, since by making the cortical group involved coherent it helps it to elevate itself above the ambient neuronal noise. The consolidation of memory, however, may be achieved via synaptic–membrane–genome interactions e.g., by stimulating genes in the neuronal genome which genes give rise to proteins that renew ('recoat') the postsynaptic membrane sites of the population involved, thereby ensuring a long-term 'engram' of the particular memory. When now stimulation comes from the environment via the peripheral nervous system, the degree of arousal of the ascending branch of the reticular formation increases and the specific thalamic nuclei, responsible for the generation of the initial oscillatory activity, get polarized by amounts of time which are roughly proportional to the speed of information pumped from the peripheral nervous system or proportional to the intensity of the impinging external stimuli on the peripheral receptors. The result is that the simple semi-Markov sequence so far describing the intermittent processing by the thalamo-cortical pacemaker turns via a phase transition as it were to a composite semi-Markov process.

This means that during the time interval the specific thalamic nuclei are polarized, the oscillatory activity of the pacemaker 'freezes', and the system gets stuck at the attractor-memory involved *in excess* of the usual residence time that one should expect from simple intermittency. After selecting the next attractor ruin but before moving to it the pacemaker holds at the previous one by an amount of time equal to the interruption interval of the specific thalamo-nucleic activity.

The time interval of the polarization of the specific thalamic nuclei — which determines the residence + holding times of the processor at the modality involved — equals of course the relaxation time it takes for the initial condition (external stimulus) to fall from the basin of attraction to the attractor involved. After that the polarization of the specific thalamic nuclei is lifted and the thalamocortical pacemaker is free to go on with its intermittent scanning activity.

So now the scanner works on a different modus, namely the modus of *metastable chaos* — that is a regime possessing a *highly inhomogeneous invariant measure* (see e.g. Fig. 3a,b). Few attractor-memories claim the 'lion's share' from the point of view of residence time; the rest are summarily scanned. Under the execution of mental tasks the thalamocortical pacemaker is a highly inhomogeneous attractor, thus it may give rise to chaotic itinerancy. How do we investigate the characteristics of such an attractor?

The issue has been addressed by collecting EEG time series during epilepsy, deep sleep, awaken regime, etc., and then by trying to infer the correlation dimension of the underlying dynamics. A basic difficulty in such an enterprise is the essentially nonstationary character of the EEG, especially during the awaken regime. Nonetheless, people came up with low dimensionalities ~ 2.1, ~ 4.1 for the epileptic and the deep-sleep regime, respectively, and with high dimensionalities ~ 7 in the case of the awaken regime, and the regime of execution of mental tasks (Babloyantz and Destexhe, 1986; Mayer-Kress and Layne, 1987).

Thus the general model emerging from the combination of the dynamical processes going on, on the one hand, within each individual strange attractor-memory and, on the other hand, between coexisting attractor-memories is the following. For a given 'bioprocessor' a small subset of environmental stimuli is partitioned in a number Λ of coexisting attractor-memories whose formation is mediated by a nonlinear dissipative operator (map). These attractors, in general, are separated by fractal basin boundaries. The entropy of such a partition (of the messages to the attractors) and the degree of compressibility afforded by each attractor $C_l = N - D_l$ (where N is the dimensionality of the raw environmental messages and D_l is the dimensionality of the individual memory) give the two essential macroparameters characterizing the *cognitive channel* between environment and processor at a given hierarchical level. This process may not be accomplished in one single hierarchical step; in a second step the attractors (if numerous at a lower level) will play the role of the members of a hyper-basin towards a new hierarchy of fewer hyper-attractors and so on.

The attractor-memories involved establish further communication via the thalamo-cortical pacemaker within the processor. This activity has two aspects:

(a) The intramemory activity (which involves rehearsal and consolidation) refers to a Markovian process within each attractor; this amounts to a slow diffusion from one part of the attractor to the other, that is, a progressive 'smearing-out' (mixing) of the specific initial stimulus. The memory of the *basin as a whole*, though, remains intact (see Fig. 3a,b). For example, when you partition the wavelengths of the visual electromagnetic spectrum you *see colours* (*attractors*), *not* individual frequencies.

(b) The intermemory activity refers to establishing an intermittent connectivity between the individual stable memories which allows an itinerant behaviour among memories (Tsuda, 1992; Freeman, 1995a,b,c; Kay *et al.*, 1995; Kay *et al.*, 1996), and in the absence of external excitation this is a simple semi-Markov process; in the presence of stimulation, though, it turns into a composite semi-Markov process with total holding time distribution depending not only on the intrinsic residence times but also on the statistics of the *external* stimulus (i.e. the holding times) modulating the pacemaker's policy (Fig. 2). The *context* generated in this way is a prerequisite of *meaning* (a necessary, but by no means a sufficient condition).

In this paper, we addressed a dynamical model of biological processors which act for the formation of dynamic memory, scanning of memory space, and the propagation of information via dynamic transition among attractor ruins, depending on the thalamo-cortical inputs. The necessity of changeover from the interpretation with a single attractor-dynamics on cognition to the one with high-dimensional itinerant dynamics was emphasized.

References

Arnold, V.I. and Avez, A. (1967), *Problèmes Ergodiques De La Mécanique Classique* (Gauthier-Villars).

Babloyantz, A. and Destexhe A. (1986), 'Low dimensional chaos in an instance of epilepsy', *Nat. Acad. Sci.*, **83**, pp. 3513–17.

Freeman, W.J. (1987), 'Simulation of chaotic EEG patterns with a dynamic model of the olfactory system', *Biological Cybernetics*, **56**, pp. 139–50.

Freeman, W.J. (1994), 'Neural mechanisms underlying destabilization of cortex by sensory input', *Physica D*, **75**, pp. 151–64.

Freeman, W.J. (1995a), *Societies of Brains — A Study in the Neuroscience of Love and Hate* (Hillsdale, NJ: Lawrence Erlbaum Associates).

Freeman, W.J. (1995b), 'Chaos in the brain: Possible roles in biological intelligence', *International Journal of Intelligent Systems*, **10**, pp. 71–88.

Freeman, W.J. (1995c), 'The creation of perceptual meanings in cortex through chaotic itinerancy and sequential state transitions induced by sensory stimuli', in *Ambiguity in Mind and Nature*, ed. P. Kruse and M. Stadler (Springer-Verlag).

Grebogi, C., Kostelich, E., Ott, E. and Yorke, J.A. (1987), 'Multi-dimensional intertwined basin boundaries: basin structure of the kicked double rotor', *Physica*, **25D**, pp. 347–60.

Hirsch, M.W. and Smale, S. (1974), *Differential Equations, Dynamical Systems, and Linear Algebra* (New York: Academic Press).

Iasemidis, L.D. and Sackellares, J.C. (1996), 'Chaos theory and epilepsy', *The Neuroscientist*, **2** (2), March, pp. 118–25.

Ikeda, K., Otsuka, K. and Matsumoto, K. (1989), 'Maxwell-Bloch turbulence', *Progress of Theoretical Physics*, Supplement, **99**, pp. 295–324.

Kaneko, K. (1990), 'Clustering, coding, switching, hierarchical ordering, and control in network of chaotic elements', *Physica D*, **41**, pp. 137–72.

Kaneko, K. (1997), 'Dominant of Milnor attractors and noise-induced selection in a multiattractor system', *Physical Review Letters*, **78**, pp. 2736–9.

Kay, L.M., Lancaster, L.R., and Freeman, W.J. (1996), 'Reafferance and attractors in the olfactory system during odor recognition', *Int. J. of Neural Systems*, **7**, pp. 489–95.

Kay, L., Shimoide, K. and Freeman, W.J. (1995), 'Comparison of EEG time series from rat olfactory system with model composed of nonlinear coupled oscillators', *International Journal of Bifurcation and Chaos*, **5**, pp. 849–58.

Matsumoto, K. and Tsuda, I. (1985), 'Information theoretical approach to noisy dynamics', *Journal of Physics A*, **18**, pp. 3561–6.

Matsumoto, K. and Tsuda, I. (1987), 'Extended information in one-dimensional maps', *Physica D*, **26**, pp. 347–57.

Matsumoto, K. and Tsuda, I. (1988), 'Calculation of information flow rate from mutual information', *Journal of Physics A*, **21**, pp. 1405–14.

Mayer-Kress, G. and Layne, S.P. (1987), 'Dimensionality of the human electroencephalogram', *An. N.Y. Acad. Sci.*, **504**, pp. 62–87.

Moser, J. (1969), 'On a theorem of Anosov', *J. of Diff. Eq.*, **5**, pp. 411–40.

Nicolis, J.S. (1982), 'Should a reliable information processor be chaotic?', *Kybernets*, **11**, pp. 269–74.

Nicolis, J.S. (1986), *Dynamics of Hierarchical systems: An evolutionary approach* (Springer Verlag); also transl. into Russian (Mir, 1990).

Nicolis, J.S. (1991), *Chaos and Information Processing* (World Scientific).

Nicolis, J.S. and Katsikas, A. (1994), 'Chaotic Dynamics of Linguistic processes at the syntactical and the semantic level(s): In pursuit of a multifractal attractor', in: *Cooperation and Conflict in General Evolutionary Processes*, ed. John L. Casti and Anders Karlquist (John Wiley).

Nicolis, J.S., Mayer-Kress, G. and Haubs, G. (1983), 'Non-uniform chaotic dynamics with implications to information processing', *Zeit. für Naturforch.*, **38a**, pp. 1157–69.

Nicolis, J.S. and Tsuda, I. (1985), 'Chaotic dynamics of information processing: The "magic number seven plus-minus two" revisited', *Bulletin of Mathematical Biology*, **47**, pp. 343–65.

Oono, Y. (1978), 'Kolmogorov-Sinai entropy as disorder parameter for chaos', *Prog. Theor. Phys.*, **60**, pp. 1944–6.

Ott, E. (1993), *Chaos in Dynamical Systems* (Cambridge: Cambridge University Press).

Rössler, O.E., Hudson, J.L., Kundsen, C. and Tsuda, I. (1995), 'Nowhere-differentiable attractors', *Int. J. of Intell. Sys.*, **10**, pp. 15–23.

Rössler, O.E., Wais, R. and Rössler, R. (1992), 'Singular-continuous Weierstrass function attractors', in *Proc. of the 2nd Int. Conf. on Fuzzy Logic and Neural Networks* (Iizuka, Japan, pp.909–12).

Ruelle, D. (1989), *Chaotic Evolution and Strange Attractors* (Cambridge: Cambridge University Press).

Shaw, R. (1981), 'Strange attractors, chaotic behaviour, and information flow', *Zeitschrift für Naturforschung*, **36a**, pp. 80–97.

Tsuda, I. (1991a), 'Chaotic itinerancy as a dynamical basis of hermeneutics of brain and mind', *World Futures*, **32**, pp. 167–85.

Tsuda, I. (1991b), 'Chaotic neural networks and thesaurus', in *Neurocomputers and Attention I*, ed. A.V. Holden and V.I. Kryukov (Manchester: Manchester University Press).

Tsuda, I. (1992), 'Dynamic link of memories — chaotic memory map in non-equilibrium neural networks', *Neural Networks*, **5**, pp. 313–26.

Tsuda, I. (1996), 'A new type of self-organization associated with chaotic dynamics in neural systems', *Int. J. of Neural Sys.*, **7**, pp. 451–9.

Tsuda, I., Körner, E. and Shimizu, H. (1987), 'Memory dynamics in asynchronous neural networks', *Progress of Theoretical Physics*, **78**, pp. 51–71.

Tsuda, I. and Yamaguchi, A. (1998), 'Singular-continuous nowhere-differentiable attractors in neural systems', *Neural Networks*, **11**, pp. 927–37.

Brian Goodwin

Reclaiming a Life of Quality

Introduction

The disappearance of organisms from contemporary biology and the absence of mind from neuroscience are, I believe, both connected with a deep conceptual and method-ological feature of Western Science. Cartesian dualism and a reductionist method-ology contribute to the replacement of organisms by genetic networks and minds by neural networks. However, these divide-and-conquer strategies that are so effective at revealing the component parts of complex systems are themselves related to a more profound axiom that is often not even recognized as an assumption. This relates to the status of subjective experience in the study of natural processes. Galileo assumed that reliable data for scientific statements about natural phenomena are restricted to mea-surable quantities such as mass, velocity, temperature, volume, and so on. Such 'primary qualities', as John Locke was later to call them, contrast with 'secondary qualities' such as the experience of colour, odour, pleasure or pain, which were considered to be purely subjective aspects of human experience, arbitrarily variable between individuals and therefore unsuitable as descriptors of real natural process.

However, primary qualities originate in human experience of force, weight, motion, etc., and so are also initially subjective. They become 'objective' only by a process of intersubjective consensus whereby subjects compare systematically the results of specific observations which become known as measurement. Once such a methodology has become established within a community of practitioners, the role of subjective experience tends to recede into the background, replaced by measuring devices which substitute for human judgement and turn observation into something regarded as real and reliable. Experience is thus withdrawn from the objectively real and the world of scientific enquiry takes on the characteristics of non-sentient matter in motion, defined as activity without experience. The result is the real world posited in modern science.

The resulting metaphysics and methodology work well in the study of non-living processes, up to a point. However, they run into severe difficulties in the study of life. Simply put, we know that we humans experience qualities such as pleasure and pain, or the colour and perfume of a flower. We have such experiences through our bodies and are consciously aware through our minds. These are two aspects of one unity, the organism. But we assume that life has evolved from non-sentient matter in motion. The result is a logical conundrum: How can experiencing subjects arise from

Journal of Consciousness Studies, **6**, No. 11–12, 1999, pp. 229–35

non-sentient matter? This question has no logically consistent answer except to deny the reality of experience, a very high price to pay for particular assumptions about '-reality'. Is there not another way in which we can simultaneously preserve the deep insights that have come from modern science and save our experience as organisms with body–minds that give us feelings and awareness?

One way of approaching a resolution to this dilemma is to go back to the distinction made in science between primary and secondary qualities, the former real, the latter in some sense illusory. The argument that I shall pursue here will take the following form. Organisms are wholes that are centres of agency. To live is to act intentionally, to discriminate and to experience. To accommodate within science an understanding of the life with which we as organisms are familiar, it is necessary to acknowledge the reality of qualitative experience. This leads to an expanded conception of science that preserves all that is of value in our tradition of exploring reality but avoids the unfortunate conclusion that some of our deepest experiences are in some sense unreal.

Organisms as Causally Efficacious Wholes with Agency

Organisms have disappeared as fundamental entities, as basic unities, from contemporary biology because they have no real status as centres of causal agency. Organisms are now considered to be generated by the genes they contain. These genes have been selected by the external forces of natural selection acting on the functional properties, or characters, that allow the organism to survive and reproduce more of its kind in a particular habitat. Thus organisms are arbitrary aggregates of characters, generated by genes, which collectively pass the survival test in a particular environment. The characters clearly cohere within the physical body which they define, but there is no causally efficacious unity that transcends the properties of the interacting parts. This is the sense in which organisms have disappeared from biology.

What would it mean for organisms to have causal efficacy above and beyond that of their interacting parts? A definition of this concept is given by Silberstein (1998) in his discussion of emergent properties: 'qualitatively new properties of systems or wholes that possess causal capacities that are not reducible to *any* of the causal capacities of the parts'. One approach to the question of such properties in organisms is to provide a systematic account of the relationships between parts and whole during the development of the adult form of an organism from a zygote (a fertilized egg). It can be shown that organisms are more than functional unities in which the parts exist for one another in the performance of a particular function or set of functions, as in a machine. They are also structural unities in which the parts exist for and by means of one another, to use Kant's descriptive phrase. That is to say, the component parts of an organism arise from an undifferentiated unity, the zygote, by the progressive emergence of distinct structures during the course of embryonic development (morphogenesis). The initial unity of the organism is maintained throughout this process and into the adult form as a condition of dynamic coherence. The traditional literature on embryonic development conforms to this view (see, e.g., Waddington, 1956; Berrill, 1972). A detailed description of morphogenesis as the emergence of integrated wholes, articulated for a variety of different types of organism and different aspects of embryonic development, is given in Webster and Goodwin (1996). I will not present details of the argument here, but simply point to this evidence that organisms are generated as causally efficacious unities, and the type of theory that is required to account for it.

What about the claim that organisms are intentional agents? A detailed argument elaborating on this concept can be found in Kauffman (1999). His position has two aspects. First, organisms are autonomous agents; that is, they are organized systems with the property that they produce more of the same organization. The biological term for this is reproduction. They are therefore logically closed systems which are open to a flow of matter and energy across their boundaries, on which they depend. Hence they are coupled to their environments but not determined by them. Their autonomy results from the self-defining logical closure which perpetuates their distinctive type of organisation. Maturana and Varela (1987) defined this as autopoiesis.

The second aspect of Kauffman's argument concerns the nature of living agency. His phrase is: organisms take action on their own behalf. They do so not by computing the set of possible actions and optimizing according to some criterion, because the set of possibilities cannot be finitely described in advance. Organisms live their lives, they do not compute them. But what does it mean to live your life rather than compute it? It means to make choices in some manner that does not depend on algorithmic prespecification and selection. That is, organisms function in ways that go beyond mechanical causality and computation. How this can be articulated in terms that are consistent with current science (including quantum mechanics), or whether new principles of action are required, is a question that cannot yet be answered with any certainty. However, it seems clear that if we are to have a concept of organisms that is consistent with our own experience of intentionality and agency, and which accommodates the observed properties and behaviour of living beings, it is necessary to recognize that life embodies a quality of sentience and experience that allows organisms to act spontaneously and appropriately, to take action on their own behalf. This is reflected in the coherence and integrity of organisms, which we perceive through qualities. To elaborate further on this, I shall now explore a particular quality of whole organisms that we describe as health.

Dynamic Indicators of Wholeness and Health

I take the position that there is a property of health of the whole organism that cannot be described in terms of the functioning and interactions of the constituent organs or tissues or molecules — whatever level of parts one wishes to consider. Furthermore, this property of the whole influences the functioning of the parts in identifiable ways; that is, it has causal efficacy. The absence of such a conception from mainstream biology and medicine is evident from the fact that there is no theory and practice of health taught to medical students that develops systematically such an emergent property of the whole organism with which one can work methodically. Health in the medical model is absence of disease, not presence of a coherent state that can be recognized and facilitated by an appropriate therapeutic relationship.

Let me describe a recent development in the study of health and disease that provides evidence of a dynamic condition of the whole that transcends the properties of parts in interaction. This comes from work on the complex dynamics of the heartbeat. The mean heart rate of an individual is reliably constant for any particular activity, such as sitting still or lying or walking. However, it turns out that if one examines a series of heartbeats for any one of these conditions, as recorded in an electrocardio-gram, there is considerable variability in the interval between successive heartbeats. What came as something of a surprise was that this variability is significantly greater in healthy individuals than in people with various types of heart condition, such as cardiac arrhythmias or congestive heart disease. In the latter cases there is more regu-larity and order in the heart rate than in healthy persons. This is a case in which too much order, or the wrong kind of order, is a sign of danger!

It is possible that the irregularity of the interbeat intervals in healthy individuals is a kind of 'noise' resulting from the sum of influences exerted on the heart by other systems of the body — the nervous, respiratory, endocrine, muscular and other sys-tems whose activities modulate heart rate. On the other hand, healthy variability might carry within it some signature of a subtle dynamic order that transcends the col-lective influences of these other parts of the organism. Poon and Merrill (1997) claim that the variability of the interbeat interval does not have the characteristics of noise, but of deterministic chaos. The order manifested by chaos is indeed subtle, the dynamics being characterized by irregularity that is unpredictable but mathemati-cally determined by the properties of strange attractors, which constrain the trajecto-ries of motion within bounds. The functional interpretation of this unexpected physiological behaviour is as follows. The healthy heart maintains continuous sensi-tivity to unpredictable demands on it from the rest of the body by continuously chang-ing its rate so that it never gets stuck in a particular pattern of dynamic order. A diseased heart, on the other hand, does tend to fall into patterns of order which fail to respond to the body's constantly changing needs. We thus get the notion of dynamic disease, and inappropriate order is indicative of danger.

Do healthy people all share the same dynamic signature of health, or are they healthy in distinctive ways? This question was addressed by Ivanov *et al.* (1996) in a study of people suffering from sleep apnoea (interrupted breathing during sleep) compared with matched healthy controls. They found that while each healthy individ-ual has a distinct pattern of variability, they all share the same generic signature of subtle dynamic order that is characteristic of chaotic systems, characterized by self-similarity and the occurrence of a well-defined scaling law of variations. Individ-uals with sleep apnoea do not have this pattern. The property in question can be char-acterized as a type of long-range order or coherence that maintains a subtle balance of activity in the heart such that a series of short interbeat intervals tends to be followed by longer intervals. The origin of this behaviour is not clear. It appears to reflect a property of the whole organism that transcends the behaviour of its parts. This points to a holistic aspect of the organism with causal efficacy; i.e., the observed dynamic is an emergent property of the whole that affects the parts, maintaining a condition of coherence throughout the organism. These studies are of considerable interest and importance in indicating ways of diagnosing different conditions of the body by a detailed dynamic analysis of particular physiological variables. Traditional diagnos-tic procedures use a similar approach, but the condition of the whole is observed through a different aspect of dynamic behaviour of the organism. To illustrate this, consider next an example that indicates the procedure in a context that extends the notion of health to include behaviour generally.

Reclaiming Qualities in Science

What type of theory and praxis go with the recognition of organisms as causally efficacious, emergent wholes? The argument that I shall now develop is logically independent of whether or not one accepts the case that organisms have whole emergent
properties, though there is logical consistency between them. How might we
approach the question of assessing the quality of life that an animal has experienced
in the past from observation of its current behaviour? We actually do this frequently.
On the whole, people have little difficulty in choosing a dog from a rescue home that
exhibits behaviour indicative of a life without serious deprivation or cruelty, which
elicits fear and aggression. However, we also make mistakes. That is, our individual
evaluations can be unreliable. Is there a way of being systematic about such evaluations? One approach is to develop a method of intersubjective consensus applicable
to this problem. This involves systematic comparison of the evaluations made independently by different individuals observing the same animal. I present here an example of this type of study carried out by Wemelsfelder *et al.* (1999) on farm animals.

The study was carried out on two groups of pigs, one of which had been living in
barren conditions (a small pen with a bare concrete floor) and the other in an enriched
environment (a large pen with straw and various objects to play with, such as fresh
branches, car tyres and metal chains). People were asked to observe the pigs behaving
in standard conditions and to assess their behaviour using qualitative descriptors of
their choice to describe the pigs' style of behaviour.

This procedure is known as 'Free Choice Profiling' and is widely used in food science and sensory research. A multivariate statistical technique called Generalized
Procrustes Analysis was used to assess consensus between different observers in their
evaluations. This identifies the degree of clustering of observer scoring patterns in a
multidimensional space using transformations that identify mathematical invariants
in the data. Analytical details are presented in the paper by Wemelsfelder *et al.* (1999).

The results of the pig study were very striking. There was a high degree of consistency in the evaluations between different people, of pigs from the two groups, barren
and enriched. Evidently human beings are pretty good at qualitative judgements of
this kind. This is not surprising; we live our lives primarily in terms of such judgements, of one another and of situations generally. Where it can be carried out, quantitative assessment is a very useful addition to qualitative judgement, but often it is not
possible or convenient. In science, however, it is regarded as the *sine qua non* of data
acquisition.

The pig study employed an analytical procedure to evaluate consensus between
different observers. This involves an effective blend of qualitative and quantitative
procedures. However, it is reasonable to suggest that a group of practitioners who are
focussed on the qualitative assessment of animal behaviour could reach consensus
without this analytical step, after systematically cultivating the development of
evaluative skills. With or without the analytical procedure, the evaluators would be
practising a systematic science of qualities. They would be using their capacity for
evaluation of the quality of life exhibited by animals through observation of their
behaviour. The primary data used in this evaluation is not measurable with an instru-
ment; it requires a human subject as the observer, assessing quality. This is not to
argue that some purely quantitative measure of behaviour might not subsequently be

found that correlates with the qualitative assessment. However, the qualitative evaluation is necessarily primary and would probably remain more reliable and effective for this type of evaluation.

Doctors and therapists do something similar to this in evaluating the health of the people that come to them for healing. They pay attention to posture, tone of voice, complexion, and other aspects of the person that reflect the condition of the whole in ways that cannot be measured by instruments. Quantitative data on body temperature, heart rate, blood pressure, blood cell counts, etc., can add significantly to a diagnosis, but qualitative evaluation of the condition of health remains a very important aspect of diagnostic skill which is developed through practice and experience. It could be cultivated more systematically during training by some type of intersubjective consensual procedure of the kind described above in the pig study. This would extend scientific data to include both quantitative and qualitative information, without losing the essential scientific principles of comparison of results within a community of persons using agreed procedures of assessment. Qualitative experience would then be recognized as a potentially reliable indicator of real situations, subject to consensus among trained practitioners.

There are many communities of investigators into qualitative methodology that are already pursuing such procedures. However, they work under the shadow of a science that has honed the quantitative study of natural process to a very fine art, while qualitative procedures, though by no means new within science, are still being explored and developed. Furthermore, the metaphysical assumptions about reality that have emerged within conventional science exclude qualities from the real and locate them within subjective, hence idiosyncratic and objectively unreliable, experience. A science of qualities requires a fundamental reappraisal of the very nature of real process, because it recognizes experience as real and primary. But this is also required if we are to accept the reality of our own experience as feeling, intending, conscious organisms. If these properties are real, then they can only arise from a reality that embodies some form of sentience as the precursor of this condition; otherwise they can be construed only as unintelligible miracles of emergence from dead matter. It seems better to extend our basic description of reality than to have to believe in this type of miracle.

Qualities Require a New Science

The change required in our conception of 'reality' to accommodate subjective experience has been the subject of many articles in this journal and I cannot add significantly to what has already been said by others. However, I can indicate which lines of argument I think will provide a metaphysical basis for a science of qualities of the type sketched above. A foundation for the requisite rethinking comes from the writings of Bergson (1911) and Whitehead (1929), with subsequent developments by Hartshorne (1972) and, most recently, by Griffin (1998). The essentials of the position are that 'matter' has sentience and 'mind' exists only as an aspect of 'matter'. What resolves these apparent antinomies is process, in which present mind gives rise to past matter as spent experience, to use the useful and evocative phrases of de Quincey (1999). There is a rough analogy here with electromagnetic waves as described in Maxwell's equations in which the electric field gives way to the magnetic field which in turn generates the electric field in a never-ending cycle of unfolding. Likewise 'mind' and 'matter' transform one into the other, mind (experience, sentience) being the creative pole that incorporates past matter into a new unfolding involving a degree of freedom and choice, this creative act then expiring in matter which produces the conditions for a new creative emergence. Working out the details of this new cosmology is a task that will occupy many a philosopher and scientist, the two areas of enquiry necessarily joining forces to define a new conception of reality. But this new conception involves a union much more extensive than philosophy and science. With qualities and feelings as essential aspects of science, the door is open to a rethinking of the relation between the arts and the sciences in our culture. The move will be beyond holistic science to a holistic culture. However, there is a great deal of work to be done if we are to get there in an effective way. As the Sufi poet, Rumi, put it:

> This talk is like stamping new coins. They pile up,
> While the real work is done outside
> By someone digging in the ground.

Acknowledgement
I am grateful to Stuart and Elisabeth Kauffman for inspiration, assistance and hospitality during the writing of this essay, and to Françoise Wemelsfelder for useful comments.

References.

Bergson, H. (1911), *Creative Evolution*, trans. A. Mitchell (New York: Henry Holt and Co).
Berrill, N.J. (1972), *Developmental Biology* (New York: Sinauer Associates).
Griffin, D.R. (1998), *Unsnarling the World Knot: Consciousness, Freedom, and the Mind–Body Problem* (Berkeley: University of California Press).
Hartshorne, C. (1972), *Whitehead's Philosophy: Selected Essays, 1935–1970* (Lincoln: University of Nebraska Press).
Ivanov, P. Ch., Rosenblum, M.G., Peng, C-K, Mietus, J., Havlin, S., Stanley, H.E. and Goldberger, A.L. (1996), 'Scaling behaviour of heartbeat intervals obtained by wavelet-based time-series analysis', *Nature*, **383**, pp. 323–7.
Kauffman, S.A. (1999), *Investigations* (Oxford: Oxford University Press) (in press).
Maturana, H., and Varela, F. (1987), *The Tree of Knowledge* (Boston: Shambala).
Poon, C-S., and Merrill, C.K. (1997), 'Decrease of cardiac chaos in congestive heart failure', *Nature*, **389**, pp. 492–5.
de Quincey, C. (1999), 'Past matter, present mind; a convergence of worldviews', *Journal of Consciousness Studies*, **6** (1), pp. 91–106.
Silberstein, M. (1998), 'Emergence and the mind–body problem', *Journal of Consciousness Studies*, **5** (4), pp. 464–82.
Waddington, C.H. (1956), *The Principles of Embryology* (London: Allen and Unwin).
Webster, G. and Goodwin, B. (1996), *Form and Transformation; Generative and Relational Principles in Biology* (Cambridge: Cambridge University Press).
Wemelsfelder, F., Hunter, E.A., Mendl, M.T., and Lawrence, A.B. (1999) 'The spontaneous qualitative assessment of behavioural expressions in pigs: First exploration of a novel methodology for integrative animal welfare measurement',*Applied Animal Behaviour Science*, in press.
Whitehead, A.N. (1929), *Process and Reality* (Cambridge: Cambridge University Press).

Valerie Gray Hardcastle

It's O.K. to be Complicated
The Case of Emotion

Since at least the time of Darwin, we have recognized that our human emotional life is very similar to the emotional life of other creatures. We all react in characteristic ways to emotionally valenced stimuli. Though other animals may not blush or cry, we all have prototypical ways of expressing anger, disgust, fear, sadness, happiness, and curiosity. In assuming that the neural circuits underlying these reactions are homologous or at least analogous across species, neurophysiologists and neuropsychologists have been able to construct impressive and substantial research programmes studying the neural correlates for emotion. They are to be applauded, for we now know quite a lot about where and how basic emotions are processed in the brain.

At the same time, there is a dangerous trend developing in the study of emotion in neurophysiology and neuropsychology, a trend toward oversimplifying and reducing emotional responses to the point of distortion. We all know that scientists must abstract away from much of what is going on in order to produce quantitative and unambiguous data. We also know that scientists operate using several basic methodological, technological, and theoretical assumptions. The question I wish to address here is whether, in the case of emotions, scientists haven't gone too far in their tendency to modularize brain processes and to reduce reactions down to their simplest components.

I: The Amygdala on Centre Stage

Studies in rats, monkeys, and people are coalescing around the idea that the anatomical circuits for emotion are relatively straightforward: prefrontal cortex, ventral striatum, and, most importantly, the amygdala (cf. Mlot, 1998). Imaging studies confirm that the amygdala lights up when we feel stress, fear, disgust, or happiness (Breiter *et al.*, 1996; LeBar *et al.*, 1998; Phillips *et al.*, 1997; though see Reiman *et al.*, 1997 for complications to this story). Those with lesions in the amygdala cannot process negative expressions on the faces of others or verbally expressed fear and anger; and they experience no fear conditioning (Adolph *et al.*, 1995; LeBar *et al.*, 1995; Scott *et al.*, 1997). In addition, mood disorders show abnormalities in the amygdala as well as prefrontal and cingulate cortex (cf. Kennedy *et al.*, 1997).

The amygdala lies at the heart of our limbic system, our so-called 'emotional brain' (see Sitoh and Tien, 1997, for review). This system has been actively investigated

Journal of Consciousness Studies, **6**, No. 11–12, 1999, pp. 237–49

since James Papez (1937; 1939) first recognized it in the late 1930s. Though our understanding of the emotional circuitry has changed since then, the fundamental hypothesis has remained intact. In brief, subcortical areas are exquisitely sensitive to emotion-laden stimuli. They tag incoming data with a valence and then send a message explaining what they have done to the rest of the brain (cf. LeDoux, 1996). More specifically, sensory information comes in via whatever sensory pathway is appropriate. This information then feeds into the thalamus, which processes the stimuli in a rough and ready way. The thalamus projects to the lateral nucleus of the amygdala, which distributes the information to other regions of the amygdala, including the central nucleus. Information then travels out of the amygdala via the central nucleus to various subcortical areas. Each area controls a different autonomic response — for example, freezing, changes in blood pressure, releasing hormones (see LeDoux, 1996, chapter 6, for a good summary).

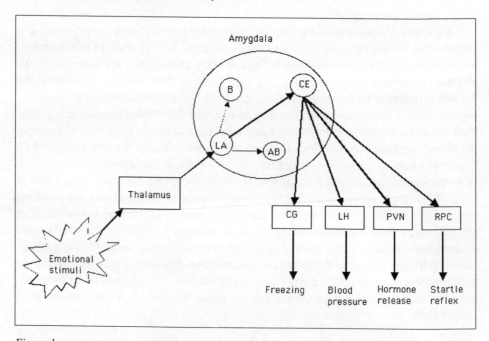

Figure 1

Sensory information travels to the lateral nucleus (LA) in the amygdala via the thalamus. LA then sends the information it has processed to other regions in the amygdala, including the basal nucleus (B), the accessory basal nucleus (AB), and the central nucleus (CE). The central nucleus then forwards the information to the central gray (CG), the lateral hypothalamus (LH), the paraventricular hypothalamus (PVN), and the reticulopontis caudalis (RPC), among other areas. These regions then initiate individual behavioural responses, such as freezing by CG, changes in blood pressure by LH, hormonal release by PVN, and a startle reflex by RPC [derived from LeDoux, 1996].

Many hold that this early and primitive valence-tagging forms the core of our emotional responses. Whatever happens later in cortex is merely in reaction to what the amygdala and other lower areas are doing. 'The neocortex can come to be influenced by emotions, and influences them through various appraisal processes, but it is not a fundamental neural substrate for the generation of affective experience'

(Panksepp, 1998, p. 42; see also Watt, 1998). Moreover, our emotional amygdala–thalamic pathway is fundamentally different from our higher level cortical processors. Joe LeDoux puts this position starkly: 'Emotion and cognition are . . . separate but interacting mental functions mediated by separate but interacting brain systems' (LeDoux, 1996, p. 69).

Perhaps the best sort of evidence for this view is data showing that the amygdala can be activated apart from cortical responses. For example, the amygdala is activated during implicit priming tasks containing affective stimuli (Whalen *et al.*, 1998). It responds to emotionally salient inputs, regardless of whether we are aware of them or can consciously respond to them. (We can and do show autonomic responses with amygdala activation; these occur outside awareness.) Even if we completely decorticate an animal, we can still elicit emotional reactions and learning (Kaada, 1960; 1967). It appears that we do not need our higher order thought processes in order to have a rich emotional life.

Conversely, if the amygdala and surrounding areas are damaged, we lose any emotional significance we would normally attach to stimuli. Henrich Klüver and Paul Bucy first described this blindness to the emotional world at the same time as Papez outlined what he took to be the limbic system (Klüver and Bucy, 1937; 1939; see also LeDoux, 1996, chapter 6, for discussion and further references). People with Klüver-Bucy syndrome, as it is now known, show little fear, anger, or anxiety. Apparently, without our amygdala, we can still perceive and understand our world; we simply no longer have any particular feelings associated with it.

The difficult question is how to go from these sorts of facts to understanding human emotions in all their complexity. Most studies, in both rats and humans, concern very simple and basic emotional responses. Most, in fact, concern fear and stress. Our emotional life, however, is quite complicated, subtle, and nuanced. Many of our human emotions have obvious developmental and cognitive components — shame and guilt are two commonly studied examples of our 'self-conscious' emotional reactions. Though there is much to say about such states, and much has been said in the psychological and anthropological literature, in this article I focus on the so-called basic emotions. I do this for simplicity's sake, but I also do it for rhetorical purposes, because even if we leave aside the more complex feelings, such as righteous indignation or guilty pleasure, we still experience a wide range of emotional phenomena. (See my forthcoming for a fuller treatment of complex emotional states.)

Should we think of these affective states merely as extensions or elaborations of some more primitive unit? Is it correct to say that the amygdala sets the valence prior to higher level interpretation or adjudication? I hold that the answer to both of these questions is no. Cortical activity not only extends and expands our feelings, but it also can determine them in the first place. Better: subcortical and cortical areas working together as a complex dynamical system produce our emotions. To focus on one area to the exclusion of the other is a mistake, both conceptually and empirically. Though I appreciate Douglas Watt's charge that neuroscientists often are 'cortico-centric' (personal communication), I believe that in the case of studying emotion, the trend runs in the opposite direction and that not enough attention is being paid to what cortex does for us emotionally. We are being amygdaloid-centric. More to the point: we are ignoring the complex feedback loops in affective processing, and we are doing this at our peril.

II: The Complexity of Our Emotional Life

We can get a flavour of what I am talking about by reviewing some individual differences in our emotions. It is well known that abused or neglected children can later
interpret what would normally be emotionally unpleasant experiences (jealousy,
anger, fear) as something with a positive valence. Not only do they seek out situations
that elicit these reactions, but they claim that they feel better for doing so. Masochists
make similar assertions. Gender influences both how we interpret the facial expressions of others and how we react in negative situations. Women react with sadness,
shame, and guilt, while men react with anger (Hess, 1998a). Social context influences
how emotions are experienced and expressed; we feel more strongly and express our
feelings more passionately when we are in a social situation than when we are alone
(Hess, 1998b). In short, personal histories and environmental contexts shape emotional responses, down to and including the basic valence attached to stimuli.

I am not claiming that emotions are socially constructed, however. Certainly we find
deep similarities across cultures, individuals, even species, when we look at affective
responses. We do appear to exhibit certain core emotions: joy, fear, distress, anger, surprise, lust, disgust. Nevertheless, how we interpret the world around us has much to do
with how stimuli inputs are experienced. Amygdala processing cannot be the first processing step in having an emotion, for its reactions are fed by previous cortical analysis.

In other words, there is something right about psychologists' appraisal theories of
emotion. Appraisal theories began with Magna Arnold's theory of emotion (for
example, Arnold 1945; 1960; as discussed in Strongman, 1996). She claims that
(excepting taste, pleasure, and pain) we always and immediately evaluate incoming
stimuli with respect to memories of our past experiences. We appraise the things in
our world as either good, bad, or indifferent, seeking what is good, avoiding what is
bad, and ignoring what is indifferent. Our current experiences later become our memories; hence our current emotional reactions will taint future experiences and our
judgements about them.

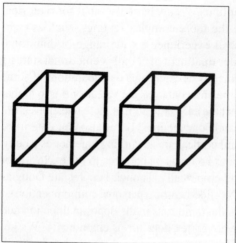

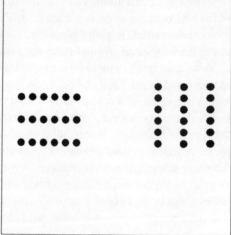

Figure 2. Necker cubes.   *Figure 3*. Rows and columns of dots.

Stanley Schacter (1959; 1964; 1970) is perhaps the best known advocate of an appraisal theory of emotion. On the basis of several ingenious but highly contentious experiments, he concluded that we label physiological arousal depending upon what we happen to be thinking about at the moment. Any one state could be labelled and thus experienced in many different ways, depending on the context. Our emotional states, he believes, are determined mainly or entirely by higher level cognitive interpretations.

We do know that humans are designed to assign singular meanings to incoming sensory data. Gestalt psychology trades heavily on this fact. We see the pair of Necker cubes in Figure 2 as both pointing in or out, but never either way at the same time or no direction at all. We can't help but see rows and columns in the dots in Figure 3.

But our interpretive skills are not limited to filling out, or filling in, visual space. We group these 'meanings' together internally into schemas which we use to parse our world. Which interpretation of some set of incoming stimuli we choose or use depends upon our particular cognitive histories, which individualize and personalize our schemas, and the particular environmental circumstances surrounding the event. In my house, because I have lots of small children, there is often a pitcher of apple juice in the refrigerator. I don't like apple juice; I like iced tea. Unfortunately, apple juice and iced tea look very similar to one another in opaque containers. Sometimes, when I am in a hurry, I grab a pitcher out of the fridge, thinking that it is tea, when it is in fact apple juice. After I pour myself a glassful and take a swallow, my immediate reaction is not, 'Ugh, apple juice.' It is, 'Ugh, I didn't know tea could go bad.' My expectation of drinking tea has shaped how I experience the world.

And I am not alone in this capacity. If male college students look at pictures of beautiful women after exercising, they will interpret their increased heart rate, etc., as sexual arousal. If they exercise without the benefit of the pictures, they will interpret the same changes in heart rate and their autonomic system as physical exertion (Cantor et al., 1975). Similarly, if male subjects happen upon a woman while crossing an unsteady bridge, they will rank her as more attractive than the same woman met while crossing a quiet street. The subjects interpret the arousal brought on by the instability of the bridge and consequent danger as sexual attraction.

People selectively look for evidence to confirm their active schemas or current beliefs about the world (Snyder, 1979). Undergraduate subjects who signed consent forms that suggested they might feel pain reported a neutral vibration as 'stinging' or 'burning' and indicated that the sensation was indeed painful. Many believed that they had been shocked electrically. Subjects who signed consent forms intimating pleasure reported that the same neutral vibration was 'tingling' and ranked their sensations as pleasurable. Subjects who received no indication of what the experience might be like reported the sensation accurately as 'vibrating', and ranked the experience as neutral (Pennebaker, 1975). The subjects' responses differed significantly, varying as their expectations did. If we believe that pain is imminent, then we are more likely to feel pain and report stimuli as painful (Blitz and Dinnerstein, 1971; Gelfand, 1964; McKenna, 1958; Neufield and Davidson, 1971). If we are expecting something pleasurable, then we are more likely to experience pleasure.

Psychologists argue that we interpret emotional changes in our bodies in the same way. We need to analyse and interpret putative affective stimuli before we get a full-blown emotion. In its most extreme versions, appraisal theories would mean that an activated amygdala–frontal lobe circuit under circumstances of duress could be interpreted, and hence experienced, as either fear or anger or even as surprise or joy, depending upon which cognitive schema was active at the time.

We already know that different emotions are instantiated in different ways in the brain so that we actually have several affective systems instead of just one (LeDoux, 1996; Panksepp, 1998). For example, happiness elicits greater activity in the ventral mesial frontal cortex than sadness (Lane et al., 1997). This fact would seem to suggest that the appraisal view must be wrong, at least in its strongest versions. Insofar as different parts of the brain are concerned with processing different emotions, then it seems that we should be able to tell our emotions apart. At the least, our brains would register the difference implicitly, even if we could not experience the difference consciously. Is there any evidence at all that indicates that some strong version of the appraisal view is correct?

Interestingly enough, Joe LeDoux himself provides some in his extended discussion of conditioned fear (1996). He reports an interesting heart conditioning study in rabbits from the mid-1970s (Schneidermann et al., 1974). When rabbits were exposed to two tones, only one of which was paired with an electrical shock, they eventually learned which tone preceded the shock and exhibited fear conditioning to that tone. However, if the auditory cortex were lesioned, then the rabbits exhibited fear to *both* tones. Without cortex, they could not distinguish the tones, though they still remembered that they should be afraid of them.

LeDoux (1996) argues that the subcortical system from the thalamus to the amygdala is the central pathway for fear conditioning. According to him, the advantage in keeping things out of cortex is one of time. It takes only 12 msec for information to travel from our auditory transducers to our amygdala via the thalamus; it also takes about that long for the information to reach auditory cortex. Hence, it would take about twice as long for the information to travel up to cortex and then back down to the amygdala. By that time, the organism could already be reacting and avoiding a potentially deadly situation. The thalamic–amygdala connection can deliver only crude interpretations of stimuli compared to what our cortex can do, but it can do it much faster.

However, his interpretation does not fit all the facts. If the thalamic–amygdala pathway were the primary source for our affective reactions, then the intact rabbits should have exhibited some sort of fear response to both tones first, and then have one of the responses (the one to the unconditioned sound) damped down by later cortical input. That didn't happen. The intact rabbits reacted only to the appropriate sound. The thalamic–amygdala tract may be crude and fast, but it also must be tied somehow into cortical activity more strongly that LeDoux suggests.

It takes both the cortex and the amygdala, working together, to create fear in intact rabbits. The amygdala might be able to do it alone in lesioned animals, but that does not show that it can do it alone in whole ones. When I bench-press weights in the gym, I don't engage my leg muscles. However, when I pick up my son to greet him after school, I do use muscles in my legs. In the gym, I lie such that I prevent certain muscles from contracting. At home, I do not take such precautions. If given the chance, I will use more muscles rather than fewer when lifting something heavy. That I don't use my leg muscles in the gym doesn't show that they aren't important for picking items up; they are. Moreover, when I can use my legs, the entire structure of my movements changes; it isn't the same lift anymore.

I maintain that rabbit emotions work the same way. Without cortex, you get one neuronal configuration; with cortex, you get another. It is unlikely that humans function all that differently. Relative to other animals, our neocortex is quite large and it has more and stronger connections with the subcortical areas. I would expect, therefore, that (if given the chance) our cortex has much influence over what the amygdala does and which emotions are present.

There are now two views of emotions on the table. Cognitive appraisal theories claim that cognitive interpretations are an integral part of emotion and either precede or co-occur with physiological responses. In contrast, the basic emotion theorists propose that physiological arousal precedes cognitive appraisal and is the more fundamental process. Definitive data regarding which view is correct is not yet forthcoming. At best, the waters are murky here (see, for example, Tomaka *et al.*, 1997), with some studies showing that appraisal precedes arousal and others indicating that events occur the other way round. Nevertheless, and issues of timing aside, there is enough evidence to warrant the conclusion that our higher cortical areas are importantly and fundamentally involved in our emotional responses.

We do know that essentially the whole brain lights up with emotion in fMRI studies (Lang *et al.*, 1998); both cortical and subcortical areas are strongly active. PET studies have shown that happiness, sadness, and disgust are all associated with increases in activity in the thalamus and medial prefrontal cortex, as well as with activation of anterior and posterior temporal structures. Emotional visual stimuli also activate the visual pathways, including primary and secondary visual cortex, especially the regions tied to object and spatial recognition. Richard Lane and his colleagues conclude, and I concur, that even though different specific emotions can activate different subcortical areas, emotional responses still require complex sensory, association, and mnemonic circuitry (Lane *et al.*, 1997; see also Paradiso *et al.*, 1997).

III: Emotions on the Fly

What should we take away from this discussion? How should we understand our emotions? Paul Griffiths (1998) argues for a dynamical systems view of emotion. He believes that we do have several basic 'affect programs' with stereotypical response profiles (Griffiths, 1997). These are complemented by social and psychological variables, which can alter or expand our more innate reactions. In this way, he does not differ very much from LeDoux, Panksepp, or Watt, who also hold that we have several primitive affective modules which react apart from cortical influences (see also Damasio, 1994). However, he also thinks that a model of epigenetic inheritance best describes our emotional development. Genes and culture acting together determine our emotional phenotypes; there is a many–many relationship between genetics and neurology, with lots of external factors contributing. Our emotions depend upon a psychoneural development that requires a social and cultural environment. We might have a genetic blueprint laid down for some affect programs, but the context in which we are raised determines our emotional outcome. He concludes that organisms embedded in an environment form a complex dynamical system with great sensitivity to initial conditions.

I am sensitive to his metaphor. However, he also believes that underlying the multifarious emotional responses humans have, we can still find some common core of properties attached to a set of differences. In other words, basic emotions, the affect programs, are homologous across cultures and closely related species. These various emotions have a common evolutionary origin, and we can use that fact to pick out underlying identities for each of them. Cultural influences entail non-linear models, but the basic affect programs remain more or less intact across all creatures with a common ancestor.

Here I diverge from the way he understands emotional categories. Neurophysiologists and neuropsychologists have traditionally individuated their natural kind categories *functionally* and not evolutionarily. Structures that perform the same duty across humans or primates or mammals or whatever, are grouped together as one sort of thing, regardless of whether these structures are homologous or analogous, or completely unrelated to each other. Conversely, even if some anatomical structures are homologous across whatever domain interests the scientists, this does not entail that the structures should be thought of as a single kind of thing from the perspective of neuroscience. Though Griffiths believes otherwise, individuating physiological and psychological categories functionally does strike me as an appropriate way to proceed, given that neurophysiology and neuropsychology are both fields interested in studying functional units. They want to know what various structures are doing now in an organism, not what led to its being there over evolutionary time. (Both, of course, are entirely legitimate questions. One set just happens to interest one group; the other set, another group.) Neurophysiologists and neuropsychologists want to know what roles and responsibilities isolated brain structures have, and not so much why we have them in the first place. Though an evolutionary story can certainly be useful in thinking about functions and in individuating psychological categories, the buck doesn't stop there.

I grant that not all physiological or psychological categories are well formed. Indeed, many of them probably should be seriously revised or outright eliminated. Nevertheless, this fact does not entail that the general approach to individuation in neurophysiology or neuropsychology is incorrect. Insofar as these fields remain primarily interested in the role a structure plays, it does not make methodological sense to seek to alter their program of functional individuation in favour of a historically-minded genetic approach.

Though I do not have the space to develop a complete argument for my position over Griffith's, for the purposes of this essay, let us assume my perspective on this matter. That contemporary neuroscience has already (rightly or wrongly) decreed that neurophysiology and neuropsychology are functional sciences surely puts the theoretical presumption in my favour. Let us agree that if we have two different inputs passing through two different neural pathways in two different organisms resulting in two different sets of responses, we have little reason to assume that the two neural pathways are both token instances that fall under a single natural kind category. Consequently, that we have (or might have) a genetic specification for certain affect programs does not mean that we can isolate those programs in the organism functionally, nor does it mean that they exist phenotypically in any interesting sense (even from the perspective of evolutionary biology).

It is true that anger or grief or happiness does result in similar physiological changes across humans, regardless of their individualized schemas or their cultural backgrounds. It is also true that we can relate these changes to those seen in other primates. I do not want to claim that our emotions spring *sui generis* from nowhere. Certainly, our biological heritage explains much about our affective reactions. Nevertheless, we should categorize these emotions based on their physiological similarities and not on their common ancestors.

Still, I think a dynamical systems approach gives us a useful metaphor for thinking about our emotions. We can think of our brains as resonating circuits, constantly active as we are always interpreting, reacting to, and predicting the world around us. Neural firing patterns in the amygdala forms part of this distributed activity, as do the firing patterns in cortex. These circuits will naturally differ across individuals and within the same individual over time. We can map the firing patterns of these circuits over time to a trajectory in a multi-dimensional phase space. (Each dimension would represent each variable that determines how the firing pattern might go; the phase space then would be a region that includes all and only the possible activity patterns in the circuit.)

My conjecture is that, given the dependence of the firing patterns on previous activity and the dependence of previous activity on the organisms' complicated and unique environmental history, the trajectories in phase space are most likely to be well-behaved but chaotic. That is, no activity pattern will precisely duplicate any other pattern, but the patterns will settle into attractor basins in the phase space. (I believe that we will find attractor basins instead of virtually random trajectories because organisms are relatively predictable. The behavioural regularities have to be caused by something other than near random events.)

The chaotic attractors would be a natural way to individuate our physiological or psychological events, for it would identify common trends across firing patterns without entailing that these trends contain isolable defining properties. Assuming we have identified the relevant variables for physiology or psychology, it will indicate how, speaking broadly, the brain interprets incoming stimuli. Incoming stimuli would be incorporated into the current activity pattern and it would either force the trajectory of the circuit into another chaotic region, indicating that the brain has changed its mind, as it were, or it will not, indicating that the brain is continuing to believe and react as it did before. In short, the chaotic attractors will show how the brain individuates and understands its world.

When the brain receives some emotion-laden stimulus, it will seek to interpret that, just as it would any other stimuli. Amygdala activation is part of the interpretation; so is cortical activation. That is, both the amygdala and the cortex are part of the resonating brain circuit relevant to emotion. These two regions work together to establish coherent and cohesive responses to input, to maintain the circuit's trajectory in an established attractor basin. The stimulus drops in, so to speak, onto ongoing interpretive efforts against a background of experiences, onto an activity pattern sensitive to initial conditions. How the brain has carved up its phase space will shape how the emotion-laden stimulus is perceived and reacted to.

No two emotional responses are ever going to be exactly the same, since they are determined by the brain's current resonating activity. Still, family resemblances will exist among them, as seen in the established attractor basins. Though we may not be able to identify certain core reactions common to any particular 'basic' emotion, our emotions partake in related activity configurations. That is, if we create a phase space of possible affective responses by the brain, we should be able to identify regions that correspond to our various emotional categories.

As an analogy and perhaps a simplistic example for what I am describing, consider Margaret Bradley and Peter Lang's marvellous work in plotting physiological response to affective stimuli along the dimensions of arousal and valence (Bradley, 1998; Lang, 1998). The first thing one should notice when looking at the data is that the physiological, linguistic, and behavioural responses do not correlate well with one another. There is tremendous individual variation for each stimulus. At the same time, we can still see general trends developing across groups, though what these trends mean exactly is unclear.

In normal male adults, we find negative valence and significant arousal in response to unpleasant stimuli. In contrast, psychopaths show little arousal toward the same stimuli. One might be tempted to conclude that negative valence and significant arousal are two of the core reactions in negative emotion (as Griffiths argues we should) and that psychopaths present an abnormal case. However, normal little boys also show little arousal in response to unpleasant stimuli. Instead of concluding that little boys are psychopaths, or — probably more accurately — that psychopaths are little boys, we should be sensitive to the range of possible reactions to the same inputs. Our model of emotion should seek family resemblances, basins of attraction, instead of a list of necessary or sufficient conditions.

Anxiety patients and older women show little arousal to pleasant stimuli, and both show increased arousal to unpleasant ones. We can see with these simple measures that a whole range of responses — ranging from significant arousal down to no arousal at all — are possible given the same stimuli. Our models of emotion should recognize that fact. Trying to find a common core of responses distorts the data, even when we are plotting along two dimensions. Imagine the distortion if we included many more variables. *Pace* Griffiths, I conclude that a dynamical systems approach will be useful when thinking about our emotions, for we need a way to model human emotions in all their complexity.

IV: The Bane of Reductionism

Most neurophysiologists operate under the reductionist assumptions that if they probe deeply and carefully enough into the brain, then they will be able to isolate anatomical or physiological differences delineating the different emotional states. In contrast, appraisal theorists claim that this sort of reductionist approach cannot work, for part of what determines what the emotion is, is a higher-order cognitive interpretation. The 'feel' of the feeling depends upon what we believe our emotional state should be. Any complete theory of emotion should reflect our best theories of cognition as well as our best neurophysiological data.

For now, it appears that both sides are at least partially correct. The best data suggest that the emotions of humans, and most likely other creatures as well, are fundamentally and inescapably tied to our interpretive abilities. Hence, a purely reductive approach which focuses on single areas in the brain to analyse them will miss important aspects of our emotional experiences and will be less well suited to predict behaviour than a more encompassing dynamic model. Nevertheless, claiming that emotions are partially interpretive does not make them mysterious or exempt from detailed neuroscientific scrutiny. As our appreciation for brains being complex dynamical systems embedded in a world increases, and we are better able to connect activation patterns in neural circuits with meaningful attractor basins, then our theories of emotions will become better as well.

In sum: emotions are wonderfully recursive and complicated phenomena, and it is okay to devise theories that reflect this fact. Our emotions are fundamentally tied to our cognitive processing (and vice versa) and both shape future reactions. At the moment, neuroscience can gesture towards the areas in the brain in which these events occur, but it cannot tell the full story, for that would require a much richer theory concerning what our resonating brain circuits are doing as they receive input from the world. Our best theory of emotion will be a non-reductive, but biological one.

Acknowledgements

I owe thanks to Geoffrey Sayre-MacCord for a delightful conversation regarding Hume, the amygdala, and fear conditioning. This interaction solidified my views regarding the connection between emotion and cognition. Embryonic versions of these ideas appeared in the *AAAI Proceedings*, 1998, and were presented at the Workshop on Emotions, Consciousness, and Qualia, in Ischia, Italy, 1998. I owe much to the participants in that workshop, for I learned a great deal from them all.

References

Adolph, R., Tranel, D., Damasio, H. and Damasio, A.R. (1995), 'Fear and the human amygdala', *Journal of Neuroscience*, **15**, pp. 5879–91.

Anderson, D. and Pennebaker, J. (1980), 'Pain and pleasure: Alternative interpretations of identical stimulations', *European Journal of Social Psychology*, **10**, pp. 207–10.

Arnold, M.B. (1945), 'Physiological differentiation of emotional states', *Psychological Review*, **52**, pp. 35–48.

Arnold, M.B. (1960), *Emotion and Personality* (2 volumes) (New York: Columbia University Press).

Blitz, B. and Dinnerstein, A. (1971), 'Role of attentional focus in pain perception: Manipulation of response to noxious stimulation by instruction', *Journal of Abnormal Psychology*, **77**, pp. 42–5.

Bradley, M. (1998), 'Psychophysiology of perception', Presentation at the Workshop on Emotions, Qualia and Consciousness, International School of Biocybernetics, Ischia, Italy.

Breiter, H.C., Etcoff, N.L., Whalen, P.J., Kennedy, W.A., Rauch, S.L., Buckner, R.L., Strauss, M.M., Hyman, S.E. and Rosen, B.R. (1996), 'Response and habituation of the human amygdala during visual processing of facial expression', *Neuron*, **17**, pp. 875–87.

Cantor, J.R., Zillman, D. and Bryant, J. (1975), 'Enhancement of experienced arousal in response to erotic stimuli through misattribution of unrelated residual arousal', *Journal of Personality and Social Psychology*, **32**, pp. 69–75.

Damasio, A. (1994), *Descartes' Error: Emotions, Reason and the Human Brain* (New York: Grosset/Putnam).

Gelfand, S. (1964), 'The relationship of experimental pain tolerance to pain threshold', *Canadian Journal of Psychology*, **18**, pp. 36–42.

Griffiths, P. (1997), *What Emotions Really Are: The Problem of Psychological Categories* (Chicago: University of Chicago Press).

Griffiths, P. (1998), 'What emotions really are: The problem of psychological categories', Presentation at the Workshop on Emotions, Qualia and Consciousness, International School of Biocybernetics, Ischia, Italy.

Hardcastle, V.G. (forthcoming), 'Dissolving differences: How to understand the competing approaches to human emotion', in *The Caldron of Consciousness: Desire and Motivation*, ed. N. Newton and R. Ellis (Amsterdam: John Benjamins Press).

Hess, U. (1998a), 'The experience of emotion: Situational influences on the elicitation and experience of emotions', Presentation at the Workshop on Emotions, Qualia and Consciousness, Inernational School of Biocybernetics, Ischia, Italy.

Hess, U. (1998b), 'The communication of emotion', Presentation at the Workshop on Emotions, Qualia and Consciousness, Inernational School of Biocybernetics, Ischia, Italy.

Kaada, B.R. (1960), 'Cingulate, posterior orbital, anterior insula and temporal pole cortex', in *Handbook of Physiology, Volume 2*, ed. J. Field, H.J. Mogoun and V.E. Hall (Washington: American Physiological Society).

Kaada, B.R. (1967), 'Brain mechanisms related to aggressive behavior', in *Aggression and Defense — Neural Mechanisms and Social Patterns*, ed. C. Clemente and D.B. Lindsley (Berkeley: University of California Press).

Kennedy, S.H., Javanmard, M. and Vaccarino, F.J. (1997), 'A review of functional neuroimaging in mood disorders: Positron emission tomography and depression', *Canadian Journal of Psychiatry*, **42**, pp. 467–75.

Klüver, H. and Bucy, P.C. (1937), '"Psychic blindness" and other symptoms following bilateral temporal lobectomy in rhesus monkeys', *American Journal of Physiology*, **119**, pp. 352–3.

Klüver, H. and Bucy, P.C. (1939), 'Preliminary analysis of functions of the temporal lobes of monkeys', *Archives of Neurology and Psychiatry*, **42**, pp. 979–1000.

Lane, R.D., Reiman, E.M., Ahern, G.L., Schwartz, G.E. and Davidson, R.J. (1997), 'Neuroanatomical correlates of happiness, sadness and disgust', *American Journal of Psychiatry*, **154**, pp. 926–33.

Lang, P.J., Bradley, M.M., Fitzsimmons, J.R., Cuthbert, B.N., Scott, J.D., Moulder, B. and Nangia, V. (1998), 'Emotional arousal and activation of the visual cortex: An fMRI analysis', *Psychophysiology*, **35**, pp. 199–210.

Lang, P. (1998), 'Imagery and emotion: Information networks in the brain', Presentation at the Workshop on Emotions, Qualia and Consciousness, Inernational School of Biocybernetics, Ischia, Italy.

LeBar, K.S., Gatenby, J.C., Gore, J.C., LeDoux, J.E. and Phelps, E.A. (1998), 'Human amygdala activation during conditioned fear acquisition and extinction: A mixed-trial fMRI study', *Neuron*, **20**, pp. 937–45.

LeBar, K.S., LeDoux, J.E., Spencer, D.D. and Phelps, E.A. (1995), 'Impaired fear conditioning following unilateral temporal lobectomy in humans', *Journal of Neuroscience*, **15**, pp. 6846–55.

LeDoux, J. (1996), *The Emotional Brain: The Mysterious Underpinnings of Emotional Life* (New York: Simon and Schuster).

Mandler, G. (1984), *Mind and Body: Psychology of Emotion and Stress* (New York: W.W. Norton & Company).

McKenna, A. (1958), 'The experimental approach to pain', *Journal of Applied Psychology*, **13**, pp. 449–56.

Mlot, C. (1998), 'Probing the biology of emotion', *Science*, **208**, pp. 105–7.

Neufield, R. and Davidson, P. (1971), 'The effect of vicarious and cognitive rehearsal on pain tolerance', *Journal of Psychosomatic Research*, **15**, pp. 329–35.

Panksepp, J. (1998), *Affective Neuroscience: The Foundations of Human and Animal Emotions* (New York: Oxford University Press).

Papez, J.W. (1937), 'A proposed mechanism of emotion', *Archives of Neurological Psychiatry*, **38**, pp. 725–43.

Papez, J.W. (1939), 'Cerebral mechanisms', *Association for Research in Nervous and Mental Disorders*, **89**, pp. 145–59.

Paradiso, S., Robinson, R.G. andreasen, N.C., Downhill, J.E., Davidson, R.J., Kirchner, P.T., Watkins, G.L., Ponto, L.L. and Hichwa, R.D. (1997), 'Emotional activation of limbic circuitry in elderly normal subjects in a PET study', *American Journal of Psychiatry*, **154**, pp. 384–9.

Phillips, M.L., Young, A.W., Senior, C., Brammer, M., Andrew, C., Calder, A.J., Bullmore, E.T., Perrett, D.I., Rowland, D., Williams, S.C., Gray, J.A. and David, A.S. (1997), 'A specific neural substrate for perceiving facial expressions of disgust', *Nature*, **389**, pp. 495–8.

Reiman, E.M., Lane, R.D., Ahern, G.L. and Schwartz, G.E. (1997), 'Neuroanatomical correlates of externally and internally generated human emotion', *American Journal of Psychiatry*, **154**, pp. 918–25.

Robbins, T.W. (1998), 'Neurobiology of positive emotions and reward mechanisms', Paper presented at the Fourth Annual Wisconsin Symposium on Emotion: Affective Neuroscience, Madison, WI, 17–18 April.

Schacter, S. (1959), *The Psychology of Affiliation* (Stanford: Stanford University Press).

Schacter, S. (1964), 'The interaction of cognitive and physiological determinants of emotional state', in *Mental Social Psychology, Vol. 1*, ed. L. Berkowitz (New York: Academic Press).

Schacter, S. (1970), 'The assumption of identity and peripheralist–centralist controversies in motivation and emotion', in *Feelings and Emotions: The Loyola Symposium*, ed. M.B. Arnold (New York: Academic Press).

Schneiderman, N., Francis, J., Sampson, L.D. and Schwaber, J.S. (1974), 'CNS integration of learned cardiovascular behavior', in *Limbic and Autonomic Nervous System Research*, ed. V.C. DiCara (New York: Plenum).

Scott, S.K., Young, A.W., Calder, A.J., Hellawell, D.J., Aggleton, J.P. and Johnson, M. (1997), 'Impaired auditory recognition of fear and anger following bilateral amygdala lesions', *Nature*, **385**, pp. 254–7.

Sitoh, Y.Y and Tien, R.D. (1997), 'The limbic system. An overview of the anatomy and its development', *Neuroimaging Clinics of North America*, **7**, pp. 1–10.

Snyder, M. (1979), 'Self-monitoring processes', in *Advances in Social Psychology, Volume 2*, ed. L. Berkowitz (New York: Academic).

Strongman, K.T. (1996), *The Psychology of Emotion: Theories of Emotion in Perspective, Fourth Edition* (New York: John Wiley & Sons).

Tomaka, J., Blascovich, J., Kibler, J. and Ernst, J.M. (1997), 'Cognitive and physiological antecedents of threat and challenge appraisal', *Journal of Personality and Social Psychology*, **73**, pp. 63–72.

Watt, D. (1998), 'Implications of Affective Neuroscience for ERTAS (Extended REticular Thalamic Activating System) Theories of Consciousness', Paper presented at Emotions, Consciousness, Qualia Workshop, International School of Biocybernetics, Ischia, Italy.

Whalen, P.J,. Rauch, S.L., Etcoff, N.L., McInerney, S.C., Lee, M.B. and Jenike, M.A. (1998), 'Masked presentations of emotional facial expressions modulate amygdala activity without explicit knowledge', *The Journal of Neuroscience*, **18**, pp. 411–8.

Hilary Rose

Changing Constructions
of Consciousness

No fresh-minted concept like the fluid genome or indeed sexual harassment (neither concept being available thirty years ago), consciousness has become immensely fashionable, but this time round as part of the new found cultural popularity of the natural sciences. However, what is immediately noticeable about the proliferation over the past decade of books and journals with 'consciousness' in their titles or invoked in their texts is that they seem to be drawn to the cultural glamour of the concept, but with little sense that the concept of consciousness has an entirely other history. Consciousness seems to lie around in the culture like a sparkling jewel, irresistible to the neuro-theorists. There seems to be no recognition amongst the many biologists, artificial intelligencers, physicists and philosophers who have played in print with their new toy that consciousness is part of another discourse and has an entirely other history. Above all, I want to underline that while for these neuro-theorists, consciousness is located within the individual human organism (and sometimes just the brain within that), the older tradition, coming from the humanities and social theory, sees consciousness as located in subjectivity and inter-subjectivity in historical context. The methodological individualism expressed in the objectivist language of the natural sciences erases both 'me' and 'you'; by contrast, in social theory, both agency and structure are crucial. For social theory there can be no development of individual consciousness without a social context.

My intention in this paper is to recall this other history, as I fear that much current theorizing seems to have thrown the consciousness baby, in all its subjectivity and sociality, out with the ineffable bath water. My hope is to foster a situation where we acknowledge these entirely different disciplines and epistemologies and so become more aware of the limits of our own discourses. Recognizing limits may help friendly and productive conversations and offer the best possibilities for good theorizing. Feeling a bit like Dickens with his 'It was the best of times; it was the worst of times', I believe that there are good reasons, despite the Science Wars and the splenetic displays of deliberate mutual incomprehension, to believe that the possibilities for such friendly conversations are opening up. The Science Wars of the nineties originally broke out (Gross and Levitt, 1994; Ross, 1996) in response to the pervasive challenge ranged against the old hegemony of natural science as privileged above all other forms of knowledge. This challenge has two main strands: first, from the new social

Journal of Consciousness Studies, **6**, No. 11–12, 1999, pp. 251–8

movements, most of all environmentalism and feminism, and second, from developments within the philosophy, history and sociology of science. What we have seen emerging over the last three decades is a breach in the erstwhile happy marriage between the social studies of science and science itself.

But while the natural science community has become fascinated by consciousness, social sciences and the humanities have become equally fascinated with both nature and the body. While it is evident that there are very different consciousnesses, natures and bodies in play, the new commonalities of concern can be read positively. Others, as I suggest below, such as Varela (1996), Freeman (1999) and, even more, Damasio (1994;1999) are contributing to making constructive conversations possible in these contested times.

From Marx and Freud to Feminism and Black Consciousness

The Shorter Oxford English Dictionary defines consciousness philosophically, as 'the state or faculty of being conscious, as a concomitant of all thought, feeling and volition' (earliest usage, 1678). As this fusion of reason, emotion and intentionality, consciousness was discussed only by those interested in the systematic exploration of subjectivity and intersubjectivity. Those in pursuit of positive objective knowledge of either nature or society necessarily either disregarded or dismissed the topic. Consciousness was the ineffable, perhaps to be graciously acknowledged as with qualia and the discussion of the redness of red, or Searle's ignorant translator of Chinese locked in a sealed room, but not as the subjective and intersubjective, to be seen as researchable by positivistic science. Thus, for much of this century, the intellectual giants dominating the discussion of consciousness were Freud and Marx. Regardless of what Popper had to say about either of their claims to be doing science, it is unquestionably clear that both understood themselves to be true heirs of the Enlightenment, believing that their researches were scientific, universal and progressive. However, their views of the gains to be achieved through enhanced consciousness could scarcely be more different. Freud's utopian vision was for individuals to exchange life-deforming neuroses for modest unhappiness. Marx's utopian vision was for nothing less than world revolution, to be attained by the collective and correct consciousness of proletariat which would enable them to transform capitalism into socialism and thence communism.

As I sketch out below, consciousness as a phenomenon only accessible through the exploration of subjectivity and intersubjectivity (that 'me' and 'you', where 'you' may be one or several), changeable, embodied, recognizably composed of feeling, cognition and intentionality, has long escaped the disciplined texts of Freud and Marx and their disciples. 1989 dealt a fatal blow to the claims of actually existing socialism, whilst arguably selective serotonin re-uptake inhibitors, whether in herbal or allopathic form, modify depression, if not neuroses, more speedily than analysis. Nonetheless, consciousness as subjectivity and intersubjectivity within historical context has both entered our everyday culture and been the focus of a tremendous amount of research in both the humanities and the social sciences, not least through the memoir and autobiography (Stanley, 1992).

That nineteenth century project of the young Marx was specifically to overthrow false theories of consciousness, above all Hegel's 'idealism' and Feuerbach's 'materialism', and to discover what he spoke of as the 'true role of consciousness' in history. Consciousness for Marx did not lie outside history but was immanent within it, thus for historical materialism, 'consciousness did not determine being but being determined consciousness'. It followed that the task of philosophers like himself was simply to explain to the world 'what its own struggles are about'. Above all, this meant helping the proletariat advance from its necessary stance of collective self-defence (trade unionism), which he spoke of as limited to 'class in itself', to that conscious unity of revolutionary theory and practice of 'class for itself'. The powerful elaboration of History and Class Consciousness by George Lukacs (1971) still makes astonishing reading eighty years on, not least in his reflections on the relations between political economy and violence.

From the present point in our own tumultuous and often savage century, few are left able to share in the Marxist proposition that the longed for 'leap to the realm of freedom' depends on the capitalist world dividing into two great classes, with one of these, the proletariat, carrying the hopes of all humanity. It was not just that the theory failed to grapple with the harsh grip of nationalism, so that even while the Russian revolution was taking place, the European working class was letting itself be slaughtered by the millions in a nationalistic and imperialist war, but that even despite Mao's ingenuity in recruiting the peasants as the necessary allies of, or substitutes for, the proletariat, somehow still that universal symbol of humanity, the proletariat in the old capitalist countries, remained deeply gendered and raced.

But what is central within both these struggles and the attempts to theorize them is the role of consciousness; as always a dynamic, never a static concept. Chinese peasants speaking bitterness, debated 'does the landlord need the peasant or does the peasant need the landlord?' in order to break through to a new level of consciousness. The capacity to overthrow the landlords required that the peasants became conscious of them as not merely exploitative but unnecessary. Shared discussion was the key to the transformation of shared consciousness. Franz Fanon's The Wretched of the Earth (1965) drew an inescapable connection between the development of black consciousness and the necessity of revolutionary violence — an analysis which powerfully influenced US African Americans, not least Malcolm X. Biko's Black Consciousness movement in South Africa, although committed to non-violence, was seen by the Apartheid state as challenging white hegemony and hence requiring his murder. Women could only begin to grasp the specificity of their oppression within gender relations through closed 'consciousness-raising groups' in which experience was shared and analysed in order to find new words, new concepts, for 'the problem that had no name'. All these transformative movements have required a shared cognitive and emotional understanding of exploitation and oppression and the shared will to change it.

Feminist consciousness-raising, like black consciousness-raising, was carried out by the oppressed themselves. In this they were spared an old contradiction within the communist movement, a proletarian movement led by Marxist intellectuals, who, like Marx himself, were drawn from the bourgeoisie. Unlike the black and feminist intellectuals, these communist leaders were morally but not necessarily committed by what Marx termed 'species being', to social transformation. Initially, the project of global sisterhood seemed to offer the realm of freedom with every bit as much promise as the proletariat, but the problem of just exactly 'who' this sisterhood was appeared rather quickly. Challenges not only from black feminists but also from post-colonial, older, younger, lesbian, disabled women etc., argued against a monolithic construction that the needs and interests of white, middle-class, young, Northern, able-bodied, heterosexual women were equated with those of all women everywhere. Although feminists have grappled with the complexities of difference in the context of hypermodernity, it has been an extraordinarily difficult process. Those dreams of a shared universal consciousness through the common species being of women have faded. Alliances between social groups including women cannot be read off from gender; instead they have to be built.

The usual clarity of hindsight enables us to see that such social movements could not succeed in their own Enlightenment terms. The social and economic gains have been modest. Poverty is still all too present in the lives of women and their children, whether in the world's richest or poorest societies. But despite this, the cultural gains over the last three decades have been huge. Of course there are unreconstructed social groups and individuals, but what is extraordinary is the immense change in gender relations since the mid-century. Those notions of femininity and masculinity once seen as 'natural' in EuroAmerica (typically forever frozen in some appalling 1950s' domestic play) are now widely seen as historically constructed. A changed shared consciousness, thought, feeling and intentionality works to expose and name phenomena flowing from such gendered concepts of human nature. New words come into existence, old words are given new social meaning: 'sexual harassment', 'male violence', 'rape', ' racism', 'child abuse'; phenomena which had hidden themselves in nature — and legally in the private domain — are pulled into culture and the public court room. Putting such concepts into culture calls the perpetrators to account. Rape is no longer discounted as part of masculine nature, but located in the intentionality of particular men. Even in war, that historic legitimator, not all men rape. But our failures to prevent these ills as swiftly as many of us hoped has also made us understand that changing subjectivity, changing the consciousness of abuse, is no simple matter for either a perpetrator or even a victim.

Nor has the huge shift in, say, gender or race consciousness been plain cultural sailing, not least because of sociobiology's immense efforts in the seventies and beyond to reassert that biology is destiny. What a changed cultural and political consciousness names as intentional crimes, are for this renewed Social Darwinism simply the working through of genes. In the nineties, sociobiology has been rebranded and reconstituted as evolutionary psychology, and individual genes have ceded in explanatory power to evolutionary imperatives fixed in a universal but gendered human nature laid down in the Pleistocene; but the consequences for assumed universal human nature are not dissimilar (Rose and Rose, in press).

For most of this century, Marx's other great insight, that nature also has a history and both shapes and is shaped by humanity, was lost, relegated to a subordinate cultural current. Projects of human liberation were for many years cut off from the need to conceptualize or care for the ecosystem of which human beings are an integral part. Thus, until the past three decades, social movements — classically the working-class movement, but also movements of national liberation — have excluded any serious analysis of either nature or the natural sciences as knowledge about nature. The movements both constituted and reflected Snow's 'two cultures'. Nonetheless, what is clear and encouraging today is the widespread concern, particularly among children, for nature. A new environmental consciousness, however often it gets pushed onto the back burner because inflation or unemployment means that the urgent pushes out the important, is an integral if uneven presence within our present political and cultural consciousness.

Not only has green nature 'out there' entered our consciousness, but so have our bodies. Today social theory understands bodies not as some fixed entity, but as embodied experience. Increasingly we recognize that bodies are dieted, exercised, pierced, handed over to cosmetic surgeons, fretted over in an unprecedented way in order to achieve conformity to new aesthetic visions of lean muscularity. Nature and the body are back, both in everyday cultural life and also in the academic discourse of the humanities and the social sciences (Schatzki and Natter, 1996) . No longer are the constructions of nature and the body seen as the automatic and exclusive responsibility of biomedicine. Other discourses have a great deal to say — and this is perhaps what is making conversations difficult.

Internal Phrenology

While natural scientists not infrequently rage against the (mis)appropriation of their concepts, by, for example, new age philosophising as in 'the quantum body' (a horror incidentally that I managed to persuade one of my favourite graduate students to take out of her thesis), or by the Lacanian feminist philosopher Irigary (Sokal and Bricmont, 1997), they seem to feel that their own (mis)appropriations are entirely justified. How else are we to understand the way in which, after a decade of vast funding for the neurosciences generating thick empirical descriptions of brain function with so far fairly modest theorizing (Horgan, 1999), the past decade has designated 'consciousness' as the hot theoretical topic. The topic has certain instantly notable characteristics, in that, unlike normal hot natural science topics, it is not pursued by brilliant young scientists, whether doctoral students or even post doctoral fellows as in, say, the DNA story, but by the already famous. Career-minded young neuroscientists have been heard to describe an interest in consciousness — even while giving papers at a meeting of that name — as a CLM (a career limiting move).

The main contributors to the consciousness debates have rarely been drawn from within neurobiology itself but are established figures in other related areas: biology (Francis Crick), philosophy (Daniel Dennett), linguistic psychology (Steven Pinker) and physics/mathematics (Roger Penrose). Does this rather peculiar demographic and professional profile of middle-aged and more consciousness theorists reflect what experimental natural scientists self-mockingly call the 'philosopause', that change of life when experimental fertility runs out? Anthropology and literature suggest that it is unwise to dismiss jokes as either trivial or treacherous. A joke is typically a rich source of dangerous comment on matters that might be socially difficult to discuss head on. It was Lear's fool who stayed with him.

Let me briefly explore some influential examples of the new science-based constructions of consciousness. First, the Cartesian inheritance of the dominant tradition in the natural sciences means that most, philosophers and scientists alike, equate consciousness with cognition (Dennett, Flanagan, the Churchlands, even Searle among the philosophers, Penrose and the artificial intelligencer Aleksander). Daniel Dennett (1991) provides an almost classically one-dimensional concept of consciousness, thus what cannot be subsumed under his personal area of expertise, namely 'cognition', is excluded from consideration. It has been left to Damasio (1994; 1999) to insist that 'Descartes' error' is precisely that of focussing on cognition at the expense of feeling in terms of theorizing brain function and conscious experience.

Numbers of the new objectivist theories of consciousness seem to lose the consciousness plot, for the theorists intellectually evade the philosophical challenge of subjectivity. Crucially, where is the 'I' of agency? Jerry Fodor (1998), in his review of Pinker's (1997) *How the Mind Works*, with its modules for this and that cognitive function, echoes this in his question, asking where is the 'I' that holds the modules together. The objectivist stance leads many natural scientists to speak of consciousness as the equivalent of not being unconscious (Stuart Hameroff, Susan Greenfield, Gerald Edelman). Crick (1994) makes a virtue of the narrow concept. Consciousness, he argues, may be a hard problem, but if we reduce it to awareness, it becomes more tractable and we can then take a segment of conscious experience, for example, visual perception, and study this as a model for the whole.

This approach assumes that human subjectivity is the equivalent of that of any other creature that is capable of being either awake, asleep or unconscious. Without assuming that there is nothing in common between the neurobiology of different species, the idea that there is a one-on-one consciousness between species borders on farcical. It suggests that my cat Hypatia and I (who get on rather well and are both experienced at being awake and asleep and sundry drowsy states between) have the same consciousness. Thus Susan Greenfield's metaphor of consciousness as a dimmer switch, in her otherwise delightful guided tour of the brain (1997), describes very well both our varying levels of sleepiness and wakefulness and their neuronal correlates, but this entirely avoids saying anything about the subjective consciousness of either Hypatia or myself. For that matter, who switches the switch? Greenfield is not alone. Koch and Crick share a not dissimilar metaphor of the 'searchlight' of consciousness (Crick, 1994). The problem of theorizing consciousness is evaded by the simple-minded acid of objectivism.

Elsewhere, the explanations demonstrate a bizarre reprise of nineteenth century phrenology. The partially localized but also internally compensatory brain functions shown up in imaging studies of the brain (Freeman, 1999), which, while not unfriendly to the localizing materialistic endeavour of phrenology, emphasize process. By contrast, claims by the new internalist phrenology are as fixed as amativeness or philoprogenitiveness. This new phrenology is not limited to knobs and whorls on the surface of the skull, but goes deep within the organ. Thus, Crick explains free will as located precisely in the anterior cingulate sulcus. Yet this insistence that 'free will' can be located in a specific brain region reads as a curious echo of nineteenth century phrenology with its materialist longing similarly to locate highly complex behaviours, feelings and dispositions. The cultural reprise taking place within the discourse of neurobiology finds itself once more able to map into the brain highly complex social phenomena, whether 'homosexuality' (LeVay, 1993) or 'free will', rather than the 'spirituality' or 'amativeness' of our Victorian ancestors. As for philoprogenitiveness, unequivocally the sociobiologists (notably Richard Dawkins) are convinced that it lies in brain processes switched on by our selfish genes. Such reductionist materialism simply eschews context, not least the social and cultural.

While Crick positions himself within that *cogito ergo sum* assertion of Cartesian identity, he also wants to find space for agency, or volition — in his terms, 'free will'. This preoccupation with 'free will' stems from the self-imposed problems which arise from his and other molecular biologists' belief in the determinism of DNA as the vaunted 'master molecule'. A similar problem besets others of the 'biology as destiny' school, (notably sociobiologists and fundamentalist Darwinians) who, having come to certain biologically inevitabilist conclusions, find themselves faced with highly negative social futures for human beings. Unwilling to accept these gloomy scenarios, they invoke free will, typically in the closing pages of their books, to escape the iron cage they have themselves constructed. Pinker's assertion that if he didn't like what his genes programmed him to do, he would tell them 'to jump in the lake' (1997, p. 52) is a classic in this skyhook approach to free will.

Amo Ergo Sum

From the perspectives of, say, the history or sociology of ideas, such propositions sound extremely strange, for they assume that concepts such as 'free will' or 'homosexuality' exist outside history and culture and can therefore have a physiological location in the individual's brain. To historically informed discourses, concepts such as 'free will' are integral to specific cultures at specific historical periods, and indeed are typically attributed only to particular kinds of people within those cultures. Only as we move towards the twenty-first century does it even begin to be possible to think of free will as something to be found equally among both genders, rich and poor, slave and freeborn, and all 'races'. Without such a historical sensitivity, a theoretical biologist must argue either that those Others had the specific brain area but that there was no expression of the attribute, or alternatively that evolutionarily the Others developed without that bit of brain.

Thus, in the public space occupied by men of a certain social class, Descartes claimed reason for himself and others like him in the universal name of Man. It was left to Rousseau to spell out the destiny of female Others, stressing the affinity of women for emotion and to the private. In his utopian educational project, women were supposed to claim (that is if Rousseau had not excluded the classics from Sophie's education) *amo ergo sum*. My hunch is that Sophie's claim could make a theoretically better starting point for a less bio-gendered approach to consciousness. 'I love, therefore I am' offers feeling — with and for others — as the condition of identity and consciousness. The first-person perspective is placed on an ontological level with an ethical awareness of others.

Recovering Feeling

Despite this general drift towards reductionism and objectivism, there is currently a modest counter current within this natural science-based discourse of consciousness, an attempt to put emotion back in. From the animal psychologist Nick Humphrey's tender if confused set of essays called *Consciousness Regained* (1982), to the terrific title if rather schmaltzy contents of Daniel Goleman's *Emotional Intelligence* (1995), or Joseph LeDoux's *The Emotional Brain* (1996), and most recently Antonio Damasio's *The Feeling of What Happens* (1999) and Walter Freeman's *How Brains*

Make up their Minds (1999), there have been a number of attempts from within the natural sciences to put feeling — that which was seen as the business of Sophie and her sisters — back into thinking.

Regrettably, even the best of these attempts are carried out with little direct reference to feminist scholarship. While few can escape the feminist challenge, whether in our most intimate lives or in the most public of spaces, it remains possible for masculinist academic discourse to exclude feminist theorizing. Yet part of the immense intellectual energy of feminism has entailed a critical revisioning of reason — rationality itself. Such a project is surely an ally of the attempt to expand consciousness beyond cognition. I will not even begin to trace that research programme, but go straight to its conclusion — one incidentally shared by many theorists of the environmental movement — which sees the asocial construction of rationality embedded in modern western science as indifferent to the needs of people and nature alike (Rose, 1994). Both movements and their theorists seek to re-vision the concept of rationality, to make rationality socially and environmentally responsible. Bringing in the social, bringing in an ethical concern with people and nature is of course to admit subjectivity, caring rationality, and the shared will to do things differently. Taking feelings seriously helps an interdisciplinary discussion of consciousness to richer and more responsible places.

References

Crick, Francis (1994), *The Astonishing Hypothesis* (New York: Scribner).

Damasio, Antonio (1994), *Descartes' Error* (New York: Grosset).

Damasio, A. (1999), *The Feeling of What Happens: Body, Emotion and the Making of Consciousness* (London: Heinemann).

Dennett, D. (1991), *Consciousness Explained* (Boston, Toronto & London: Little, Brown and Co.).

Fanon, Franz (1965), *The Wretched of the Earth* (Harmondsworth: Penguin).

Fodor, J. (1998), 'The trouble with psychological Darwinism', *London Review of Books*, **20** (2), January 1998.

Freeman, W. (1999), *How Brains Make up their Minds* (London: Weidenfeld and Nicolson).

Goleman, Daniel (1995), *Emotional Intelligence* (New York: Bantam).

Greenfield, Susan (1997), *The Human Brain: A Guided Tour* (London: Weidenfeld and Nicolson).

Gross, P. and Levitt, N. (1994), *Higher Superstition: The Academic Left and its Quarrels with Science* (Baltimore: Johns Hopkins).

Horgan, John (1999), *The Undiscovered Mind: How the Brain Defies Explanation* (London: Weidenfeld and Nicolson).

Humphrey, Nicholas (1982), *Consciousness Regained: Chapters in the Development of Mind* (Oxford: Oxford University Press).

Le Doux, Joseph (1996), *The Emotional Brain* (New York: Simon & Schuster).

LeVay, Simon (1993), *The Sexual Brain* (Cambridge, MA: MIT Press).

Lukacs, G. (1971), *History and Class Consciousness* (London: Merlin; first published 1920).

Pinker, Stephen (1997), *How the Mind Works* (Cambridge, MA: MIT Press).

Rose, Hilary and Rose, Steven (In Press), *Alas Poor Darwin: Arguments against evolutionary psychology*.

Rose, Hilary (1994), *Love Power and Knowledge: Towards a Feminist Transformation of the Sciences* (Cambridge: Polity).

Ross, A. (ed. 1996), *Science Wars* (Durham, NC, and London: Duke University Press).

Sokal, A. and Bricmont, J. (1997), *Intellectual Impostures* (London: Profile).

Schatzki, T.R. and Natter, W. (ed. 1996), *The Social and Political Body* (New York and London: Guildford).

Stanley, Liz (1992), *The Auto/Biographical I: The Theory and Practice of Feminist Auto Biography* (Manchester: Manchester University Press).

Varela, F. (1996), 'Neurophenomenology: A methodological remedy for the Hard Problem', *Journal of Consciousness Studies*, **3** (4), pp. 330–49.

Maxine Sheets-Johnstone

Emotion and Movement

A Beginning Empirical-Phenomenological Analysis of Their Relationship

I: Introduction

In his discussion of time and of 'how many ways we speak of the "now"', Aristotle unwittingly highlights in a striking way the nature of a qualitative dynamics. He says that '"now" is the link of time' referenced in expressions such as 'at some time', 'lately', 'just now', 'long ago', and 'suddenly' (*Physics* 222b27–29). Something radically different is conveyed by the last example: 'suddenly' has a decisively dynamic aspect wholly distinct from the other terms or phrases. Aristotle says simply that '"Suddenly" refers to what has departed from its former condition in a time imperceptible because of its smallness' (*Physics* 222b15–16). He is obviously taking 'suddenly' as a quantitative term parallel to the other quantitative terms. But 'suddenly' is basically something both more and other than an interval of time 'imperceptible because of its smallness'. It is a *qualitatively* experienced temporality, just as rushed, prolonged, and creeping are *qualitatively* experienced temporalities. In brief, the distinctive dynamic that defines 'suddenly' derives from felt experience. It is fundamentally not a quantitative term but an experienced kinetic quale. As such, it has a certain affective aura: 'suddenly' may describe an earthquake, a fall, an ardent kiss, an urge or inspiration, or one of multiple other possible experiences, each of which has a certain affective resonance. What is kinetic is affective, or potentially affective; by the same qualitative measure, what is affective is kinetic, or potentially kinetic.

Recognition of the everyday qualitative character of *suddenly* opens up an intricate and challenging domain of experience emblematic of the intimate bond between emotions and movement. In what follows, I offer a beginning sketch of the relationship, concentrating first on empirical research that preceded the rise of cognitivist science with its prominencing of an information-processing brain (Bruner, 1990) and its correlative dislocation of movement. I summarize three empirical studies of emotion [1] that carry forward the work of Darwin, and that vindicate in

---

[1] A reviewer of this essay stated that 'the three investigators the author selects . . . have not really produced the types of rigorous studies that most scientists would currently deem to be of sufficient quality to

Journal of Consciousness Studies, **6**, No. 11–12, 1999, pp. 259–77

different ways the work of physiological psychologist Roger Sperry on perception
and his principle thesis that the brain is an organ of and for movement (Sperry,
1952).[2] The summaries make evident the theoretics that bind the studies together and
reveal the tactile-kinaesthetic body that is in each case their foundation. I turn then to
a summary phenomenological analysis of movement, showing how the dynamic
character of movement gives rise to kinetic qualia. The analysis exemplifies how
empirical studies may be epistemologically deepened through phenomenology, in
this instance through a phenomenological elucidation of the fundamentally qualitative structure of movement, a structure that grounds the relationship between movement and emotion in a qualitative dynamics and formal dynamic congruency. In
virtue of that congruency, motion and emotion — kinetic and affective bodies — are
of a dynamic piece.[3] Methodological consequences follow from this exposition. So
also do implications for cognitivism, which range from the observation that movement is not behaviour and that the term 'embodied' is a lexical band-aid to the observation that animate forms are not machines and that a kinetic, qualitative
(meta)physics follows naturally from the study of animation and animate form.

II: Empirical Studies of Emotion

The first research that warrants our attention is the lifelong experimental work,
empirical methodology, and related clinical practice of Edmund Jacobson. A close
friend of Karl Lashley, Jacobson was a medical doctor and neuropsychiatrist with a
doctorate in psychology.[4] Jacobson developed and honed a form of introspection, a
practice he called 'auto-sensory observation', which he taught to his patients, enabling them to monitor and ultimately dissipate excessive, unproductive bodily

---

constitute essential and unambiguous empirical progress in the area'. It is important to point out that the
research of the three investigators has not been critically shown to be lacking in rigour, to be of inferior
quality, and so on, but has only been ignored, and this most probably because the research is not currently
popular: it deals with *experience*, not with behaviour, and it deals with intact living humans, not with
brains. In this respect, it should be noted, on the one hand, that the positive value of an empirical study
holds until specifically shown to be indefensible — e.g., the study is invalidated on procedural grounds, it
is shown to be unreliable because unreplicable, and so on; and on the other hand, that science progresses
as much by discovery in arrears — e.g., Mendel, Wegner — as by advancing discoveries. In evidential
support of both hands, we may readily look to Karl Pribram's commending citations of Nina Bull's
research (e.g. in Pribram, 1980, pp. 246, 256) and to Manfred Clynes' positive citation of the same (in
Clynes, 1980, p. 281).

[2] Sperry's later groundbreaking experimental research involving brain commissurotomies eclipsed his
earlier groundbreaking experimental research on perception and movement. To be noted in this context is that however much present-day textbooks veer off into a pre-eminently information-processing
view of brains, in their sections on movement, some of them contradict the view and clearly support
Sperry's thesis, e.g., 'The brain is the organ that moves the muscles. It does many other things, but all
of them are secondary to making our bodies move.' (Carlson, 1992, p. 214.)

[3] The dynamic congruency is elegant in a way analogous to the way in which mathematical formulations
and scientific explanations are said to be elegant.

[4] For information on his background, see Jacobson (1970), pp. 11–21. In a paper on the
electrophysiology of mental activities, Jacobson mentions '[a] rather amusing comment' made by
Lashley: 'Lashley told me with a chuckle that when he and Watson would spend an evening together,
working out principles of behaviourism, much of the time would be devoted to introspection' (Jacobson, 1973, p. 14; see also Jacobson, 1967, p. 16). Because of its omission of introspection, Jacobson
regarded behaviourism 'only half a science' (ibid., p. 17).

tensions, and in consequence to decrease felt anxieties and other debilitating feelings. In this way, they were able to take personal responsibility for their problems (Jacobson 1929; 1967; 1970). Jacobson's technique of self-observation was learned and taught by other physicians and psychiatrists, and by other persons as well. During World War II, for example, his technique was taught to U.S. Navy Air Cadets — 15,300 men — who suffered '[a]nxiety states accompanied by fatigue, restlessness and insomnia, including what were called *breakdowns*' (Jacobson, 1967, p. 171).

The self-observational technique that Jacobson developed centres on a tactile-kinaesthetic awareness of the tension level of one's specific and overall bodily musculature. Jacobson validated the technique by electroneuromyometry, i.e., the measurement of neuromuscular action potentials. He is in fact credited with being 'the first to record the action potentials in the muscles and to show that they vary in a predictable way with mental activity and especially with feelings of tension' (Fishbein in Jacobson, 1967, p. viii). A basic principle of the theory emanating from his experimental findings and clinical practice is quite simple: neuromuscular tension is emotionally laden; 'neuromuscular acts participate in mental activities . . . including emotions' (Jacobson, 1970, p. 34). It is notable that Jacobson pointedly contrasts his theory with the traditional view of the brain, the view 'that all mental activity occurs in the brain alone; that the brain does our thinking, e.g., as the alimentary tract does our digestion,' or, as he later says, with the view of those who regard neuromuscular activity 'as the tail wagged by the dog' (ibid., p. 32). He calls our attention as well to the error of those who, hearing of the practice of 'auto-sensory observation' equate it to 'suggestion' by the instructor (ibid.). A number of Jacobson's findings are of particular interest, such as

> [T]he trained observer (not the tyro) identifies and locates signals of neuromuscular activity as integral parts of the mental act [of 'attention, imagination, recall, fantasy, emotion, or any other mental phenomena']. He does not discern two acts, one so-called 'mental' and the other 'neuromuscular', but one act only (ibid., p. 35); and 'objective and subjective data indicate conclusively that when the trained observer relaxes the neuromuscular elements apparently specific in any mental activity, the mental activity as such disappears accordingly' (ibid.).[5]

In sum, Jacobson's fundamental experimental finding — and hence the significance of auto-sensory observation — is that what happens in a brain does not happen apart from muscular innervations. 'Those who would do homage to the brain with its ten billion cell-amplifiers can well continue to do so', Jacobson says, but they must also not overlook empirical evidence: that 'muscles and brain proceed together in one effort-circuit, active or relaxed' (ibid., pp. 36, 34).[6]

---

[5] To assure clarity, I add the following annotation: Jacobson does not say *all* mental activity disappears; he says that 'the mental activity as such disappears'. The *as such* qualifies the particular mental activity that disappears, i.e., the mental activity ongoing before the onset of relaxation. With all due attention to Jacobson's emphasis upon the necessity of developing capacities in auto-sensory observation and differential relaxation — of being a trained observer, not a tyro — readers might nevertheless try consulting their own experience to corroborate the disappearance of a specific mental activity upon neuromuscular relaxation.

[6] That muscle-brain constitute a unitary circuit is a key insight supported and emphasized by notable contemporary investigators as well as by Roger Sperry in his initial and highly influential work on perception, which led him to identify the brain as an organ of and for movement (Sperry, 1952). Dynamic systems theorist J. A. Scott Kelso, for example, writes, 'It is important to keep in mind . . . that the brain

Empirical evidence of a singular muscle–brain 'effort-circuit' confirms the basic premise implicit in Darwin's *The Expression of the Emotions in Man and Animals*: movement and emotion proceed hand in hand. The fundamental concordance between the two phenomena lies in the fact that bodily movement is expressive. What Darwin sought to explain in his book was the origin of the concordance on the basis of serviceable habits, the principle of antithesis, and the phenomenon of 'nerve-force'; that is, certain movements arise because they are of benefit to the animal, or because they are called forth in opposition to innate kinetic practices, or because of a spontaneity or excess of 'nerve-force'. Throughout the book, what Darwin basically describes is *movement*. For example, with respect to joy and vivid pleasure, he writes that 'there is a strong tendency to various purposeless movements' (Darwin, 1965[1872], p. 76), and several sentences later remarks,

> Now with animals of all kinds, the acquirement of almost all their pleasures, with the exception of those of warmth and rest, are associated with active movements, as in the hunting or search for food, and in their courtship. Moreover, the mere exertion of the muscles after long rest or confinement is in itself a pleasure, as we ourselves feel, and as we see in the play of young animals. Therefore on this latter principle alone [the principle of the action of the nervous system] we might perhaps expect that vivid pleasure would be apt to show itself conversely [that is, in contrast with long rest and confinement] in muscular movements (p. 77).

The implicit premise is furthermore explicitly and succinctly attested to in his remark concerning the variable relationship of movement and emotion: 'I need hardly premise that movements or changes in any part of the body, . . . may all equally well serve for expression' (ibid., p. 28). In short, the *expression* of emotion in man and animals is a kinetic phenomenon, a neuromuscular dynamic that, as we will presently see, has a certain spatiality, temporality, intensity, and manner of execution. This complex kinetic structure is essentially demonstrated in movement notation analyses by ethologists who thereby capture the dynamics of animal behaviour. The ethological studies of mammalian pre-copulatory interactions (Golani, 1976) and of the dynamics of wolves fighting (Moran *et al.*, 1981) are classics in this respect.

The import of Jacobson's work to Darwin's evolutionary studies of emotion, and to movement-oriented ethological studies as well, lies in the strong empirical data it presents showing that emotions are grounded in a neuromuscular dynamic. The dynamic is delineated along further empirical lines in the experimental research of psychiatrist Nina Bull. Bull's work shows that emotions are shaped by motor attitudes, that 'a basic neuromuscular sequence is essential to the production of affect' (Bull, 1951, p. 79).[7] It demonstrates, and in a striking way, that there is a *generative* as

---

did not evolve merely to register representations of the world; rather, it evolved for adaptive action and behavior. Musculoskeletal structures coevolved with appropriate brain structures so that the entire unit functions together . . .' (Kelso, 1995, p. 268). He goes on to say that 'Edelman arrived at a similar conclusion,' i.e., 'For him, like me, it is the entire system of muscles, joints, and proprioceptive and kinesthetic functions plus appropriate parts of the brain that evolves and functions together in a unitary way' (ibid.).

[7] Ginsburg and Harrington (1996), in their review of research on bodily states and emotions, thoroughly misrepresent Bull's monograph and the experimental work that it details when they characterize her view of feeling as a 'pause' between '"motor attitude" and instrumental action' (p. 249). Bull is at pains to describe emotions as a *process*, and a process that includes thinking. Toward the end of her first chapter, with respect to one aspect of that process, 'attitude-affect', the aspect with which she is

well as *expressive* relationship between movement and emotion. Her work is in this respect a significant amplification of Darwin's. A summary account follows.

In a first group of experimental studies showing how a preparatory postural attitude is vital to the feeling of emotion, subjects were hypnotized, then told '[that] a word denoting a certain emotion would be uttered, that they would then experience this emotion, that they would show this in outward behavior in a natural manner', and that they would afterward be asked to describe what happened (ibid., p. 78). Six emotions were investigated in this manner: fear, anger, disgust, depression, joy, triumph. The subjects' reports validate Bull's thesis that a certain neuromuscular attitude is necessary to, and coincident with, each particular emotion. With respect to fear, for example, one subject reported 'First my jaws tightened, and then my legs and feet . . . my toes bunched up until it hurt . . . and . . . well, I was just afraid of something' (ibid., p. 59). With respect to anger, 'subjects mentioned wanting to throw, pound, tear, smash and hit' — and what restrained them was 'always the same, *clenching the hands*' or making some similar restraining movement (ibid., p. 65).[8] It is important to emphasize that the preparatory postural attitude is in all instances a spontaneously arrived at attitude; what subjects are reporting in each case is *how they were moved*. In the succeeding set of experimental studies, hypnotized subjects were read a particular description from one of their own experiential reports, the description beginning with phrases such as 'Your jaws are tightening' (fear), or 'You feel heavy all over' (depression), or 'There is a feeling of relaxation and lightness in your whole body' (joy), or 'You can feel your chest expanding' (triumph), and so on. Following this initial descriptive reading, the subject was told 'You are now locked in this physical position. There will be no changes in your body — no new bodily sensations — until I specifically unlock you.' The experimenter then told the subject, 'When I count to five I shall utter a word denoting a certain emotion. When you hear the word you will feel this emotion — feel it naturally — and will be able to tell us about it afterward' (ibid., pp. 79–80). The emotion the experimenter named was antithetical to the one coincident with the position in which the subject was locked. What the experiment showed is that subjects were unable to have any other feeling than the one into which they were locked. In other words, they were unable to feel the designated contrasting emotion, and this because any change in affect required a change in postural set or

---

particularly concerned, she states that while it may seem to be a *state* or a 'static quality', attitude-affect is actually 'a moving series of neuromuscular events, a process which, for want of any better name, we must continue to call *emotion*' (Bull, 1951, p. 13).

[8] Of particular significance is Bull's attention to kinetic detail in the form of identifying conflicting motor attitudes. In anger, there is 'a primary compulsion toward aggression or attack, and a secondary powerful restraint, or holding back, which was always muscular and attitudinal' (1951, pp. 62–3). Equally divergent but different attitudes are found in fear and in disgust. Bull speaks of '[t]he jointed character' of disgust (p. 48), the one distinct reaction being a felt nausea and a preparation for vomiting and the other a turning away or avertive attitude of the body. Thus, one reaction was 'predominantly visceral and the other predominantly skeletal' (p. 48). Proportions were different in each case so that the overall experience varied, but '[the] two reactions [were] so closely interwoven as to be apparently inseparable' (pp. 48–9). Again, in fear, 'two separate incompatible reactions [were] going on at once, but in this case the conflict was between posture and movement within the same muscular system', both reactions being skeletal rather than skeletal and visceral in nature. The '*desire to get away* [was] opposed by the *inability to move*' (p. 58).

bodily attitude. As one subject said, 'I reached for joy — but couldn't get it — so tense'; and as another said, 'I feel light — can't feel depression' (ibid., pp. 84, 85).

From a methodological viewpoint, what makes Bulls' study of particular interest is that it utilizes hypnosis to access the experience of emotions. Introspective reports so obtained do not require time-intensive observational training as, for example, Jacobson's introspective auto-sensory observational studies do. Most important, however, are two facts: first, experiences of emotion reported by hypnotized subjects are near indisputable, i.e., there is no reasonable basis for challenging their authenticity; second, experiences so obtained are readily and incontrovertibly detailed as pre-eminently experiences of the tactile-kinaesthetic body. The avertive pattern of disgust, for example, is described by one subject as 'I tried to back away — pushed back on the chair — straight back. All the muscles seemed to push straight back. I could feel that rather strong'; the dual character of fear is described by another subject as 'I wanted to turn away in the beginning . . . I couldn't . . . I was too afraid to move . . . [my legs were] made of lead . . . I couldn't move my hands either. It was as if they were nailed to the chair'; the expansive and powerful character of triumph is described by another subject as 'I had an urge to stand on my toes in order to look down on people at a more acute angle' (ibid., pp. 53, 58, 73). With respect to the evidential pre-eminence of the tactile-kinaesthetic body, and to the origin of emotion in a qualitatively felt neuromuscular dynamics, Bull's comment about the subjects' general lack of distinction between bodily feelings and the feeling of an emotion is significant. Although she also remarks that subjects 'seemed always aware of a difference', she concludes by saying that '[t]his important matter requires further investigation, and no exact definition of emotional feeling or affect, as distinct from organic sensation, will be attempted at the present time' (ibid., p. 47). In effect, she leaves the question of the relationship between bodily feelings and emotional feelings in mid-air. Yet if having a feeling in an emotional sense depends on a certain postural set, a certain tactile-kinaesthetic attitude and thus a certain tactile-kinaesthetic feel[9], and if one must get out of this tactile-kinaesthetic attitude and feel in order to have a different emotion, then clearly, definitions and distinctions are less important than the recognition and descriptive analysis of a basic corporeal matter of fact: *affective feelings and tactile-kinaesthetic feelings are experientially intertwined*. That subjects generally do not distinguish between the two feelings is testimony to the fact that they are regularly experienced holistically, not as piecemeal parts that become progressively apparent, and not as causally sequenced phenomena, but integrally. It thus suggests that bodily feelings and feelings of emotion are divisible only reflectively, after

---

[9] Usual counterexamples offered to this line of reasoning concern paraplegics and paralysed persons. What is not customarily recognized, however, is that persons so afflicted were once not so afflicted. As experimental subjects, there is no doubt but that their testimony is conditioned by previous experience: they know what it is like to feel anger, fear, and so on, in a full bodily felt sense. (It is telling that most people do not realize — or they commonly forget — that Helen Keller was not blind and deaf from birth, but became blind and deaf when she was nineteen months old. Fundamental experiences and learnings in the first months and years of life can be neither ignored nor discounted.) It is furthermore apparent from the empirical studies cited in the present text that some kind of preparation is necessary to obtaining veridical reports on the tactile-kinaesthetic body, and this because adults, especially Western ones, are notoriously afflicted with Cartesian disease. In other words, adults need to be trained to attend to their bodies and to be meticulous observers. This applies to all persons involved in introspecting tactile-kinaesthetic experience.

the experience. Further, as the experimental evidence shows, affective feelings are consistently true to tactile-kinaesthetic dynamics; the two sets of feelings are mutually congruent. Their congruency defines the character or nature of their intertwinement. The summary phenomenological analysis of movement that will presently follow lays the groundwork for elucidating the foundational dynamics undergirding the congruency.

Now a postural attitude is defined by Bull as a readiness to do something, a corporeal readiness to act in some way or other, and it is this postural attitude that is the generative source of emotion (cf. Varela, 1999, pp. 132–3 on 'ontological readiness' and 'readiness potential'; Sheets-Johnstone, 1999, Chapter 9, on 'readiness toward meaning').[10] The postural attitude is thus coincident with what might be designated the onset of emotion: *with a felt urge to do something* — approach something, strike something, touch something, run from something, and so on. Emotion, then, is not *identical* to kicking, embracing, running away, and so on, but is, from the beginning by way of the postural attitude, the motivational-affective source of such actions.[11] As such, it might be conceived within Bull's analysis as the necessary substrate or foundation of action. An observation by Darwin succinctly illustrates this point. Darwin writes that '[W]hen we start at any sudden sound or sight, almost all the muscles of the body are involuntarily and momentarily thrown into strong action, for the sake of guarding ourselves against or jumping away from the danger, which we habitually associate with anything unexpected' (Darwin, 1965, p. 284). The action itself, that is, the 'guarding ourselves against' or 'jumping away from' is not the feeling nor does it generate the feeling; the guarding or the jumping is its expression. By the same token, the 'strong action' of the postural attitude — 'all the muscles of the body are involuntarily and momentarily thrown into strong action' — is what makes the guarding or jumping possible. Without the readiness to act in a certain way, without certain *corporeal tonicities*, a certain feeling would not, and indeed, could not be felt, and a certain action would not, and indeed, could not be taken, since the postural dynamics of the body are what make the feeling and the action possible.

Psychologist Joseph de Rivera's 'geometry of emotions' (Dahl in de Rivera, 1977, p. 4) provides further documentation of the essential relationship between emotion and movement, and in ways that both corroborate and extend Bull's experimental studies. His 'geometry' or structural theory of emotions rests on two fundamental observations: when we experience emotion, 'we experience ourselves . . . as *being*

---

[10] Specific attention should be called to the fact that the postural readiness to act is a *spontaneous* bodily happening, not a voluntary cultivated one. Attention too should be called to the fact that *readiness* is a phenomenon in dire need of recognition and study by cognitivists and researchers generally in the area of cognition and semantics. Readiness is obviously related to attention — one of the 'mental powers' itemized by Darwin (Darwin, 1981 [1871]), and to receptivity, a fundamental dimension of experience analyzed by Husserl (1973b) in terms of *turning toward*. Implicit in both Darwin's and Husserl's accounts is a recognition of living bodies, i.e., readiness is a phenomenon that is anchored in living bodies and, being so anchored, is a phenomenon that necessarily requires the study and understanding of animate form.

[11] The movements of grief and of joy are not actions but precisely movements. In effect, there is no less a distinctive postural readiness to the having of these and other such emotions. The body folds heavily inward in grief, for example, in contrast to its expansive lightness in joy. Hence, 'preparation for action' may in some instances be a certain postural readiness and corporeal tonicity tied not to action but to a purely qualitative kinetics or kinetic form — a way of being a body.

moved' (ibid., p. 11, italics in original; see also, among others,[12] Sartre, 1948, p. 15:
'[T]he phenomenologist will interrogate emotion. . . . He will ask it not only what it is
but what it has to teach us about a being, one of whose characteristics is exactly that
he is capable of *being moved*' [italics added]); and when we examine our experience,
we discover 'different movements of the emotions' and in turn can specify 'the nature
of the movement that each [emotion] manifests' (de Rivera 1977, pp. 35, 38). De Rivera elaborates the first observation when he writes that 'the paradox of emotional
experience' is that 'we are passively being moved rather than acting and yet this
movement seems to be coming from *within* us' (ibid., p. 12). He does not inquire specifically into the provenience of this *coming from within*; he does not trace its roots to
Jacobson's felt bodily tensions, to Bull's felt neuromuscular dynamics, or to what I
have identified phenomenologically as the tactile-kinaesthetic body, but it is clear
that he recognizes this *generative* source of emotions even as he focuses on what he
defines as the transformative nature of emotion (they transform our relation to the
world [ibid., p. 35]), and even as he fuses, or perhaps better, prematurely fuses and
thus confuses 'the movement of emotions' and emotional transformations. His recognition of a tactile-kinaesthetic dynamics — of the coming from within as a postural
attitude that engenders an urge to move in certain ways — is evident in the corporeal
illustration he gives of the four basic differential movements of emotion. The illustration implicitly specifies, and in concrete kinetic terms, the coming from within. Presented in the chapter 'The Movements of the Emotions', the illustration names four
fundamental kinetic relations — what we might designate four basic *kinetic forms* —
that can obtain between subject and object and that are instanced in the feelings of
anger, fear, affection, and desire. De Rivera's illustration of the differences between
and among these forms is firmly anchored in common, everyday bodily movement
experiences and warrants full quotation:

> It is intriguing that the distinction between these four basic relations [of anger, fear,
> affection, and desire, which he delineates in terms of moving against or away from an
> object in the first two instances and in terms of moving toward an object in the second two
> instances] may be captured by different bodily movements of extension and contraction.
> If the arms are held out in a circle so that the fingertips almost touch, they may either be
> brought toward the body (a movement of contraction) or moved out in an extension. The
> entire trunk may follow these movements. [So also, we might add, may one's legs, and
> thus one's whole body.] Now if the palms are facing in, the extension movement corre-
> sponds to a moving toward the other — a giving — as in tenderness, while the contraction
> movement suggests a movement toward the self — a getting — as in longing. If the palms
> are rotated out, the extension movement corresponds to the thrusting against of anger,
> while the contraction intimates the withdrawal away of fear. . . . If one allows oneself to
> become involved in the movement and imagines an object, one may experience the corre-
> sponding emotion (ibid., p. 40).

On the basis of these 'four basic emotional movements' (ibid., p. 41), de Rivera
elaborates a complex structure of emotions that includes consideration of a subject's
emotions toward him/herself (emotions such as shame and pride), of emotions as
fluid or fixed, of movement from one emotion to another, and so on. The point of
moment here is not the complex interrelated structure that de Rivera progressively
builds, but the basic kinetic structure underlying the whole: *all emotions resolve*

[12] See below in this text: emotions 'happen' to us (Ekman *et al.*, 1990).

themselves into extensional or contractive movement, movement that goes either toward or against or away from an object, including the object that is oneself. The simple self-demonstration that de Rivera describes aptly captures this basic kinetic structure and with it, the quintessential kinetic dynamics of emotion. It does so through a recognition of the spatiality inherent in the generative kinetic form of emotions: we are moved to move toward or against or away; we are moved basically to extend or to contract ourselves. The correspondence between the spatiality of these basic movements and the spatiality expressed in statements of Bull's subjects is transparent: they say not only that 'I reached for joy — but couldn't get it — so tense', but 'My chest was expanded and held out'; 'I wanted to pound the table or throw something, but I clasped my hand instead'; 'I tried to back away — pushed back on the chair — straight back. All the muscles seemed to push straight back' (Bull, 1951, pp. 143, 146, 153); and so on. The spatial dimension of movement is thematic and palpably evident in these statements. But spatiality is only one dimension of movement; temporality, intensity, and the projectional character of movement are basic dimensions as well. The global phenomenon of movement is compounded of dynamically interrelated elements that together constitute the fundamental dynamic congruency of emotion and motion. Indeed, emotions are from this perspective *possible kinetic forms of the tactile-kinaesthetic body*. This is the direction in which all of the empirical research points. A phenomenological analysis of movement will elucidate the dynamic structure underlying these possible kinetic forms.

III: The Phenomenology of Movement:
A Summary Account

When we bracket our natural attitude toward movement,[13] which includes suspending the object-tethered, dynamically empty, and in turn epistemologically and metaphysically skewed definition of movement as 'a change of position',[14] and turn our attention to a phenomenological analysis of the *experience* of movement, we find a complex of four basic qualities: tensional, linear, amplitudinal, and projectional (Sheets-Johnstone, 1966/1980; 1999). These qualities, separable only analytically, inhere in the global experience of any movement, including most prominently our experience of self-movement. Any time we care to notice them, there they are. We shall take an everyday experience of moving ourselves — walking — as a 'transcendental clue' (Husserl, 1973a; see also Sheets-Johnstone, 1999), that is, as a point of departure for a summary phenomenological analysis of movement.

Walking is a dynamic phenomenon whose varying qualities are easily and plainly observable by us: we walk in a determined manner, with firm, unswerving, measured steps; we walk in a jaunty manner with light, cambering, exaggerated steps; we walk in a disturbed manner with tense, erratic steps that go off now in this direction, now in that, and that are now tightly-concentrated, now dispersed; we walk in a regular walking-to-get-some-place manner with easy, flowing, striding steps. Tensional, linear, amplitudinal, and projectional qualities of movement are present in each instance

---

[13] For a detailed account of bracketing (the phenomenological *epoché*), see Sheets-Johnstone (1999), Chapter 4: Husserl and Von Helmholtz — and the Possibility of a Trans-Disciplinary Communal Task.

[14] What changes position are objects in motion, not movement. Movement is thus not equivalent to objects in motion (see Sheets-Johnstone, 1979; 1999).

and in each instance define a particular dynamic.[15] With the recognition of these qualities comes a beginning appreciation of their complexity and of their seemingly limitless interrelationships — and an appreciation as well of the fiction and vacuity of defining movement as 'a change of position' much less of conceiving it as output. As the examples of walking indicate, movement is a variable phenomenon because it is an inherently complex dynamic phenomenon. Motor physiologists have long recognized this fact in what they term 'the degrees of freedom problem' (Bernstein, 1984). The problem is aptly designated phenomenologically 'the kinaesthetic motivation problem' (Sheets-Johnstone, 1999): we can raise our arm from the wrist, from the elbow, from the shoulder, for example, with different possible tensions and amplitudes, different possible speeds, in different possible directions, and so on. Regarded in the phenomenological attitude, movement is both a variable-because-complex and complex-because-variable dynamic happening, an experience which, as indicated, is there any time we care to notice it.[16]

By the very nature of its spatio-temporal-energic dynamic, bodily movement is a *formal* happening. Even a sneeze has a certain formal dynamic in which certain suddennesses and suspensions of movement are felt aspects of the experience. Form is the result of the qualities of movement and of the way in which they modulate and play out dynamically. In a very general sense, tensional quality has to do with our felt effort in moving; linear quality with both the felt linear contour of our moving body and the linear paths we describe in the process of moving, thus, with the directional aspect of our movement; amplitudinal quality with both the felt expansiveness or contractiveness of our moving body and the spatial extensiveness or constrictedness of our movement, thus, with the magnitude of our movement; projectional quality with the manner in which we release force or energy — in a sustained manner, for example, in an explosive manner, in a ballistic manner, in a punctuated manner, and so on. Linear and amplitudinal qualities obviously constitute spatial aspects of movement; temporal aspects of movement are a complex of projectional and tensional qualities. It is of singular moment to note that movement *creates* the qualities it embodies and that we experience. In effect, movement does not simply take place *in* space and *in* time. We qualitatively create a certain spatial character by the very nature of our movement — a large open space or a tight resistant space, for example, a spatial difference readily suggestive of the distinctive spatialities of joy and fear.

---

[15] Languaging the dynamics of movement is a challenging task, perhaps more so than languaging any other phenomenon one investigates phenomenologically. Pinpointing the exact character of a kinetic experience is not a truth-in-packaging matter; the process of moving is not reducible to a set of ingredients. The challenge derives in part from an object-tethered English language that easily misses or falls short of the temporal, spatial, and energic qualitative dynamics of movement.

[16] A reviewer called my attention to a paper by Georgieff and Jeannerod in connection with his concern that 'it is not obvious that kinaesthesia ALONE could be responsible for awareness of movement as self-initiated.' The paper by Georgieff and Jeannerod (http://www.isc.cnrs.fr/wp/wpjea9805.htm) in part concludes that 'normal subjects appear to be unable to consciously monitor the signals generated by their own movement' (p. 4). The experiment on which the conclusion is based, however, assumes a key element that needs to be investigated and taken into account, namely, attention (cf. Darwin, 1981 [1871]). What one attends to is what one is conscious of: if one's attention is visually tethered to a visual desired result (and given 'the well-known dominance of visual information over information from other modalities,' a point that Georgieff and Jeannerod themselves make [ibid.]), kinaesthetic awareness will be proportionately lessened. The conclusion, in effect, is vitiated by oversight of a key 'mental power' (Darwin, 1981 [1871], pp. 44–45).

Analogous relationships hold with respect to the created temporal character of movement — a hurried and staccato flow of movement, for example, or a leisurely and relatively unpunctuated flow, temporal differences readily suggestive of the distinctive temporalities of agitation and calmness. In sum, particular energies, spatialities, and temporalities come into play with self-movement and together articulate a particular qualitative dynamic.

IV: The Dynamic Congruency

As the examples of walking show, the formal dynamics of movement are articulated in and through the qualities of movement as they are created in the act of moving. The challenge now is to demonstrate concretely how dynamic kinetic forms are congruent with dynamic forms of feeling — how motion and emotion, each formally distinctive experiences, are of a dynamic piece. Because it is a common and well-researched emotion, I will use *fear* to illustrate the dynamic congruency.

Phenomenologically, it is sufficient to imagine oneself fearing (Husserl, 1983, Section 4) — as in being pursued by an unknown assailant at night in a deserted area of a city[17] — in order to begin studying the kinetic dynamics of fear. A beginning phenomenological account of the kinetic experience might run as follows:

> An intense and unceasing whole-body tension drives the body forward. It is quite unlike the tension one feels in a jogging run, for instance, or in a run to greet someone. There is a hardness to the whole body that congeals it into a singularly tight mass; the driving speed of the movement condenses airborne and impact moments into a singular continuum of motion. The head-on movement is at times erratic; there are sudden changes of direction. With these changes, the legs move suddenly apart, momentarily widening the base of support and bending at the knee, so that the whole body is lowered. The movement is each time abrupt. It breaks the otherwise unrelenting and propulsive speed of movement. The body may suddenly swerve, dodge, twist, duck, or crouch, and the head may swivel about before the forward plunging run with its acutely concentrated and unbroken energies continues.

Compare this brief phenomenological description to the description of Martina's fear on experiencing a change in accustomed habit. Ethologist Konrad Lorenz writes:

> One evening I forgot to let Martina [a greylag goose] in . . . and when I finally remembered . . . I ran to the front door, and as I opened it she thrust herself hurriedly and anxiously through, ran between my legs into the hall and . . . to the stairs. . . . [A]rriving at the fifth step, she suddenly stopped . . . and spread her wings as for flight. Then she uttered a warning cry and very nearly took off. Now she hesitated a moment, turned around, ran hurriedly down the five steps and set forth resolutely . . . ' (Lorenz, 1967, pp. 65, 66–67)

Compare it to the fear of Temple in novelist William Faulkner's *Sanctuary*:

> She surged and plunged, grinding the woman's hand against the door jamb until she was free. She sprang from the porch and ran towards the barn and into the hallway and climbed the ladder and scrambled through the trap and to her feet again, running towards the pile of rotting hay. Then suddenly she ran upside-down in a rushing interval; she

---

[17] Obviously, this is only one possible example. A complete phenomenological analysis requires 'free variations' (Husserl, 1973a; 1977), or in other words, consideration of multiple experiences of fear in order to identify invariants. A complete analysis would thus entail, for example, consideration of instances in which one is paralyzed with fear as well as mobilized by it.

could see her legs running in space, and she struck lightly and solidly on her back and lay still . . . ' (1953, pp. 75–76).

Descriptions of the dynamics of fear illustrate in each instance how the four basic qualities of movement inhere in an ongoing kinetic dynamic and how that dynamic is through and through congruent with the dynamics of fear: its felt urgency, clutchedness, stops and starts, desire for escape, sense of sudden impending disaster coming from everywhere and nowhere, and so on. In short, movement qualities can be described (and both more finely and more extensively than in the brief sketches above); and *fear* movement can in turn be distinctively detailed, and in different species as well as different instances. This is essentially because movement is movement — it is analytically the same in all instances — and because fear moves us — living creatures, animate forms — as all emotions move us: to move in ways coincident with its felt dynamics. Dynamics vary because fear itself varies: the clutchedness of fear may predominate over the desire for escape; urgency may be extreme at one moment or in one situation and far less pressing in another; and so on. Moreover, each particular experience unfolds in a particular way, articulating a particular overall formal dynamic that begins in a certain way from a certain here-now other emotion,[18] that waxes and wanes, or is attenuated, heightened, reinforced, compounded, intensified, or unexpectedly calmed. Whatever the particular instance, when fear 'happens' to us (Ekman *et al.*, 1990), i.e., when it moves us, we move in ways qualitatively congruent with the way(s) in which we are moved to move; spatial, temporal, and energic qualities of our movement carry us forward in an ongoing kinetic form that is dynamically congruent with the form of our ongoing feelings. Unified by a single dynamics, the two modes of experience happen at once; simultaneity of affect and movement is made possible by a shared dynamics.

It is evident, then, that a particular kinetic form of an emotion is not identical with the emotion but is dynamically congruent with it. Because there is a *formal* congruency, one can separate out the emotion — the felt affective aspect and the postural attitude that generates it, or allows it to generate — from the kinetic form that expresses it. An emotion may thus be corporeally experienced, on the one hand, even though it is not carried forth into movement, and it may be mimed, on the other hand, but not actually experienced.[19] In other words, one can inhibit the movement dynamics toward which one feels inclined — opening one's arms, moving quickly forward, and hugging; or throwing one's arms upward, wheeling about, and pacing; and equally, one can go through the motions of emotion — opening one's arms, moving quickly forward, and hugging; or throwing one's arms upward, wheeling about, and pacing — without experiencing the emotion itself.[20] The dual possibilities testify unmistakably to the dynamic congruency of emotion and motion. Corporeal tonicities are congruent with specific emotions from the beginning, as Bull's research shows. Whether and how one gives kinetic form to these tonicities is a matter of

---

[18] Limited space precludes showing how emotion is continuous rather than a set of neatly packaged states that descend on us individually every so often.

[19] One can precipitate autonomic nervous system activity, however, merely by 'putting on a face'. See Levenson *et al.* (1990); Ekman *et al.* (1983).

[20] The striking power of movement in dance to present us with the semblance of emotion (Langer, 1953) through the choreographic formalization of a kinetic dynamics is testimonial to the latter possibility (see also Sheets-Johnstone, 1966/1980). Martha Graham's *Lamentation* is a classic example.

choice.[21] The two options appear to have different origins. With respect to inhibition, one ordinarily learns in childhood that to avoid certain unwanted consequences, self-restraint is desirable. However, precisely in these circumstances, one may learn to simulate — to go through the motions of — what parents or other adults deem proper. While inhibition is actively learned, that is, a child is taught to restrain him/herself from, for example, hitting, no one teaches a child to dissemble or simulate, e.g., to move compliantly when she/he feels like hitting. A child learns this from her/his own experience and intuitively practices the art of movement deception.[22] The dual possibilities not only testify to dynamic congruence; they underscore the fact that what is affective is kinetic or potentially kinetic, and that what is kinetic is affective or potentially affective. Restraining movement and simulating emotion attest to each fact respectively.

V: Methodological Significance of a Whole-Body Dynamics

The kinetic dynamics of emotion may be studied objectively through the use of a movement notation system. The possibility is not entirely new (see review of non-verbal behaviour studies in Rosenfeld, 1982), but its methodological significance for empirical studies of emotion has not been recognized, in large measure because the fundamental congruity of emotion and motion has been neither acknowledged nor examined. To appreciate the significance, consider first some well-known empirical data.

Fear 'is the dominant component of anxiety'; it measures the highest tensional mean of all emotional situations; it 'brings about a tensing and tightening of muscles and other motor mechanisms, and in terror the individual may 'freeze' and become immobile' (Izard, 1977, pp. 378, 366, 365). Psychologist Carroll Izard amplifies these basic empirical findings, stating that 'Intense fear is the most dangerous of all emotion conditions' and that 'The innate releasors or natural clues for fear include being alone, strangeness, height, sudden approach, sudden change of stimuli, and pain' (p. 382).[23] The data bear out and broaden Jacobson's studies of anxiety. With respect to felt bodily experiences of fear, however, they fall far short of what Bull's subjects offer. While facial expression is described, and extensively so, the body is not, except to say that 'The person feels a high degree of tension and a moderate degree of impulsiveness' (p. 383).

---

[21] Whether and how one moves (or, in highly simplified third-person behavioural terms, whether and how a person acts, e.g., aggressively, friendly, or disgustedly, for example, and what actions a person performs, e.g., pounding, patting, or turning away) is something over and above the corporeal tonicities themselves in that whether and how one moves are both volitional. However much one is moved to move, and however much one is a creature of habit, one can elect to move — angrily or compassionately, for example — or not to move, e.g., to be indifferent, or uninvolved, which kinetically means turning away in some manner, averting one's eyes, and so on. One is, in short, always responsible for one's movement (behaviour).

[22] See von Helmholtz (1971 [1870]) and Husserl (1980) on intuition and its distinction from reasoned processes of thought.

[23] Todd (1937, p. 267): 'The terrified cat at the top of the elm, his muscular strength greatly enhanced by his adrenalin secretion, stops digesting because of his more pressing needs. Rescue him, and he curls up in his corner and is soon fast asleep, recovering his equilibrium. Man, however, being the only animal that can be afraid all the time, prolongs his conflicts even after the danger is past. Proust died of introspection long before he died of pneumonia, burned out by the chemistry of seven volumes of *Remembrance of Things Past* (p. 274). See also Averill (1996, p. 218): '[F]ear: no animal has as many as man, not only of concrete, earthly dangers, but also of a whole pantheon of spirits and imaginary evils as well.'

Bodily movements coincident with emotion are different from both facial expression and autonomic nervous system activity, these phenomena being the prime focus of empirical studies of emotion. Studies of the former present emotion in the form of visual stills and deduced facial muscle involvement; studies of the latter measure physiological responses.[24] Neither focus on the *whole-body experience* of emotion, which means neither focus on the felt experience of being moved and moving. This is *not* to minimize the far-reaching epistemological value and significance of studies of facial expression or of autonomic nervous system activity *vis à vis* emotions. It is rather to call attention to the near complete lack of attendance to the felt bodily experience of emotion as in Bull's studies, to the felt kinetic unfolding or bodily process of emotion as adumbrated in de Rivera's 'movement of emotion,' and to the twin formal dynamics of being moved and moving as evident both in the above descriptive accounts and in diverse literatures generally.[25] In fact, by itself, our immediate and untroubled comprehension of descriptions of emotion — descriptions regularly given in primatological studies, ethological studies, and in all manner of literature on humans — calls attention to the foundational grounding of emotion in motion, which is to say in the experience of our own kinetic/tactile-kinaesthetic bodies. How else explain our untutored understanding of a tightly tensed running body that suddenly stops, turns, swivels, then pitches on, or of a goose's 'hurried' and 'anxious' stop-and-start movements, or of a character's 'surging', 'plunging', 'grinding', 'springing', 'scrambling' movements? We know immediately — in our muscles and bones — what it is to be pursued, to experience sudden and disruptive change, to be trapped; we know in a bodily felt sense what it is to be — in a word — fearful. Indeed, we recognize fear in these purely kinetic descriptions in the same way that experimental subjects recognize fear on being shown composite photographs of faces with widely opened eyes, raised and pulled together brows, and drawn-back lips, and who furthermore recognize their own facial expression of fear on being asked to make these composite gestures themselves (Ekman *et al.*, 1983; Levenson *et al.*, 1990). *We recognize the kinetics of fear on the basis of our own kinetic/tactile-kinaesthetic bodily experiences of fear.* Primatologist Jane Goodall documents this fact straightaway and more broadly when, in describing a variety of intraspecific whole-body emotional comportments in a chimpanzee society, she states, 'We make these judgements [about how a chimpanzee is feeling] because the similarity of so much of a chimpanzee's behaviour to our own permits us to empathize' (Goodall, 1990, p. 17).

---

[24] With respect to physiological studies, it is worthwhile pointing out that various researchers localize emotions in the (primitive) brain, especially the limbic system, and that the practice of localization is not without criticism. After considering various localization scenarios, Averill (1996, p. 221) comments that 'As Von Holst and Saint-Paul (1960) have emphasized, questions of "how" and "why" are too frequently turned into the seemingly more simple problem of "where".' He goes on to remark, 'The recent past has been a period of great neuroanatomical progress, made possible by advances in electronic recording and stimulating devices; unfortunately there is little sign of corresponding progress in the conceptualization of psychophysiological relationships. The macroscopic phrenology of Gall and Spurzheim may be dead, but a kind of microscopic phrenology is alive and well in many a neurophysiological laboratory.'

[25] The lack is occasionally recognized: e.g., Ginsburg and Harrington (1996, p. 245): 'There is a relative dearth of systematic research on the relationships between (sic) emotions and movements and postures.'

To omit attention to a *whole-body* dynamics is to reduce the dynamics of emotion — and more particularly, the dynamic form of an emotion as it unfolds — to a single expressive moment or to isolated internal bodily happenings. It is to de-temporalize what is by nature temporal or processual. Correlatively, it is to skew the evolutionary significance of emotion, which is basically not to communicate, but *to motivate action*. Sperry's principal finding — that the brain is an organ of and for movement — is central to this evolutionary understanding. Not only is the social significance of emotion, i.e., the value of letting others know how one feels and of knowing how others feel, contingent on being social animals, a comparatively late evolutionary development, but knowledge of the feelings of others is itself tied to movement. 'Fearful behaviour' — a 'display' of emotion or what primatologist Stuart Altmann more generally and rightfully terms a 'comsign' (Altmann, 1967; see also Sheets-Johnstone, 1990) — is articulated in bodily movement. Being articulated in bodily movement, it has a distinctive kinetic form recognizable by others. Indeed, like all communicative emotional behaviours, 'fear behaviour' *originates in movement*, movement that is communal in the sense of being performed or performable by conspecifics, movement that thus falls within the 'I cans' or movement possibilities of the species and on that basis is immediately meaningful to all — *a comsign*. In short, emotional behaviours are fundamentally kinetic bodily happenings that originate in experiences of being moved to move and that evolve kinetically. Their communicative value is an evolutionary outgrowth of what is already there: motivations (from Latin *movere*, to move) are felt dispositions or urges to move in certain ways — to strike or to back away, or to peer, stalk, touch, snatch, or squeeze. To say that the social derives from what is evolutionarily given is to say that it derives from species-specific kinetic/tactile-kinaesthetic bodies (Sheets-Johnstone, 1994; see also Ekman, 1994; Ekman and Davidson, 1994).

In sum, emotions are prime motivators: animate creatures 'behave' because they feel themselves moved to move. Short of this motivation, the social significance of emotion would be nil. What would be the value of knowing another's feelings or of another knowing one's own if in each instance the knowledge was kinetically and affectively sterile, generating nothing in the way of interest, curiosity, flight, excitement, amicability, fear, agitation, and so on?[26]

Movement notation systems allow empirical study of a whole-body kinetic process in ways that would provide insight into the *differential dynamics of emotions*. In Labananalysis and Labanotation especially, both the *what* and the *how* of movement is notated, thus not merely a flexing of the knee or a twisting of the torso (Labananalysis), for example, but the manner in which the knee is flexed or the torso is twisted (Labanotation or Effort/Shape). In effect, one could specify both the qualitative dynamics of movement and the formal dynamics of emotion as they are simultaneously played out. One could thereby demonstrate empirically the dynamic congruency of movement and emotion in real life. It bears noting that, through the use of movement notation systems, dynamic congruency can be elucidated in species-specific ways that draw our attention to kinetic domains (Sheets-Johnstone, 1983; 1999), thus to similarities and differences among and between species. Moreover,

---

[26] Limited space precludes showing that interest, curiosity, excitement, and other such feelings are no less emotions than fear and anger.

dynamic congruency can be elucidated in culture-specific ways, allowing one to distinguish what is evolutionarily given from what is culturally transformed — exaggerated, suppressed, neglected, or distorted (Sheets-Johnstone, 1994). Insofar as one can find only what one's methodology allows one to find and to know only what one's methodology allows one to know, the value of movement notation systems to the empirical study of emotion is self-evident: the systems offer a methodology proper to *dynamical* studies of emotion, emotion as it is actually experienced in the throes, trials, and pleasures of everyday life. In this respect, they offer the possibility of a complete empirical science of emotion, a science that, not incidentally, is capable of addressing evolutionary and cultural questions on the basis of detailed pan-species and pan-human empirical evidence.

VI: Implications for Cognitivism

Emotions move us, and in moving us are quintessentially linked to kinetic/tactile-kinaesthetic bodies. Preceding sections have shown that they are clearly tied to animation and to kinetic possibilities of animate life. Broad but conceptually fundamental implications follow from this beginning analysis. The characterization of living organisms as information processors or algorithmic machines and in turn as things whose various mechanisms can be thoroughly explained by studies of brains and behaviour — i.e., *in exclusion of experience, which means in exclusion of phenomenological and empirically-focused investigations and analyses of experience* — skews an understanding of animate life. Calling attention to this experientially deficient understanding, the foregoing analysis has the following implications for cognitivism:

(1) Movement is not behaviour; experience is not physiological activity, and a brain is not a body. What emerges and evolves — ontogenetically and phylogenetically — is not behaviour but movement, movement that is neatly partitioned and classified as behaviour by observers, but that is in its own right the basic phenomenon to be profitably studied; what is of moment to living creatures is not physiology *per se* but real-life bodily happenings that resonate tactilely and kinaesthetically, which is to say experientially; what feels and is moved to move is not a brain but a living organism.

(2) A movement-deficient understanding of emotion is an impoverished understanding of emotion. Being whole-body phenomena, emotions require a methodology capable of capturing kinetic form. When serious attention is turned to kinetic form and to the qualitative complexities of movement, emotions are properly recognized as dynamic forms of feeling, kinaesthesia is properly recognized as a dimension of cognition, cognition is properly recognized as a dimension of animation, and animation is no longer regarded mere output but the proper point of departure for the study of life.

(3) Movement notation systems provide real-life as opposed to computational or engineering conceptions and mappings of animal movement. Modelled movement is no match for a real-life kinetics, which alone can provide detailed understandings of the spatio-temporal-energic dimensions of movement itself and of the dynamics of kinetic relationships and contexts.

(4) The penchant to talk about and to explain ourselves and/or aspects of ourselves as embodied — as in 'embodied connectionism' (Bechtel, 1997), and even as in 'embodied mind' (Varela *et al.*, 1991; Lakoff and Johnson, 1999), 'embodied schema' (Johnson, 1987), 'embodied agents', 'embodied actions' (Varela, 1999), and 'phenomenological embodiment' (Lakoff and Johnson, 1999) — evokes not simply the possibility of a disembodied relationship and of near or outright tautologies as in 'embodied agents', 'embodied actions' and 'the embodied mind is part of the living body' (Lakoff and Johnson, 1999, p. 565), but the spectre of Cartesianism. In this sense, the term 'embodied' is a lexical band-aid covering a 350-year-old wound generated and kept suppurating by a schizoid metaphysics. It evades the arduous and (by human lifetime standards) infinite task of clarifying and elucidating the nature of living nature from the ground up. Animate forms are the starting point of biological evolution. They are where life begins. They are where animation begins. They are where concepts begin. They are where emotions are rooted, not in something that might be termed 'mental life' (e.g., Cabanac, 1999, p. 184: 'emotion is a mental feeling'), a 'mental' that is or might be embodied in some form or other, but in animate form to begin with. *Embodiment* deflects our attention from the task of understanding animate form by conceptual default, by conveniently packaging beforehand something already labelled 'the mental' or 'mind' and something already labelled 'the physical' or 'body' without explaining — to paraphrase Edelman (1992, p. 15) — 'how "the package" got there in the first place' (cf. Sheets-Johnstone, 1998; 1999).

(5) Machines are sessile systems/devices anchored in one place as animate creatures are precisely *not* anchored. Robots are not forms of life to whom emotions happen but remote-control puppets to which signals are sent; they are not *moved* to move, but are *programmed* to move. Zombies are even more remote, being mere intellectual figments plumped with sound and fury but signifying nothing pertinent to understandings of animate life. In this respect, the hard problem is to forego thought experiments and to listen assiduously to our bodies, and to observe phenomenologically and empirically what is going on. The hard problem is to give animate form and the qualitative character of life their due. More broadly, the hard problem is to see ourselves and all forms of life as intact organisms, living bodies, rather than as brains or machines. We come into the world moving; moving and feeling moved to move are what are gone when we die. Surely when we lament or fear our own death, we do not lament or fear that we will have no more information to process. We lament or fear that we will no longer be *animate* beings but merely material stuff — *lifeless, unmoved, and unmoving*. Nature is 'a principle of motion', as Aristotle recognized, and kinetic form is its natural expression.

References

Altmann, Stuart (1967), 'The structure of primate social communication,' in *Social Communication Among Primates*, ed. Stuart A. Altmann (Chicago: University of Chicago Press).

Aristotle (1984), *Physics*, trans. R.P. Hardie and R.K. Gaye, in *The Complete Works of Aristotle*, ed. Jonathan Barnes (Princeton: Princeton University Press).

Averill, James R. (1996), 'An analysis of psychophysiological symbolism and its influence on theories of emotion,' in *The Emotions*, ed. R. Harré and W. G. Parrott (London: Sage Publications).

Bechtel, William (1997), 'Embodied connectionism,' in *The Future of the Cognitive Revolution*, ed. David Martel Johnson and Christina E. Erneling (New York: Oxford University Press).

Bernstein, Nicolas (1984), *Human Motor Actions: Bernstein Reassessed*, ed. H.T.A. Whiting (New York: Elsevier Science Publishing Co).

Bruner, Jerome (1990), *Acts of Meaning* (Cambridge, MA: Harvard University Press).

Bull, Nina (1951), *The Attitude Theory of Emotion* (New York: Nervous and Mental Disease Monographs [Coolidge Foundation]).

Cabanac, Michel (1999), 'Emotion and phylogeny', *Journal of Consciousness Studies*, **6** (6–7), pp. 176–190.

Carlson, Neil R. (1992), *Foundations of Physiological Psychology* (Boston, MA: Allyn and Bacon).

Clynes, Manfred (1980), 'The communication of emotion: Theory of sentics', in *Emotion: Theory, Research, and Experience, vol. I, Theories of Emotion*, ed. R. Plutchik and H. Kellerman (New York: Academic Press).

Crick, Francis and Koch, Christof (1992), 'The problem of consciousness', *Scientific American* **267/3**: pp. 153–159.

Dahl, Hartvig (1977), 'Considerations for a theory of emotions', in de Rivera (1977).

Darwin, Charles (1965 [1872]), *The Expression of the Emotions in Man and Animals* (Chicago: University of Chicago Press).

Darwin, Charles (1981 [1871]), *The Descent of Man and Selection in Relation to Sex* (Princeton: Princeton University Press).

de Rivera, Joseph (1977), *A Structural Theory of the Emotions* (New York: International Universities Press).

Edelman, Gerald (1992), *Bright Air, Brilliant Fire* (New York: Basic Books).

Ekman, Paul (1994), 'Strong evidence for universals in facial expressions: A reply to Russell's mistaken critique', *Psychological Bulletin*, **115** (2), pp. 268–87.

Ekman, Paul and Davidson, Richard J. (Ed. 1994), *The Nature of Emotion: Fundamental Questions*, (New York: Oxford University Press).

Ekman, Paul, Davidson, R.J. and Friesen, W.V. (1990), 'The Duchenne Smile: Emotional expression and brain physiology II', *Journal of Personality and Social Psychology*, **58** (2), pp. 342–53.

Ekman, Paul, Levenson, R.W. and Friesen, W.V. (1983), 'Autonomic nervous system activity distinguishes among emotions', *Science*, **221**, pp. 1208–10.

Faulkner, William (1953), *Sanctuary* (Harmondsworth: Penguin Books).

Georgieff, Nicolas and Jeannerod, Marc (1998), 'Beyond consciousness of external reality. A "Who" system for consciousness of action and self consciousness', http://www.isc.cnrs.fr/wp/wpjea9805.htm: 1–10.

Ginsburg, G.P. and Harrington, Melanie E. (1996), 'Bodily states and context in situated lines of action', in *The Emotions*, ed. R. Harré and W. G. Parrott (London: Sage Publications).

Golani, Ilan (1976), 'Homeostatic motor processes in mammalian interactions: A choreography of display,' in *Perspectives in Ethology*, vol. 2, ed. P.P.G. Bateson and Peter H. Klopfer (New York: Plenum Publishing).

Goodall, Jane (1990), *Through a Window: My Thirty Years wih the Chimpanzees of Gombe* (Boston, MA: Houghton Mifflin).

Husserl, Edmund (1973a), *Cartesian Meditations*, tr. Dorion Cairns (The Hague: Martinus Nijhoff).

Husserl, Edmund (1973b), *Experience and Judgment*, ed. Ludwig Landgrebe, tr. James S. Churchill and Karl Ameriks (Evanston, IL: Northwestern University Press).

Husserl, Edmund (1977), *Phenomenological Psychology*, tr. John Scanlon (The Hague: Martinus Nijhoff).

Husserl, Edmund (1980), *Ideas Pertaining to a Pure Phenomenology and to a Phenomenological Philosophy*, Third Book (*Ideas III*): *Phenomenology and the Foundations of the Sciences*, tr. Ted E. Klein and William E. Pohl (The Hague: Martinus Nijhoff).

Husserl, Edmund (1983), *Ideas Pertaining to a Pure Phenomenology and to a Phenomenological Philosophy*, First Book (*Ideas I*), tr. F. Kersten (The Hague: Martinus Nijhoff).

Izard, Carroll E. (1977), *Human Emotions* (New York: Plenum Press).

Jacobson, Edmund (1929), *Progressive Relaxation* (Chicago: University of Chicago Press).

Jacobson, Edmund (1967), *Biology of Emotions* (Springfield, IL: Charles C).

Jacobson, Edmund (1970), *Modern Treatment of Tense Patients* (Springfield, IL: Charles C. Thomas).

Jacobson, Edmund (1973), 'Electrophysiology of mental activities and introduction to the psychological process of thinking', in *Psychophysiology of Thinking*, ed. F.J. McGuigan and R.A. Schoonover (New York: Academic Press).

Johnson, Mark (1987), *The Body in the Mind* (Chicago: University of Chicago Press).
Kelso, J.A. Scott (1995), *Dynamic Patterns: The Self-Organization of Brain and Behavior* (Cambridge, MA: Bradford Books/MIT Press).
Langer, Susanne K. (1953), *Feeling and Form* (New York: Charles Scribner's Sons).
Lakoff, George and Johnson, Mark (1999) *Philosophy in the Flesh* (New York: Basic Books).
Levenson, Robert W., Ekman, Paul and Friesen, W.V. (1990), 'Voluntary facial action generates emotion-specific autonomic nervous system activity', *Psychophysiology*, 27 (4), pp. 363–84.
Lorenz, Konrad (1967), *On Aggression*, tr. Marjorie Kerr Wilson (New York: Bantam Books).
Moran, Greg, Fentress, John C. and Golani, Ilan (1981), 'A description of relational patterns of movement during 'ritualized fighting' in wolves', *Animal Behavior*, 29, pp. 1146–65.
O'Connell, Alice L. and Gardner, Elizabeth B. (1972), *Understanding the Scientific Bases of Human Movement* (Philadelphia, PA: Lea & Febiger).
Pribram, Karl H. (1980), 'The biology of emotions and other feelings', in *Emotion: Theory, Research, and Experience, vol. I, Theories of Emotion*, ed. R. Plutchik and H. Kellerman (New York: Academic Press).
Rosenfeld, Howard M. (1982), 'Measurement of body motion and orientation', in *Handbook of Methods in Nonverbal Behavior Research*, ed. K.R. Scherer and P. Ekman (Cambridge: Cambridge University Press).
Sartre, Jean-Paul (1948), *The Emotions: Outline of a Theory*, tr. B. Frechtman (New York: Philosophical Library).
Sheets-Johnstone, Maxine (1966/1980), *The Phenomenology of Dance* (New York: Arno Press; London: Dance Books Ltd.).
Sheets-Johnstone, Maxine (1979), 'On movement and objects in motion: The phenomenology of the visible in dance', *Journal of Aesthetic Education*, 13 (2), pp. 33–46.
Sheets-Johnstone, Maxine (1983), 'Evolutionary residues and uniquenesses in human movement', *Evolutionary Theory*, 6, pp. 205–9.
Sheets-Johnstone, Maxine (1990), *The Roots of Thinking* (Philadelphia, PA: Temple University Press).
Sheets-Johnstone, Maxine (1994), *The Roots of Power: Animate Form and Gendered Bodies* (Chicago, IL: Open Court Publishing).
Sheets-Johnstone, Maxine (1998), 'Consciousness: A natural history,' *Journal of Consciousness Studies*, 5 (3), pp. 260–94.
Sheets-Johnstone, Maxine (1999), *The Primacy of Movement* (Amsterdam: John Benjamins Publishing).
Sperry, Roger W. (1952), 'Neurology and the mind/brain problem', *American Scientist*, 40, pp. 291–312.
Todd, Mabel Elsworth (1937), *The Thinking Body* (New York: Dance Horizons).
Varela, Francisco J. (1999), 'Present-time consciousness', *Journal of Consciousness Studies*, 6 (2–3), pp. 111–40.
Varela, Francisco J.,Thompson, Evan and Rosch, Eleanor (1991), *The Embodied Mind* (Cambridge, MA: MIT Press).
von Helmholtz, Hermann (1971 [1870]), 'The origin and meaning of geometric axioms (I)', in *Selected Writings of Hermann von Helmholtz*, ed. and tr. R. Kahl (Middletown, CT: Wesleyan University Press).

INDEX

CATALOGUE ENQUIRIES

If you have enjoyed this book, you might be interested in information
on other publications from Imprint Academic.

Name..

Address...

....

....Email.............................

Please send me your catalogue of publications:

☐ CONSCIOUSNESS STUDIES

Journals:

Journal of Consciousness Studies
Cybernetics and Human Knowing

Books:

Gallagher and Shear, *Models of the Self*
Goguen, *Art and the Brain*
Libet et al., *The Volitional Brain*
Metzinger, *Conscious Experience*
Núñez & Freeman, *Reclaiming Cognition*
Shawe-Taylor, *Consciousness at the Crossroads*
Varela & Shear *The View From Within*

☐ CULTURE, POLITICS & SOCIETY

Journals:

History of Political Thought
*Polis: The journal of the society for the study
of Greek political thought*

Books:

Coleman, *Scholastics, Liberties and Radicals*
Edwards, *Gangræna*
Goldsmith and Horne, *The Politics of Fallen Man*
Glyn-Jones: *Holding up a Mirror*
Sutherland, *The Rape of the Constitution?*
Sutherland, *Vox Populi: Culture and the Mass Media*
The ROTA (subscription series)

Mailing address details overleaf

Full details also on http://www.imprint.co.uk

CATALOGUE ENQUIRIES

If you have enjoyed this book, you might be interested in information
on other publications from Imprint Academic.

Name..

Address...

....

....Email.............................

Please send me your catalogue of publications:

☐ CONSCIOUSNESS STUDIES

Journals:

Journal of Consciousness Studies
Cybernetics and Human Knowing

Books:

Gallagher and Shear, *Models of the Self*
Goguen, *Art and the Brain*
Libet et al., *The Volitional Brain*
Metzinger, *Conscious Experience*
Núñez & Freeman, *Reclaiming Cognition*
Shawe-Taylor, *Consciousness at the Crossroads*
Varela & Shear *The View From Within*

☐ CULTURE, POLITICS & SOCIETY

Journals:

History of Political Thought
*Polis: The journal of the society for the study
of Greek political thought*

Books:

Coleman, *Scholastics, Liberties and Radicals*
Edwards, *Gangræna*
Goldsmith and Horne, *The Politics of Fallen Man*
Glyn-Jones: *Holding up a Mirror*
Sutherland, *The Rape of the Constitution?*
Sutherland, *Vox Populi: Culture and the Mass Media*
The ROTA (subscription series)

Mailing address details overleaf

Full details also on http://www.imprint.co.uk

Insert in envelope and send to either address:

North American continent:

Consciousness Studies
Department of Psychology
University of Arizona
Tucson   AZ 85721
USA

Phone: 520 621 9317
Fax: 520 621 9801
Email: center@u.arizona.edu

Rest of World:

Imprint Academic
PO Box 1
Thorverton
EX5 5YX
UK

Phone: +44 1392 841600
Fax: +44 1392 841478
sandra@imprint.co.uk

Insert in envelope and send to either address:

North American continent:

Consciousness Studies
Department of Psychology
University of Arizona
Tucson   AZ 85721
USA

Phone: 520 621 9317
Fax: 520 621 9801
Email: center@u.arizona.edu

Rest of World:

Imprint Academic
PO Box 1
Thorverton
EX5 5YX
UK

Phone: +44 1392 841600
Fax: +44 1392 841478
sandra@imprint.co.uk